```
QD              122285
8.5             Bottle, R. T.
.I47              Information
1992            sources in
                chemistry
```

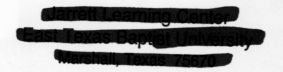

Guides to Information Sources

Information Sources in
Chemistry

Guides to Information Sources

A series under the General Editorship of
D.J. Foskett, MA, BSc, MRIC
and
M. W. Hill, MA, BSc, MRIC

This series was known previously as 'Butterworths Guides to Information Sources'.

Other titles available include:

Information Sources in Sport and Leisure
   edited by Michele Shoebridge
Information Sources in Patents
   edited by C.P. Auger
Information Sources for the Press and Broadcast Media
   edited by Selwyn Eagle
Information Sources in the Medical Sciences (Fourth edition)
   edited by L.T. Morton and S. Godbolt
Information Sources in Information Technology
   edited by David Haynes
Information Sources in Grey Literature (Second edition)
   by C.P. Auger
Information Sources in Pharmaceuticals
   edited by W.R. Pickering
Information Sources in Metallic Materials
   edited by M.N. Patten
Information Sources in the Earth Sciences (Second edition)
   edited by David N. Wood, Joan E. Hardy and Anthony P. Harvey
Information Sources in Cartography
   edited by C.R. Perkins and R.B. Barry
Information Sources in Polymers and Plastics
   edited by R.T. Adkins
Information Sources in Science and Technology (Second edition)
   edited by C.C. Parker and R.V. Turley
Information Sources in Physics (Second edition)
   edited by Dennis Shaw
Information Sources in Economics (Second edition)
   edited by John Fletcher
Information Sources in the Life Sciences (Third edition)
   edited by H.V. Wyatt
Information Sources in Engineering (Second edition)
   edited by L.J. Anthony

Guides to Information Sources

Information Sources in
Chemistry

Fourth Edition

Editors
R.T. Bottle
J.F.B. Rowland

London • Melbourne • Munich • New York

© Bowker-Saur Ltd 1993
All rights reserved. No part of this publication may be reproduced or transmitted in any form or by any means (including photocopying and recording) without the written permission of the copyright holder except in accordance with the provisions of the Copyright, Designs and Patents Act 1988 or under the terms of a licence issued by the Copyright Licensing Agency, 90 Tottenham Court Road, London, W1P 9HE. The written permission of the copyright holder must also be obtained before any part of this publication is stored in a retrieval system of any nature. Applications for the copyright holder's written permission to reproduce, transmit or store in a retrieval system any part of this publication should be addressed to the publisher.

Warning: The doing of an unauthorized act in relation to a copyright work may result in both a civil claim for damages and criminal prosecution.

British Library Cataloguing in Publication Data

Information Sources in Chemistry. -
4Rev.ed. - (Guides to Information Sources Series)
  I. Bottle, Robert Thomas   II. Rowland, J. F. B.   III. Series 540.7
  ISBN 1-85739-016-4

Library of Congress Cataloging-in-Publication Data
Information sources in chemistry / editors, R. T. Bottle, J. F. B. Rowland -- 4th ed.
        p. cm. -- (Guides to information sources)
    Rev. ed. of: Use of chemical literature. 3rd ed. 1979.
    Includes bibliographical references and index.
    ISBN 1-8...
    ...Chemical literature. I. Bottle... II. Rowland, J. F. B.
III. Use of Chemical Literature. IV. Series: Guides to information sources (London, England)
QD8.5.I47 1992
540'.72--dc20   92-17946

Bowker-Saur is a division of REED REFERENCE PUBLISHING,
60 Grosvenor Street, London W1X 9DA, UK.
Tel: +44 (0)71 493 5841. Fax: +44 (0)71 409 0932.

Cover design by Calverts Press
Printed on acid-free paper
Typeset by Typographics, Whitstable, Kent
Printed and bound in Great Britain by Antony Rowe Ltd, Chippenham, Wiltshire

Series editor's foreword

As is obvious, any human being, faced with solving a problem or with understanding a task, reacts by thinking, by applying judgement and by seeking information. The first two involve using information from the most readily available source, namely one's memory, though searching this and retrieving what is wanted may be conducted below the level of consciousness. If this source does not provide all that is needed then the information searcher may turn to all or any of three external sources: observation (which can consist simply of 'going and looking' or can involve understanding sophisticated research); other people, who may be close colleagues or distant experts; and stores of recorded information, for example a local filing system or an electronic databank held on a computer network or even a book or journal in a national library.

The order (observation; other people; recorded information), is not significant though it is often a common sequence. Certainly it is not intended to impute an order of importance. Which of the three or what combination of them one uses depends on a number of factors including the nature of the problem and one's personal circumstances. Suffice to say that all three have their place and all three are used by every literate person.

Nowadays the amount of information in any field, even if one can exclude that which has been superseded, is so large that no human being or small group of people can hope to know it all. Thus, company information systems, for example, get bigger and bigger even when there are efficient means of discarding unwanted and out-of-date information. Managing these systems is a complex and full time task.

Many factors contribute to the huge information growth and overload. Throughout the world, large amounts of research continue to be undertaken and their results published for others to use or follow up. New data pours out of the financial markets. Governments keep passing new legislation. The law courts keep generating new rulings. Each organization and everyone in it of any significance, it seems, is in the business of generating new information. Most of it is recorded and much of it is published.

Although there is a growing tendency to record information in electronic media and to leave it there for distribution via electronic networks of one sort or another, the traditional media are still in use. Even tablets of stone are still used in appropriate circumstances but, of course, it is paper that predominates. The electronics age has not yet led to any reduction in the amount of printed material being published.

The range of types of published or publicly available information sources is considerable. It includes collections of letters, monographs, reports, pamphlets, newspapers and other periodicals, patent specifications, standards, trade literature including both manufacturers' product specifications and service companies' descriptions of their services, user manuals, laws, bye-laws, regulations and all the great wealth of leaflets poured out by official, semi-official and private organizations to guide the information in other than verbal form; maps; graphs; music scores photographs; moving pictures; sound recordings; videos. Nor, although their main content is not published information, should one forget as sources of information collections of artefacts.

In an attempt to make some of the more frequently needed information more easily accessible, these sources of primary information are supplemented with the well-known range of tertiary publications, and text books, data books, reviews and encyclopedias.

To find the information source one needs, another range of publications has come into being. To find experts or organizations or products there are directories, masses of them, so many that directories of directories are published. To find a required publication there are library catalogues, publishers' list indexes and abstracting services, again a great many of them.

For librarians and information specialists in the industrialized countries, access to abstracting services is now normally achieved online, i.e. from a computer terminal over the telephone lines to a remote computerbase database host. Since 1960 the use of libraries by information and even document seekers has changed considerably and can be expected to change further as the British Library study Information 2000 indicates. More and more primary information is being stored electronically and more and more copies of printed documents are supplied via telephone or data networks. Sets of newspapers or other major publications can be acquired on optical discs for use in-house. Scientists in different universities, perhaps working on a common project, are sharing their results via the medium of electronic bulletin boards. The use of electronic messaging systems for disseminating information is now commonplace. Thus the combination of computer technology with telecommunications engineering is offering new ways of accessing and communicating information.

Nevertheless, the old ways continue to be important and will remain so for many years yet.

The huge wealth of sources of information, the great range of resources, of means of identifying them and of accessing what is wanted increase the need for well aimed guides. Not all sources are of equal value even when only those well focused on the required topic are considered. The way new journals proliferate whenever a new major topic is established, many it seems just trying to 'climb on the bandwagon' and as a consequence substantially duplicating each other, illustrates this. Even in an established field the tendency of scientists, for example, to have a definite 'pecking order' for selecting journals in which to publish their research is well known. Some journals submit offered articles to referees; some others publish anything they can get. Similar considerations apply to other publications. The degree of reliance that can be placed on reports in different newspapers is an illustration. Nor is accuracy the only measure of quality. Another is the depth to which an account of a given topic goes.

The aim of this Guides to Information Sources series is to give within each broad subject field (chemistry, architecture, politics, cartography etc.) an account of the types of external information source that exist and of the more important individual sources set in the context of the subject itself. Means of accessing the information are also given though only at the level of the publication, journal article or database, individual chapters are written by experts, each of whom specializes in the field he/she is describing, and give a view based on experience of finding and using the most appropriate sources. The volumes are intended to be readable by other experts and information seekers working outside their normal field. They are intended to help librarians concerned with problems of relevance and quality in stock selection.

Since not only the sources but also the needs and interests of users vary from one subject to another, each editor is given a free hand to produce the guide which is appropriate for his/her subject. We, the series editors, believe that this volume does just that.

Douglas Foskett
Michael Hill

About the contributors

Michael Archer
Graduated in Chemistry from Sheffield University, England and more recently from the UK's Open University in Computing and Information Technology. He worked in the cosmetics industry before specializing in information work. Since 1982 he has headed the Information Sciences Department at SmithKline Beecham's Development and Production site at Worthing.

Robert Bottle
Taught and did research in physical chemistry for twenty years, becoming increasingly interested in chemists' information problems. He ran the first Use of Chemical Literature course in 1959 (on which the first edition of this book was based). He taught Information Science at Syracuse University for two years, and at Strathclyde, before becoming Director of City University's Centre for Information Science in 1973. Professor Bottle built up the department's reputation for research and innovative courses. He has published extensively and was the initiator of this series of information guides.

Peter Carter
After graduating in Pharmacology from Chelsea College, London University, he worked in biomedical research, before completing an MSc in Information Science at the City University. He was Information Officer at the UK Chemical Industries Association for four years, and has also worked for Chemical Intelligence Services. His present post, since 1990, is Database Coordinator for *European Chemical News* magazine.

Michael Cooke
After completing his D.Phil. in Organic Chemistry at Oxford University, he filled Radiochemist and Medicinal Chemist posts for Hoechst UK. After a brief spell with the chemical information specialist ORAC Ltd., he was appointed Head of Internal Information Services within the R&D Directorate of the Wellcome Foundation Limited in 1988. During his work he has authored numerous scientific papers and patents, and has also translated Roth and Kleemann's *Pharmaceutical chemistry: drug synthesis*.

Judith Deschamps
Started her career in the analytical laboratories of a well-known pharmaceutical company. Her experience of that company's library led her to a career change. She has considerable experience of working in government libraries. Her special interests are environmental information (with an international slant) and grey literature.

Tamara Eisenschitz
Originally qualified and did research in Theoretical Physics but turned to Information Science. Before being appointed by City University, England, she worked briefly as a patent searcher. Dr. Eisenschitz has written extensively on patent issues and the problems of technology transfer and has published a book on intellectual property.

Hamish Kidd
After graduation from the University of Edinburgh in 1970, has held positions in information work for the Royal Society of Chemistry (RSC) in London, Nottingham and Cambridge, and in particular worked on pesticide databases between 1983 and 1991. He is currently Assistant General Manager (New Business) within the Secondary Services Department of the RSC.

Alexander Mullen
Obtained his doctorate at Glasgow University, Scotland, but has worked in Germany since then. He joined Bayer AG in 1982 and is responsible for information activities in the chemistry, patent and competitor activities areas at the Pharmaceutical Research Centre. He has developed a particular interest in trend and similar types of analyses based on patents and R&D-related economics information. Dr. Mullen has written and spoken frequently on chemistry and information science as well as having translated major German-language scientific textbooks into English. He is currently the President of the Pharma Dokumentationsring eV (PDR).

Alan G. Osborne
Trained as an organic chemist, gained PhD in 1979. Previously Senior Experimental Officer, Department of Chemistry, City University, London now Chief Research Officer, Department of Chemistry and Biological Chemistry, University of Essex at Colchester. Research interests involve N.M.R. spectral studies of heteroaromatic compounds.

Terry Owen
Graduated in Chemistry from Manchester University. After research in industry, he lectured in organic chemistry and was one of the contributors to the first edition of this book. In 1963 Dr. Owen became Professor of Organic Chemistry at the University of South Florida, Tampa.

Sheila Pantry
Is a qualified library and information specialist with many years experience. A varied career in many backgrounds, iron and steel, engineering, coal, research and public libraries culminated in 1977 in the Health and Safety Executive where she is Head of Library and Information Services which operates through 30 sites in the UK. Author of many papers and books, she has carried out consultancy work in Australia, Canada, Jordan, Dublin and New Zealand as well as work for the EEC Directorate V (OHS) and gives many lectures in the UK and in other countries on occupational health and safety information. At present is an information adviser to the UK Ministry of Arts and Libraries, is a member of the Advisory Committee of the British Library Document Supply Centre, is the UK representative for the International Labour Office (CIS) Health and Safety Centres, and is a member of the EEC Health and Safety Information Strategy Working Group. A computer addict and is enthusiastic about teaching information awareness to all possible users.

Roy Pickering
Qualified as a Biologist at Exeter University, England, before joining the Wellcome Foundation Ltd. He was in charge of the company's central R&D information function for some years, but is now working on the improved pharmaceutical information exchange with Developing Countries including the Library and Information Services Advisory Council (UK) and the Board of the European organization Pharma Dokumentationsring eV. He retains an interest in the problems of information management brought about by the application of technology.

Sir Raymond Rickett, CBE
Is Chairman of the CNAA having been Director of Middlesex Polytechnic from 1973–1991. After taking a BSc in 1953 he was awarded his PhD at the Illinois Institute of Technology in 1959. He has been active in the field of International Education for which he was knighted in 1990.

Fytton Rowland

Is currently a Research Fellow in the Department of Information and Library Studies at Loughborough University of Technology. He originally qualified as a biochemist before training in Information Science at The City University, and was for many years a member of the staff of the Royal Society of Chemistry, latterly as Publications Production Manager.

John Sweeney

Graduated in Chemistry in 1969 from Exeter University and obtained an MSc in Information Science in 1976 from City University. Since 1983 he has worked for the British Library where he was originally the subject specialist in chemistry and is now head of the priced information services in science and technology.

John Timney

Graduated from Newcastle University in 1976 and stayed on to complete a PhD in metal carbonyl chemistry. In 1984, after five years teaching in Northumberland, he moved to North Tyneside College. He remains an active researcher and is a reporter for the RSC's annual review of organometallic chemistry.

Gerald G. Vander Stouw

Is Projects Manager in the Research Department at Chemical Abstracts Service, with responsibility for supervising staff engaged in research efforts related to storage and search of chemical structures and reactions. Received a PhD degree in Organic Chemistry from Ohio State University in 1964 and then joined CAS as a member of the Research staff.

Contents

Section A: General Sources

1 Information, communication and libraries
   R.T. Bottle   3

2 Primary literature
   J.F.B. Rowland   17

3 Abstracting and indexing services
   R.T. Bottle and J.F.B. Rowland   31

4 Books, reviews and encyclopaedias
   J.M. Sweeney   49

5 Online searching for chemical information
   G. G. Vander Stouw   67

6 Chemical structure handling by computer
   M.D. Cooke   105

Section B: Pure Chemistry

7 Standard tables of physico-chemical data
   R.T. Bottle   119

8 Inorganic and nuclear chemistry
   J.A. Timney   139

9 Organic chemistry: the Beilstein Handbuch
   T.C. Owen, R.M.W. Rickett, R.T. Bottle   147

10 Organic chemistry: other reference works
   A.G. Osborne   163

Section C: Industrial Chemistry

11 Patents
   T.S. Eisenschitz   191

12 Technocommercial information
   P. Carter   213

13 Health and safety
    S. Pantry   231

14 The pharmaceutical industry
    M. Archer, A. Mullen, W.R. Pickering   251

15 Agrochemical and food industries
    H. Kidd and J.F.B. Rowland   279

16 National and international governmental information sources
    J.A. Deschamps   289

Section D: Conclusion

17 Practical use of chemical information sources
    R.T. Bottle and J.F.B. Rowland   305

Index-glossary of acronyms, databases, etc.   325

Index   335

Preface to the fourth edition

It is now nearly fifteen years since the third edition was written. As noted in Chapter 1, this is a time span in which the literature of science typically doubles in volume. During this period the non-print formats, which necessitated so many changes in the decade between the second and third editions, have now become commonplace. In particular, methods of handling chemical structures by computer warrant a new chapter and the material on chemical coding and WLN introduced into the third edition is now but of historical interest.

Perhaps also of historical interest is the history of this book. It started in 1959 when one of us (RTB) offered a three-day course for chemistry graduate students on the Use of Chemical Literature. Few students attended but it attracted quite a number of information scientists and technical librarians. When the course was repeated the following year, the clientele were all information workers. These courses, run in Liverpool Public Library during its Golden Age, are believed to be among the first subject-oriented information science courses offered in the UK. The lectures were edited, some additional material and exercises added and eventually published by Butterworths in 1962. This established a pattern whereby subject specialists, including many who actually used the literature for their own R&D work, produced a more authoritative literature guide than one or two authors attempting to cover the subject on their own. The second edition was written whilst the editor was taking a sabbatical from teaching physical chemistry at Bradford University and had been invited to teach information science at Syracuse University Library School during 1968–1970. The second edition became the first member of the Butterworth Series *Information Sources for Research and Development* (now the Bowker-Saur *Guides to Information Sources*). The third edition was extensively rewritten and a number of new sections added. Only the Beilstein chapter bore much resemblance to the previous editions — even in the present edition, this is the least changed chapter. With the recent establishment of the Beilstein database and the declining knowledge of German among both chemists and information scientists, this situation could well change in a fifth edition.

As always, there are changes in the list of contributors and there is an additional editor for this edition. Of the original contributors to the first edition, apart from one of the editors, only Professor Owen and Sir Raymond Rickett remain on the list of contributors to this edition. We *xvi Preface to the fourth edition*

would like to take this opportunity to thank the contributors to the third edition for their help and to welcome those contributing for the first time to this edition. We are indebted to the Controller of HMSO for permission to reproduce part of the *UK Patent Application* GB 2182658 A. We also thank Ivor Williams for information on primary journals in Chapter 2 and Shirley Darke for help with nomenclature problems.

The structure of the first part of the book remains broadly similar to the third edition. What were merely chapter subsections then on online searching and chemical structure handling have now become full chapters (Chapters 5 and 6) in the General Sources section. Then follows Section B on Pure Chemistry. This is followed by Section C on Industrial Chemistry where technocommercial information and health and safety also merit full chapters (Chapters 12 and 13) rather than subsections. Chapters on the pharmaceutical (14) and agrochemical and food industries (15), patents (11), and government publications (16) complete this section.

The Conclusion section is a brief practical guide to using chemical information sources and contains some material which does not conveniently fit elsewhere. It is hoped that separating out the acronyms and abbreviations from the main index will be helpful and also aid the reader in identifying those more commonly encountered in our field.

The reader will note that Chapter 1 contains a section on the communication process amongst chemists and that the final chapter also contains a section entitled 'Communication to others'. We see communication as the alpha and omega of the information problem in science. Apart from being a text for the information scientist, we hope that this book will also ease communication problems for the bench chemist.

RTB February 1992
JFBR

Section A: General Sources

CHAPTER ONE

Information, communication and libraries

R.T. BOTTLE

When the first edition of this book was being written over 30 years ago, computerized systems were still at the experimental stage at Chemical Abstracts Service, the US National Library of Medicine and a few other places. The proportion of information in non-print format has increased very rapidly since the last edition. Such sources have grown markedly in usefulness and sophistication, and consequently in their use. Although bibliographic databases are still predominant, the range and importance of full text and evaluated factual databases, or databanks is growing yearly. Chemical structures are particularly convenient access points to them for the chemist. Thus a new chapter on handling chemical structures in computer systems has been included in this edition (Chapter 6).

Although some information about techniques, apparatus, etc. especially in chemistry's biomedical fringes, is quite short lived, there is still much useful information in the literature published in the latter half of the nineteenth century. Some of it, where it relates to organic compounds, is now being stored electronically in the Beilstein database (Chapter 9), albeit in a very highly condensed form. There is still a vast amount of pre-1960 material still only available in printed format. Time and economics dictate that it will remain so. Whilst much of it is unlikely ever to be read, the task of selecting which might be useful sometime and which might not, probably would require more skilled effort than computerizing the lot — quite apart from prior disclosure considerations for patents (see Chapter 11).

There seems therefore a probable need to conserve this material and to know how to gain access to it until well into the twenty-first century. Whilst knowing how to get publication details of books and how to use libraries may not be quite so important today as when previous editions were published, such material cannot be omitted entirely. It, therefore, forms the second half of this chapter rather than a separate short chapter on its own. Hopefully this serves to emphasize the important role which the printed word and libraries still play in the communication of information between chemists.

Communication by chemists

Chemists can communicate directly by correspondence, by visits, at conferences and electronically (see also Chapter 17). On most occasions, however, they communicate indirectly through the literature. This communication consists of a number of well-defined pathways reminiscent of a complex reaction kinetics scheme (Bottle, 1973). The person requiring information must intercept it at one of the several stages it passes through on its path from the original author's mind to where it becomes integrated into the general fund of knowledge. This book's aim is to make the interception of information a much less random process than it normally is, and to guide the reader through the very considerable volume of conventionally printed literature, either directly or via online systems. Measured in terms of the man-years taken to produce it, the chemical literature is the most expensive tool available to the chemist. He cannot therefore afford to misuse it. About 12 million papers, reports, patents, etc. have so far been published (and this number is doubling approximately every 15 years). Making plausible estimates for the number of man-hours each item represents (and remembering the additional man-years required to abstract it and distil it into storage files such as *Chemical Abstracts, Beilstein, Gmelin,* etc.) and costing this at current levels, the chemical literature probably represents £10^{11}–£10^{12} worth of research already completed. Not all papers are of lasting value; many on experimental techniques quickly obsolesce as they are replaced by improved methods. The chemist, therefore, may require data from the past and techniques from the present. The chemist's research can be regarded as processing such information so that questions are thrown up to which answers can be found through experimental work, thus generating new information for the future.

The chemist can make this new information available to the scientific community without undue difficulty and others can then build on it through further work and produce more information — provided that it reaches the right person at the right time. It is in the last stage of this information processing scheme which we designate research that the communication links are weakest. Because of the size of the chemical literature, virtually no one now obtains all the information he or she requires by direct communication. The chemist will get access to most of it through a variety of temporary or permanent stores for the primary sources of information. The lines of scientific communication are indeed tenuous and surrounded by a vast ocean of literature. Nevertheless the means for dealing with this literature problem exist today, if only people know how to use them, and in the foreseeable future improvements in information storage and retrieval will prevent science from becoming submerged under its own literature. The tendency for information to disperse itself in the literature was recognized as long ago as 1882 when Dr

H. C. Bolton addressed the newly formed Chemistry Section of the American Association for the Advancement of Science in these words: 'Chemical literature is characterized by two opposing forces, a tendency to dispersal and an effort to collect the widely scattered fragments' (quoted by Van Patten, 1950). One thinks immediately of the analogous Le Chatelier's Principle. Le Chatelier's Principle, however, is concerned with a system in equilibrium but the above two opposing forces are never long in equilibrium. Our current literature problems are commonly ascribed to the literature 'explosion' which is often imagined to be a post-war phenomenon. Price (1963) has shown, however, that this is merely the exponential portion of a logistic growth curve which started when science was born in the seventeenth century. Any parameter of science we look at shows this growth pattern — be it number of journals or of abstracts or of the number of new compounds discovered per year (which has grown from about 3000 per year a century ago to over 400 000 per year now). Whilst the number of entries in *American Men (and Women) of Science* showed exponential growth from its first edition in 1903 to the twelfth edition in 1971, such growth has not continued. We found that the number of entries when divided by the US population figure has shown an exponential decrease since 1969, at a rate at which the percentage of scientists in the US population would halve in 30 years. We have observed a similar decline for the percentage of scientists in the UK population (Bottle, Henderson and Norbury, unpublished data). Thus in the developed countries, the proportion of scientists in the population has probably almost reached saturation (having regard for the proportion of those above a certain minimum intelligence level and the competitive demands of other professions). In the less developed countries there is no such saturation and indeed we may well see from this source (particularly China), an escalation in the growth of science which will more than compensate for any reduction in its growth rate in the West. It is, however, an interesting thought that some 80 per cent of all the scientists who have ever lived are alive today (Price, 1963).

One consequence of the exponential growth pattern appears to have been largely overlooked. If we take a conservative figure for the doubling time as 15 years, a new graduate will have x units of information available at the start of his or her career, but on retirement 45 years later, the total amount of information available will be $8x$ units. This underlines the necessity to teach students how to use the literature effectively. The main methods used to teach students how to use their subject literature were reviewed by Bottle (1967). They appear to have changed little since then, apart from introducing online searching. There is some evidence that even the little that was taught in the 1970s is now getting squeezed out of the curriculum.

Many of the older generation of chemists were shown at the start of their careers how to use the chemical literature to the best advantage by a senior and more experienced colleague. Today many factors militate against this ideal method of instruction. The need for more formal instruction in this subject results from what can be termed 'the problem of numbers', referring to both the students to be instructed and the amount of material existing, making it difficult for anyone who is not an information specialist to have much expert knowledge outside the field of his particular interest. (This is, of course, why there are so many contributors to this book.) The growth of the technical library and the emergence of the information specialist, while welcome events in themselves, have tended to place intermediaries between the chemist and one of the most important tools of his trade. While saving much time and effort in retrieving the information requested, this feather-bedding has often resulted in the chemist being insufficiently aware of what treasure actually is hidden, so that no advantage is taken of the facilities which are available. This book aims, not at making bench chemists their own information scientists, but to save researchers' and their information scientist colleagues' time and to facilitate use to the full of the excellent facilities which are within the reach of all in this country today.

The basic problem facing the inexperienced worker is firstly to be aware that the information required may be already in the literature, secondly to be aware that it can be retrieved from storage and thirdly to understand storage and retrieval mechanisms so that all types of information may readily be obtained. There are two main parameters — subject matter and time — which determine which of three main methods of information storage are most appropriate in a particular case. The oldest information storage method is the comprehensive treatise or monograph, in which material is collected from a closely defined subject field over a long time interval. The material is condensed, is often critically evaluated and processed and is retrievable from storage through the layout of the treatise. The second method involves storing the information article by article as it arrives at the abstracting centre, the month's (or week's) catch being broadly classified for current use. Specific information is retrieved from the store through the subject indexes which should preferably be cumulated over a period of years. The third main type derives from the second and its development has doubtless been stimulated by the time lags inherent in the first two types. This type covers all the self-indexing systems, including computer-sorted storage and retrieval systems. Because information is coded into the store, this effectively indexes it and the whole store is continuously and completely searchable.

Recognizing that a knowledge of the structure of chemical literature is the key which unlocks its treasures, the next step is to show how the practical problems connected with the subject are best approached. The first requirement is a clear delineation of precisely what information is being sought. Formally writing down, as concisely as possible, exactly what is the nature of the query is usually a good start. There is, however, no set formula for solving all problems. The approach adopted depends on the nature of the problem, the depth of the inquiry and, to a large extent, the experience of the searcher. Use of the imagination can lead to information from unsuspected sources such as stockbrokers' reports and company reports to shareholders (Chapter 12), patent literature (Chapter 11) and even newspapers and company chairmen's statements.

Because the chemical information needs of the pharmaceutical industry have long caused new information systems and products to be developed, Chapter 14 discusses such needs at length. This overview should be regarded as a case study in chemical information use. It starts with compound design, synthesis, property data, etc. — all areas of as much relevance to the academic as to the industrial chemist in other fields. It then progresses through development to production. Inevitably this case study approach has produced a small overlap with material elsewhere in the book. Nevertheless we have not edited out this overlap as we believe this chapter could, and should, be read with advantage by chemists in many diverse industries unconnected with pharmaceuticals.

Types of information sought

Several types of information query or problem are encountered. A rough classification is as follows:

1. *Facts or data related to a specific compound.* This is often the simplest enquiry to make, since the compounds can be easily indexed by their formulae.

2. *Examples of a specific type of reaction or measurement.* A more exhaustive search may be necessary than in the previous case since indexing tends to be of compounds rather than of reactions or measurements. Computer sorting of structural fragments of encoded product and reactant molecules is now an established method in the documentation of organic reactions. (See Chapter 6.)

3. *First footholds in a new field or reconnaissance reading.* This is the problem which faces all workers when they start a new topic or try to keep abreast of recent trends in other fields. It is especially acute for the new graduate when coming to bridge the gap between undergraduate knowledge and research work. This requirement of background information is dealt with in Chapter 4.

4. *Correlative material.* The collection from different areas of chemistry

of facts which are unrelated except for fitting into a particular theory is especially difficult and this is where initiative and imagination can yield the greatest dividends.

5. *Historical approach.* This is perhaps of greatest interest to the bibliophile. Histories of science and biographical material are very briefly discussed in Chapter 4. Skolnik (1976) wrote a succinct history of the subject matter of this book under the title 'Milestones of Chemical Information Science'.

6. *Industrial development rather than academic research problems.* Apart from information on fundamental chemistry commercial intelligence is often required as well as a knowledge of the relevant patent literature (see Chapters 12 and 11).

7. *Communication with others.* When the chemical literature is used for this purpose, as in report writing, teaching, and so forth, much is lost if the presentation is ineffective. This is why some guides to this art have been briefly mentioned (Chapter 17). A detailed discussion of this subject is, however, outside the scope of this book.

Types (1) and (2) above represent what might be termed the *everyday approach,* where one requires specific information on a topic in hand. In contrast to this is the *exhaustive approach,* where one needs to make an extensive and detailed survey and where one uses the methods outlined in Chapters 5 and 17. The third approach to the literature is the most time-consuming of them all: this is the *current approach* — i.e. keeping up to date with current progress. Although some pointers on how to tackle this problem are given in the final section of Chapter 17, in one's more despondent moments one feels the solution to this problem must await the invention of the 40 hour day! Some amelioration of this problem can doubtless be achieved through increasing one's reading speed by the methods advocated by E. and M. de Leeuw in *Read Better, Read Faster* (Penguin, 1969).

If one understands how the various types of information source arise, one can then locate and evaluate them better. For example, many reports by scientists working in government laboratories are not published in the usual scientific journals. A knowledge of government departments sponsoring such reports or the agencies which collect or catalogue them is essential for searching the literature. Thus the development of governments' responsibilities for research are traced in Chapter 16 and for similar reasons an outline of the procedure for obtaining a patent is given in Chapter 11. The structure of *Beilstein* must be known if it is to be used when the indexes are not available. This is discussed at length in Chapter 9. (One should not be put off by the fact that the printed version of *Beilstein* is in German. The language is quite straightforward and can easily be understood by the novice without having to use a dictionary too often.) To help those who do not read German easily, a translation of the contents of Landolt-Börnstein's *Tables* has been appended to Chapter 7.

More attention is paid in this book to English-language sources than to foreign-language ones, since it is thought that the former will usually be more easily obtainable, and also more easily read, by the average chemist now that translation tests for graduating chemists have long since fallen out of favour. The UK aspects of problems of technical information have been discussed more fully than the US aspects, since these are the ones with which the contributors are most familiar; furthermore a quite recent US guide is available (R.E. Maizell's *How to Find Chemical Information: A Guide for Practising Chemists, Teachers and Students* (2nd edn, Wiley, 1987).

Having confessed to this sin of nationalistic bias, we may now caution the reader about the existence of this evil in certain handbooks, abstracts, etc., produced in Germany during the Nazi regime; it has also been detected from time to time in Soviet publications. The lack of attention paid by Western reviewers to Eastern work is a criticism noted in previous editions, but one also notices the occasional US article which pays scant attention even to any relevant UK work.

One must also remember that every article is a compromise between the ideal of comprehensiveness and the practical need for conciseness. That this latter tendency is increasing is illustrated by comparing the long introductory discourses in papers published half a century ago with the short *pro forma* accounts in today's leading journals. The exponential increase in the literature referred to elsewhere has doubtless been a powerful catalyst to this foreshortening of chemical communication. (Unfortunately this has been accompanied by greatly reduced readability (Bottle *et al.*, 1983)).

Some authorities go so far as to propose that journals should publish only abstracts of articles submitted to them, so that interested workers could then write in for photocopies of the original manuscript as and when required and electronic journals have been produced experimentally. With the growth of computer typesetting, an increasing number of journals are available for full-text searching.

While the primary publication of research papers as abstracts would undoubtedly reduce the physical volume of the literature, it would not help to collect it together but would serve to make it even more difficult for the individual scientist to get at the facts required. Similar remarks apply to proposals that all journals should be issued in microform; this has the additional disadvantage that since it is more trouble to read microforms than to skim through the conventional journal, one will then have to rely even more on a literature-unifying information retrieval service such as *Chemical Abstracts* — even though this imposes an additional delay in getting one's information and cuts down the possibility of browsing in a related field (which can frequently lead to a new line of attack on the problem in hand).

On finding books and using libraries

In a library, the books and periodicals need to be arranged in some sort of order on the shelves. Normally, in order to bring books of similar context together, a library classification scheme is used. The classification most frequently encountered is the Dewey Decimal Classification (DDC). Like most schemes it separates pure and applied science. Chemists will often find it bizarre that some books are allocated to the pure chemistry class number in the 540s and other apparently similar ones are given an applied chemistry class number in the 660s. The lesson is, when browsing for suitable books, look on both sets of shelves. A close relative of DDC, the Universal Decimal Classification (UDC) is claimed to be more suitable for classifying scientific material. A number of academic libraries in the UK as well as in the USA use the Library of Congress Scheme (LC). A comparison of how these three schemes subdivide down from Pure science to pH is given in Table 1.1. It will be noted that all the notations for Hydrogen ion concentration are much longer than pH and are not at all mnemonic. A particularly absurd example of classification, which this author once saw, was a thin book titled *Holographic Index,* on the spine of which were four more characters for its UDC class mark (016:535.41:621.375.9) than there were letters in the title! Now that reasonably user-friendly online catalogues or OPACs (Online Public Access Catalogues) are available in many libraries, a lack

Table 1.1 Comparison of the three major classification schemes

DDC, UDC or LC subject headings	DDC	UDC	LC
Pure Science	500	5	Q
Chemistry	540	54	QD
Physical & theoretical chemistry	541		QD453-655
Theoretical chemistry		541	
Physical chemistry	541.3	541.1	
Electro- & magneto-chemistry	541.37		
Electrochemistry		541.13	QD552-585
Electrolytic solutions	541.372		
Electrolytic dissociation. Ions		541.132	
Electrolytic dissociation. Hyd–rogen ion concentration			QD561
Hydrogen ion concentration	541.3728	541.132.3	

of knowledge of one's library's classification scheme will not handicap the average user. With these systems, appropriate books can be located through a variety of access terms. This obviates remembering strings of numbers.

A book can only occupy one space on the shelves, yet it may often have several facets worthy of note, under which one might wish to find it. Multi-access computerized catalogues permit one to find such books more easily than did the old card catalogues. The general principle on which the librarian works is to classify the book according to its main subject emphasis, putting it where it will be most useful, and to create an index (in the form of the library catalogue) which reveals the decisions made. In this connection the chemist must remember that a book entitled *Mathematics for Chemists* will frequently (but not always) be considered as a book on mathematics, not on chemistry.

Periodicals may be arranged in classified order, or alphabetically by title, or in numerical order with each title receiving a code number. They are often in a special section of the library, and bound volumes are usually filed separately from current issues. Where part of a library's holdings are in microtext form (microfilm, microcard or microfiche), special reading apparatus is necessary, and enquiries should be made at the library counter.

Pamphlets are often stored separately from books, in special pamphlet boxes or in vertical filing cabinets. If they contain information of permanent value, they are classified and catalogued in the usual way, but those which are ephemeral may be collected in broad subject groups, with no specific catalogue entry.

Patent specifications and standard specifications are usually filed in separate sequences by their serial numbers, and are not normally entered individually in the catalogue. Practice may, however, vary. Catalogue practice varies also for government publications trade literature and theses, and here again the guidance of the librarian should be sought.

Although the main book stock is normally arranged in numerical sequence by classification symbols, it is as well to realize that the sequence may be broken or changed to suit the particular needs of individual libraries. Older material may be kept in the stack, to which access may be restricted. Oversize books may be in a separate sequence. There may be special collections which are kept together as a unit, although the books in them cover many subjects. Because of these variations, many libraries display a plan showing the exact location of the various subjects and types of literature, and this, coupled with adequate guides on the bookshelves, should make it easy for users of the library to find what they require. In some libraries a booklet describing the library, its arrangement and its services is available on request. If you have a problem do not be afraid to ask the library staff for help.

Major science libraries

Important collections are to be found in the world's major national and academic libraries. Space does not permit the inclusion of them all, and perhaps there is little need to describe here such well-known collections as those of the Bibliothèque Nationale in Paris, the Lenin Library in Moscow, the Library of Congress in Washington, the US National Libraries of Medicine, or of Agriculture. Some parts of the British Library (BL) are extremely important to the chemist. These include the Science Reference Information Service (SRIS) and the Document Supply Centre (BLDSC). BL is, however, well described in their promotional glossy, *The British Library: Past, Present, Future* (1989). This includes an artist's impression of the new building at St Pancras in London, which lacks even the curvature of a carbuncle. SRIS will be moving there shortly.

SRIS is the largest public library for science and technology in the world. Derived from the old Patent Office Library, it now contains 30 million UK and foreign patents. Its scope and depth of coverage was extended considerably during the 1960s when it was known as the National Reference Library of Science and Invention. There are many information scientists on the staff and also linguists who will assist the user to understand the gist of a document and thereby evaluate its relevance. In recent years it has developed a business information collection which now rivals or surpasses the City Business Library.

The other important division of BL from the chemist's viewpoint is the Document Supply Centre (BLDSC) at Boston Spa, near York. Since 1962 it has become the major source for inter-library lending in science, technology and the social sciences. It has built up a collection of 3 million books, 200000 journal titles, and 300000 conference proceedings in these areas. From the start, collections of Russian and other Eastern European, Japanese, etc. material has been built up, as well as half a million translations. It is a major depository for report literature, adding 130 000 per year to its holding of 7 million reports (mainly on microfiche) from USAEC, NASA, NTIS and others of European origin. BLDSC is a major source of theses, holding 50 000 US dissertations plus 75 000 UK doctoral theses. A document delivery service, ADONIS, covers 200 biomedical journals and CD-ROM disks are distributed weekly to participating centres worldwide. It should be noted that whilst individuals living within range of BLDSC can use its facilities for reference purposes, they can only borrow through their institutional or local public library. BLDSC claims a 95 per cent satisfaction rate for requests. Both BLDSC and SRIS have facilities for online searching.

Although BLDSC plays the major role in supplying literature not available locally, many libraries throughout the country still lend to each other as part of a national system of inter-library loans which has existed for many years. Books and periodicals may be borrowed from abroad when necessary. Public, university, college and many special and government libraries play their part in this scheme.

Mention should also be made of the Royal Society of Chemistry Library in Burlington House, Piccadilly, London, W1V 0BN. It claims the largest collection of material in the UK specifically devoted to chemistry. Although somewhat starved of recent material owing to the ravages of inflation, online searching and postal borrowing and photocopying services are available to members. It has a useful collection of older material in addition to 600 current serials.

Some useful directories for locating details of other libraries are:

E.M. Codlin (ed.), *Directory of Information Sources in the United Kingdom*, 6th edn, 1990, Aslib.

A. Harrold (ed.), *Libraries in the United Kingdom and the Republic of Ireland 1992*, 18th edn, 1991, Library Association (annual listing of over 600 libraries and library authorities).

P. Dale (ed.), *Guide to Libraries and Information Units in Government and Other Organizations*, 30th edn, 1992, BLSRIS, biennial.

American Library Directory, Bowker (a guide to public, academic and special libraries in the USA and Canada, indicating numbers of volumes and any special subject interests, revised biennially).

L. Ash (ed.), *Subject Collections: a Guide to Special Book Collections and Subject Emphasis as reported by University, College, Public Museum and Special Libraries in the United States and Canada*, 7th edn, 1992, Bowker (more than 45 000 entries of collections, including size and specialization).

B.T. Darney and J.A. DeMaggio (eds), *Directory of Special Libraries and Information Centers, 1990*, 13th edn, 2 vols, Gale Research (entries for 15 000 specialized libraries in the USA and Canada, including not only special libraries and information centres, but also major special collections in university and public libraries).

L. Ash (ed.), *Subject Collections in European Libraries*, 6th edn, 2 vols, 1985, Bowker.

H. Lengenfelder *et al.* (eds), *World Guide to Special Libraries*, 2nd edn, 1990, Saur.

B. Bartz, H. Opitz and E. Richter (eds), *World Guide to Libraries*, 10th edn, 1991, Saur.

The World of Learning: Directory of the World's Universities, Colleges, Learned Societies, Libraries, Museums, Art Galleries and Research Institutes, Europa Publications (published annually).

Publication details of books

Normally the chemist will ask a librarian for assistance in tracing publication details of books. Many libraries have a special section of general and subject bibliographies; in other cases subject bibliographies are shelved with books on the same subject. For current British publications on all subjects the *British National Bibliography* (1950–) is the most useful guide, and gives full details each week of newly published books, in a classified subject arrangement, with author and title indexes. It cumulates eventually into an annual volume, and finally into multi-year cumulations. Most major countries have a similar national bibliography. The US *Cumulative Book Index* attempts to include all English-language books irrespective of country of publication. The arrangement here is of authors, titles and subjects in one alphabetical sequence.

For tracing details of books which are still in the print and which can therefore still be purchased in a bookshop irrespective of date of publication, *Whitaker's Books in Print* is the best source of information for UK books. For US publications the annual *Books in Print* and *Subject Guide to Books in Print* are the standard bibliographies. Many good booksellers help their customers to keep in touch with new publications by issuing lists or sets of cards. *New Books in Chemistry* is a fortnightly bulletin from Chemical Abstracts Service in their *CA Selects* series, which commenced in 1977.

Journals such as *Chemistry in Britain*, *Journal of Chemical Education*, *Education in Chemistry*, *Nature*, *Science*, etc. are also useful for their book reviews and publishers' announcements.

Few people read a technical book for pleasure: rather do they read it for the facts which it contains. One looks up isolated facts for reference purposes and thus reference is the commonest form of technical reading. The object of education is, of course, not to instil isolated facts, since virtually no one can remember all he may ever need, but is rather to use Whitehead's words, the 'acquisition of the art of the utilization of knowledge'. Thus it is better to learn how these facts may be utilized and correlated and last, but not least, where they may be discovered. To this last objective this book is dedicated.

References

Bottle, R.T. (1967), *Progress in Library Science*, pp. 97–115. London: Butterworths.

Bottle, R.T. (1973), *Journal of Documentation*, **29**, 281–294.

Bottle, R.T., Rennie, J.S., Russ, S. and Sardar, Z. (1983), *Journal of Information Science*, **6**, 103–108.

Price, D.J. De Solla (1963), *Little Science, Big Science*. Columbia University Press.

Skolnik, H. (1976), *Journal of Chemical Information and Computer Science*, **16**, 187.

Van Patten, N. (1950), *Journal of Chemical Education*, **27**, 431.

CHAPTER TWO

Primary literature

J.F.B. ROWLAND

A primary report is one written by research workers providing a full description of a piece of their own original work, and intended for publication, either in print or in some kind of publicly accessible archive. It should be distinguished from preprints or informally circulated manuscripts, which are not available to the public, and from review articles, abstracts and textbooks, which do not represent the first full disclosure of new research findings written by those who performed the work.

For centuries, such literature has had a central position in the structure of the scholarly enterprise (Ziman, 1968; Ravetz, 1973), and it is important to remember that its function is not solely, and perhaps not even principally, the dissemination of information for current-awareness purposes. Various means exist for the disclosure of preliminary reports of work in progress, notably the many seminars, symposia and conferences that occur every year. The formal primary literature contains reports of finished pieces of work; the authors have polished them and submitted them to an editor who has arranged for peer review by referees. (Patent specifications also are primary reports, but their quality-control system is different and is described in Chapter 11.) In most cases primary reports are published in conventional print after typesetting, though direct reproduction from the typescript and archiving of the more voluminous quantities of data are both now common. The polishing, refereeing and printing processes all take time, and so a report may well not be published until a year or more after the completion of the research work which it describes. Thus other means have to be used for rapid communication; the primary literature is essentially archival, providing researchers with a foundation of accepted, tested data upon which an edifice of further research can be constructed with confidence. It is also a quality-control mechanism; a paper is accepted for publication only after qualified referees have judged the work reliable, and thus it is the mechanism by which work receives its due credit and priority. This slow, long-term approach fits uneasily alongside the very fast pace of technological development that characterizes the present era.

Development of primary journals

Although scientists in earlier times had collected together their findings and published them in book form, or had corresponded with colleagues, by the mid-seventeenth century a new mode of communication was becoming necessary. The primary scientific literature evolved from the meetings of The Royal Society in London and its equivalents in other countries, at which Fellows read papers. It became the practice to print and publish the *Transactions* of these meetings so that members not present could learn about them, and subsequently to allow the publication of papers that had not been read at meetings. (It is still the practice of The Royal Society to allow its Fellows to publish papers in its *Philosophical Transactions* and *Proceedings* journals without any peer review; it is assumed that the fact that the work was done by a Fellow of the Royal Society is a sufficient guarantee of its quality.) Gradually, over the course of the eighteenth and nineteenth centuries, the *Proceedings* journals of various learned societies became more formalized and their rules and procedures more codified.

In chemistry, Germany became the dominant country during the nineteenth century, and its two main chemical journals, *Berichte* and *Annalen*, both published in the German language, became two principal primary journals of chemistry. The explanation of this German domination lay in the early development of the PhD system in the universities of Germany allied with the early and rapid growth of the chemical industry there. In 1909, 45 per cent of the contents of *Chemical Abstracts* originated from Germany. After the defeat of Germany in the First World War, the exodus of Jewish and other anti-Nazi scientists in the 1930s, and the second defeat in the Second World War, however, the domination of Germany in chemistry ended and that of the USA began. Hence English became the major language of chemical, indeed all scientific, publication, and journals published in the USA replaced the German-language journals as the most authoritative forum for publication of the results of chemical research. It is interesting to note that the percentage of documents in English in *Chemical Abstracts* exceeded those in German about 1936 (Bottle *et al.*, 1983).

Until the 1940s, all the major journals of chemistry had been publications of learned societies in the various countries, a tradition going back to the beginnings of these societies in the 1660s. The journals were not published for profit; in most cases, payment of the membership subscription to the society brought with it an automatic subscription to the journal. Today, there is usually an additional subscription for society journals, except for some of the smallest ones, and some of the society journals make significant financial surpluses which are used to support other activities of the societies.

Society journals remain quite dominant in chemistry, more so than in many other disciplines, mainly because in most countries there is a single unified and powerful chemical society (Rowland, 1980). In contrast, the biomedical area is highly fragmented, with many small and specialized societies, which has led to a much greater involvement of commercial publishers in primary publication in that field. Although Macmillan had long published *Nature* and Akademische Verlag and other commercial publishers were active in prewar Germany, a benchmark was the foundation of Pergamon Press Ltd by Robert Maxwell in the 1940s. Since then purely commercial publishers have become increasingly active in the publication of primary scholarly journals. The main attraction of these to a commercial publisher is that a journal is sold on advance subscription, so that the publisher receives his money before spending it, in sharp contrast with the book. Primary publications also make no payment to authors, in contrast with most other kinds of periodical. Today a clear majority of all primary scientific journals are published by commercial publishers, with Academic Press (Harcourt Brace Jovanovich), Pergamon, Springer-Verlag and the Elsevier–North-Holland group the largest participants. The Elsevier group has now taken over Pergamon and this combined group is now the largest by a considerable margin.

Hand in hand with these trends have gone several others (The Royal Society, 1981) towards a much greater number of different journals, greater specialization of content, lower sales of each title, and much higher subscription prices per page in real terms.

An outstanding exception to these trends has been the American Chemical Society (ACS). Owing to the large size of the Society (over 100 000 members) and its large home market, the ACS has been able to maintain the circulations of its journals at a level which is very high by international standards, and therefore to avoid the vicious spiral of rising real price and falling circulation. It has not, however, refrained from launching specialized journals, and now owns a stable of 17 in addition to its flagship *Journal of the American Chemical Society*. Little high-quality chemistry from US authors is published anywhere else. Authors whose native language is not English face a dilemma; if they publish in their own language their work will probably be ignored by the anglophone world, but they may find it difficult to get their work accepted by the US journals. Some UK journals perform a useful service here; their traditionally fastidious standards of copy-editing, often a source of irritation to domestic authors, are helpful in improving the English of authors who are not native speakers of English. Indeed, the UK is increasingly becoming a publishing-agency country for scholarly work, publishing work mainly by non-UK authors and selling copies mainly to non-UK customers. Some people feel that the UK gains considerable economic benefit from the historical accident that it speaks a dialect of the US language! But overall this trend undoubtedly impoverishes the domestic scholarly literature of many countries, and the response in many cases has been the publication of their own journals in English, either as the original language of publication or in cover-to-cover translations.

The trend to more journals at higher prices, with greater specialization and lower circulations, is superimposed on an overall rapid growth of the scientific literature which has been the subject of much comment (Price, 1963). The traditional technology of printing ensures that a lower circulation has to mean a higher price, because the editorial and typesetting costs of preparing the work for the printing press have to be spread over the copies sold, however few these may be. The period of very rapid growth of the scientific literature pre-dated the availability of 'new technology' for the printing and dissemination of scholarly scientific literature. Thus there was a serious problem for scientists and scientific libraries; the cost of maintaining an up-to-date and truly comprehensive collection in chemistry, or any other major scientific discipline, outstripped the rate of growth of resources available to most libraries. US librarians in particular have complained bitterly at the prices charged by European publishers, since they are so much higher than those of the ACS, and for journals whose perceived academic worth to US readers is lower; European publishers have also tended to compensate for wide fluctuations in currency exchange rates by converting their journal subscription prices into dollars at a fixed exchange rate advantageous to themselves. Even so, there is little real evidence to support the allegation that the European commercial publishers, in particular, have profiteered from the academic libraries; the rate of profit on primary journals, though substantial, is not beyond the limits of normal commercial returns, and their high prices are largely due to their specialized subject matter and consequent limited market size.

Although the norms of scholarly science require full publication of research findings, the cost of this renders it impractical in some subject areas. As computerized laboratory equipment has resulted in the automation of many types of experiments, large amounts of data can sometimes be generated very quickly, and data such as X-ray crystallographical results, amino acid sequences of proteins, and routine nuclear magnetic resonance results cannot now justify the cost of full printed publication.

There have been various suggestions for innovations in the primary publication process, several of which have been put into practice. None of them, however, has yet displaced traditional publication from its dominant position. Among the innovations are: data depositories, such as that for crystallographic data at the British Library Document Supply Centre, Boston Spa; computerized databanks, such as that for nucleic acid sequences planned within the Human Genome project in the USA (see Chapter 6); synopsis journals, such as the *Journal of Chemical Research*, where the full text is available in microfiche and miniprint only, and the conventionally printed version contains synopses of the papers; organized distribution of preprints or separates; and electronic journals, using electronic mail or conferencing systems on the national and international academic computer networks.

The most successful innovation of the period since the Second World War has undoubtedly been the rapid-publication short-communication journal. Here, each paper is only one or two pages long and does not include full experimental detail; it is intended to be followed later by the publication of a traditional full paper, but in practice not every short communication is followed by a corresponding full paper and some authors seem to publish only in short-communication mode. The advantage of these communications is that their less-delayed appearance makes them more genuinely current-awareness publications than are full-text journals, and some at least of these journals (such as *Nature*) enjoy high prestige; in some cases, such as *Journal of the Chemical Society, Chemical Communications* and *Tetrahedron Letters*, their standing is higher than that of the full-length journals from the same stable. These journals have built on a prewar tradition of rapid communication via Letters to the Editor of *Nature, Chemistry and Industry*, etc. Thus several of them have 'Letters' in their titles, e.g. *Tetrahedron Letters*.

The rapid growth of the scientific literature over many years means that the active chemist needs to read widely in many different journals. There is evidence (Rowland, 1982) that simply browsing regularly in the latest issues of 15–20 journals is still the favoured means of keeping up to date. There is undoubtedly a need for more systematic methods, and abstracting and indexing publications are discussed in Chapter 3 and computer-based information-retrieval services in Chapter 5. There are also various publications designed as guides to acquisition of the primary literature; one of these is *CASSI (Chemical Abstracts Service Serials Index)*, itself available in both printed and computer-readable forms, which lists all the serials regularly scanned by the staff of the Chemical Abstracts Service of the ACS. *CASSI* contains useful information on these periodicals (see p. 35), and also lists holdings in major US libraries and a few European ones.

Also useful is *Ulrich's International Periodicals Directory 1990–1991* (29th edn, Bowker) which gives details of publisher, where abstracted, price, etc. Although covering over 100 000 periodicals and serials, it is selective. It contains under 2000 entries in its chemistry and biological chemistry sections.

Because of the dominance of BL's Document Supply Centre (referred to in Chapter 1) in the supply of periodicals for loan, UK union catalogues of periodicals, such as *BUCOP* and *The World List of Scientific Periodicals in the Years 1900–1960*, have fallen into disuse. BLDSC lists 76 000 periodicals in *Current Serials Received* (1990). This is primarily useful for checking the exact title prior to a loan request.

The *Science Citation Index*, published in both printed and machine-readable form by the Institute for Scientific Information (ISI) and described in greater detail in Chapter 3, provides a tool for the location of relevant references by means of the approach that perhaps comes most naturally to scholars, that is, the use of authors' names and the citation lists of known relevant papers.

Non-journal format

There are a number of forms of primary literature other than the journal. In chemistry, patent specifications, as mentioned above, and doctoral theses are significant sources of information. Trade literature, internal company reports and other 'grey', that is semi-published, sources have some significance, though it should be borne in mind that they have probably not been submitted to peer review in the same way that scholarly journal articles have. The same point may often apply to published conference proceedings.

Also unrefereed were most of the vast number of reports issued during the first three decades after the Second World War by the various atomic energy, space research and other government agencies. Whilst the FIAT, BIOS and other reports on wartime German chemical research have now all been distilled into appropriate monographs and treatises, adequate initial access to nuclear chemistry and aerospace reports can normally be obtained through *Nuclear Science Abstracts* (1947–1976) and *Scientific & Technical Aerospace Reports*. Report literature is today much less important than it was when earlier editions of this book were published. Most government agencies now encourage the publication of the results of their funded research in the open journal literature. The acquisition and control of reports, theses, etc. are discussed in greater detail in *Information Sources in Grey Literature* (C.P. Auger, ed., 2nd edn, Bowker-Saur, 1988).

The monograph is a significant form of primary publication in some subject areas, particularly in the humanities, but the physical sciences are in general too fast-moving for the book to be a practical alternative for first publication of research. Books in chemistry (see Chapters 1 and 4) are essentially tertiary sources, that is, they are review publications representing a thoughtful synthesis of the content of many primary papers published over several years.

An important recent development is the availability of the full text of primary chemical journals as a machine-readable database on one of the international online networks. This service, known as CHEMICAL JOURNALS ONLINE (*CJO*), is prepared by the ACS and distributed via the Scientific and Technical Network (STN), of which the ACS is part-owner. It contains the primary journals of the ACS and those of The

Royal Society of Chemistry (RSC), together with the polymer journals of John Wiley, the *Journal of the Association of Official Analytical Chemists*, the international (English-language) edition of the German journal *Angewandte Chemie* and some Elsevier journals. Other major chemical publications are likely to be added in the future. At present the database contains the main text and figure captions, but not the figures (illustrations) themselves; however, development work is in progress which should result in the addition of the artwork to the database soon.

It is unlikely that online searching of this full-text database will replace the use of the printed journal for general reading and browsing. Print on paper remains the preferred medium for comfortable prolonged reading for most people. The merit of the full-text service is different: it enables the user to locate any item of data that is in the primary journals, independent of any decision by abstractors or indexers about what to include or index. It is of particular value for locating numerical values or other pieces of raw data which would never be included in an abstract, and for methodological and experimental information. The search software enables one to use all the words, except for a small stop-list of prepositions, conjunctions and so on, as search terms, and then to display only those paragraphs within the paper that contain the search terms. Thus one can search (for example) for a specific category of chemical substance and also for 'm.p.' or 'melting point', to obtain actual melting point values direct from the primary literature.

The availability of the full texts in machine-readable form at a reasonable cost is a consequence of technological change in the printing industry. Virtually all typesetting is now done by computer methods, and thus the keystrokes used to produce the printed journal can also be the source of the machine-readable database. Although the *CJO* database is thus a by-product of the printed publication, it is nevertheless available to many users earlier than the printed version, because distribution of journals is often undertaken by slow surface mail, for cost reasons. At present the printed version remains the main source of the publishers' revenue, but if primary publications follow the same route as the secondary (abstracts and indexes) publications, the balance will gradually shift towards the electronic version (see Chapters 3 and 5).

Furthermore, the use of 'new technology' in printing also offers the possibility of cuts in the publishers' 'first-copy cost'. Most scientific authors (or their secretaries) now use word-processor programs on microcomputers to write their papers; chemical authors often use chemical structure-drawing programs as well, to produce structures and schemes. It is still the case at present that after submission of the paper to the journal, this work is often duplicated by the typesetters and artists employed by the printers; but this situation is changing rapidly, and editors of many journals are now willing to accept papers on floppy disks from authors, and to pass these on to their printers in order to avoid duplicate keyboarding. As this becomes a more widespread practice, as it surely will, the cost of production of journals should fall in real terms. It is to be hoped, however, that the ethics of science will continue to require that all papers be subjected to peer review before they can be considered properly published, as this is the main quality-control mechanism of scientific research work.

The idea of the 'electronic journal', which does not exist in printed form at all, has been discussed for some years, after initially being suggested by Senders (1977). The printed primary publication has proved surprisingly resilient, however, probably because of its archival function and the need of scientists for the professional recognition that traditional publication provides; so far, the scientific community has not been willing to award the same status to non-conventional forms of dissemination. Senders' journal was probably ahead of its time technologically as well; only in the last five years has the infrastructure of academic networks and teleconferencing software become sufficiently widely available and robust outside North America to be a trouble-free alternative to print.

Journals

It is generally understood within any particular scientific community that there is a 'pecking order' of journals, ranging from the top ones into which all authors would like to get their work, down to those whose reputation is that they will publish almost anything (though their editors would no doubt deny this). An active researcher in the field will probably carry the approximate pecking order in his or her head. It can also be measured by citation analysis, as described by Garfield (1977), but without acknowledgement to the method for selecting the key journals in a field of interest suggested in the second edition of this book (Bottle, 1969), which predated ISI's *Journal Citation Reports* by several years. Thus one can draw up a suitable list of journals either by using *Journal Citation Reports* or by the method described in Chapter 3. In chemistry, as mentioned earlier, there would be consensus that the ACS's journals rank highly, and a current list of these is now given.

Accounts of Chemical Research
Analytical Chemistry
Biochemistry
Chemistry of Materials
Energy and Fuels
Environmental Science and Technology
Industrial and Engineering Chemistry
 — *Fundamentals*
 — *Process Design and Development*

— *Product Research and Development*
— *Research*
Inorganic Chemistry
Journal of Agricultural and Food Chemistry
Journal of the American Chemical Society
Journal of Chemical Information & Computer Sciences
Journal of Physical and Chemical Research Data
Journal of Medicinal Chemistry
Journal of Organic Chemistry
Journal of Physical Chemistry
Langmuir
Macromolecules
Organometallics

The RSC does not enjoy quite such a high status, but its journals are nonetheless of accepted quality and no chemical library anywhere in the world would be complete without them. They are listed below.

The Analyst
Analytical Proceedings
Journal of Analytical Atomic Spectrometry
Journal of Chemical Research
Journal of the Chemical Society
— *Chemical Communications*
— *Dalton Transactions* (Inorganic)
— *Faraday Transactions* (Physical)
— *Perkin Transactions I* (Organic)
— *Perkin Transactions II* (Physical Organic)

There are also two new journals launched in 1991 by the RSC: *Mendeleev Communications*, a short-communications journal in English for papers from the former Soviet Union, and *Journal of Materials Chemistry*. *Journal of Chemical Research* is produced by the RSC on behalf of the Gesellschaft Deutscher Chemiker, the Société Chimique de France and the RSC, and is a synopsis plus microform journal.

The major journals from Germany, although not as dominant as they once were, remain an important source which cannot be ignored; the three most important are *Chemische Berichte, Justus Liebigs Annalen der Chemie* and *Angewandte Chemie*, all published by VCH (Verlag Chemie). More specialized German journals include the *Berichte der Bunsen-Gesellschaft für physikalische Chemie* (VCH) and the *Zeitschrift für physikalische Chemie* (Akademische Verlag). Several significant chemical journals, formerly published in East Germany, may now have a less certain future in the aftermath of the economic problems of reunification; they include the *Zeitschrift für Chemie*, the *Zeitschrift für anor-*

ganische und allgemeine Chemie and the *Journal für praktische Chemie*, all published in Leipzig. The *Monatshefte für Chemie* is the main Austrian journal; it is published by Springer-Verlag, but from their Vienna office.

There are a number of other 'national' journals of general chemistry that must be mentioned; these include *Helvetica Chimica Acta, Acta Chemica Scandinavica, Canadian Journal of Chemistry, Bulletin of the Chemical Society of Japan,* the rapid-communications journal *Chemical Letters* (Chemical Society of Japan), *Australian Journal of Chemistry, Receuil des Travaux Chimiques des Pays-Bas* (The Netherlands — in English despite its French title), *Collection of Czechoslovak Chemical Communications* (again in English), *Australian Journal of Chemistry, Indian Journal of Chemistry* and *South African Journal of Chemistry* (English with Afrikaans abstracts or vice versa).

The important French-language journals are the *Bulletin de la Société Chimique de France, Annales de Chimie,* and the Belgian *Bulletin des Sociétés Chimiques Belges.* All three now accept papers in English as an alternative to French.

The International Union of Pure and Applied Chemistry publishes (through Blackwell Scientific Publications) *Pure and Applied Chemistry,* though not all of its contents are primary papers in the true sense; some are, for example, instructions for the use of IUPAC nomenclature.

In addition to the specialized journals of the ACS and RSC listed above, there are of course a great many other journals covering specific areas of chemistry, a large proportion of them published by commercial publishers. The polymer journals of John Wiley, for example, are included in the *CJO* online full-text service; they are listed below.

Biopolymers
Journal of Applied Polymer Science
Journal of Polymer Science
  — *Polymer Chemistry*
  — *Polymer Letters*
  — *Polymer Physics*

There are also a number of other important polymer journals, among them *British Polymer Journal* from the Society of Chemical Industry, and the German *Makromoleculare Chemie.*

In organic chemistry, the most important commercial journal is undoubtedly *Tetrahedron Letters,* a short-communications journal from Pergamon Press; its full-length stablemate *Tetrahedron,* however, enjoys rather lower prestige. Within the organic area, there are many more specialized journals of which probably the most important are *Journal of Heterocyclic Chemistry* (HeteroCorp) and *Journal of Organometallic Chemistry* (Elsevier). There are also *Synthesis* (Georg Thieme Verlag) and the rapid-communication journal *Synthetic Communications* (Marcel Dekker), *Bioorganic Chemistry* (Academic Press), *Heterocycles* (Japan Institute for Heterocyclic Chemistry), *Journal of Molecular Structure* (Elsevier) and *Agricultural and Biological Chemistry* (Japan Society for Bioscience).

Analytical chemistry has several important journals in addition to the ACS and RSC ones mentioned above; these include *Analytica Chimica Acta* and *Fresenius Zeitschrift für analytische Chemie*, and more specialized ones such as the *Journal of Chromatography*, the *Journal of Electroanalytical Chemistry and Interfacial Electrochemistry* (Elsevier), and *Magnetic Resonance in Chemistry* and *Organic Mass Spectrometry* (both Wiley/Heyden).

In the physical chemistry area there are other important journals in addition to the ACS and RSC ones; these include the *Journal of Chemical Physics*, and the UK *Molecular Physics*, published by Taylor and Francis. Others in physical-chemical areas not yet mentioned include *Journal of Catalysis* (Academic Press), *Journal of Colloid and Interface Science* (Academic Press), *Journal of Photochemistry* (Elsevier), *Photochemistry and Photobiology* (Pergamon), *Spectrochimica Acta* (Pergamon), *Theoretica Chimica Acta* (Springer) and *Journal of Solution Chemistry* (Plenum).

Other significant journals of inorganic chemistry include the former *Journal of Inorganic and Nuclear Chemistry*, now renamed *Polyhedron* (published by Pergamon), *Inorganica Chimica Acta* (Elsevier) and various journals devoted to specific elements or groups of elements, such as *Journal of Fluorine Chemistry* (Elsevier) and *Journal of Less-Common Metals*.

Translations

While the increasing domination of the English language in the publication of science has already been noted, it cannot be safely assumed that all relevant material is published in English. CAS figures for the language distribution of the literature covered in *CA* in 1988 were: English, 73.4 per cent; Russian 12.0 per cent; Japanese 4.1 per cent; German 3.3 per cent; all others 6.8 per cent. In particular, until very recently virtually all science from the Soviet Union was published in that country's own journals and in the Russian language. For many years Western governments have sponsored the translation of the main Russian scientific journals from cover to cover and their republication in English, in order to provide the scientists of the West with access to this major corpus of foreign-language literature. Over 200 Russian journals in all subject areas are translated in this way. The British Library (BL) in collaboration with the RSC publishes the *Russian Journal of Physical Chemistry (Zhurnal Fizicheskoi Khimii)*, *Russian Journal of Inorganic Chemistry (Zhurnal Neorganicheskoi Khimii)* and *Russian Chemical Reviews (Uspekhi*

28 Primary literature

Khimii), and although the circulation of these translated journals is not high, sufficient copies are sold that they will generally be available in libraries in major centres of population.

Similarly, the BL collaborates with the Rubber and Plastics Research Association to produce *International Polymer Science and Technology,* based on *Soviet Rubber Technology* and *Soviet Plastics,* and with the British Carbonisation Research Association to produce *Coke and Chemistry.* The Plenum Publishing Corporation in the USA is a major publisher of Russian cover-to-cover translation journals, and among their titles are the *Soviet Journal of Bioorganic Chemistry, Soviet Journal of Coordination Chemistry* and *Soviet Journal of Glass Physics and Chemistry,* and the McElroy company of Texas translates *Soviet Chemical Industry.* Consultants Bureau, New York, translates the important chemistry section of *Doklady Akademii Nauk SSSR* (Proceedings of the Academy of Sciences of the USSR) under the title of *Doklady Chemistry,* and also publishes *Kinetics and Catalysis,* a translation of *Kinetika i Kataliz.*

There are also some cover-to-cover translation journals translated from languages other than Russian, though Russian originals account for about 90 per cent of all translation journals. If the recent changes in Eastern Europe presage a greater advance of English, in place of Russian, as the preferred first language of publication of Eastern European authors, then the emphasis in journal translation programmes may shift to Chinese. Among Chinese translation journals published in chemistry by Plenum are *Geochimica* and *Chemical Bulletin.* Translations from Japanese have lost importance since more of the main Japanese journals are themselves published in English; however, McElroy still produces *Japanese Polymer Science and Technology* in translation.

It may be that a paper is of interest which is published only in a foreign language. The British Library Document Supply Centre publishes a regular monthly bulletin entitled *British Reports, Translations and Theses received by the BLDSC,* and this should be consulted to discover whether BL already possesses a translation of the particular foreign-language paper that one requires. Another source of similar information is the *Aslib Index of Translations* (unpublished but maintained at Aslib).

Theses

Although there has sometimes been debate on the question of whether theses and dissertations submitted by candidates for higher degrees should be regarded as publications, it is now generally accepted that the information contained within them is often of value, and they are often cited as references in other publications. Although in many cases the core information of a thesis is also published in one or more journal articles, this is not invariably the case and it may be necessary to gain access to the thesis itself. Generally the only source of a thesis is the library of the institution where the degree was earned; however, within the UK, these may be obtained by other institutions of higher education on inter-library loan. A large proportion of more modern UK ones are also available on microfilm at the British Library Document Supply Centre, Boston Spa; in 1991, however, it was announced that this service might cease owing to low usage and hence high cost per user.

It may be difficult to locate information about the existence of a relevant thesis; however, some chemical ones are listed by *Chemical Abstracts*. British theses generally are listed by Aslib in its *Index to Theses (with Abstracts)*, and by the British Library Document Supply Centre in its monthly *British Reports, Translations and Theses received by BLDSC*. *Dissertation Abstracts International*, published by University Microfilms in Ann Arbor, Michigan (now part of the Bell & Howell organization), claims to list 95 per cent of all theses written in the USA and Canada plus those from a few European universities. *Dissertation Abstracts* is now available as a machine-readable database through the usual online hosts. German doctoral candidates have long been required to provide 100–300 copies of their theses. A thesis exchange programme between German universities has built up very large collections of theses. Often a thesis author has spare copies available.

References

Bottle, R.T. (1969), Abstracting and information retrieval services, Chapter 5 in *Use of Chemical Literature*, ed. R.T. Bottle, 2nd edn, pp. 71–72. London: Butterworth.

Bottle, R.T., Rennie, J.S., Russ, S. and Sardar, Z. (1983), Changes in the communication of chemical information. *Journal of Information Science*, **6**(4), 103–108.

Garfield, E. (1977), *Essays of an Information Scientist*, vol. 1, pp. 262–265. Philadelphia: ISI Press.

Price, D.J. De S. (1963), *Little Science, Big Science*. New York: Columbia University Press.

Ravetz, J.R. (1973), *Scientific Knowledge and Its Social Problems*. Harmondsworth: Penguin Books.

Rowland, J.F.B. (1980), Learned societies and associations as retailers of information, in *The Nationwide Provision and Use of Information, Proceedings of an Aslib–IIS–LA Joint Conference, 15–19 September 1980, Sheffield*, pp. 37–46. London: Library Association.

Rowland, J.F.B. (1982), The scientist's view of his information system. *Journal of Documentation*, **38** (1), 38–42.

Royal Society (1981), *A Study of the Scientific Information System in the United Kingdom*, British Library R & D Report no. 5626.

Senders, J. (1977), An online scientific journal. *Information Scientist*, **11**, 3–9.

Ziman, J.M. (1968), *Public Knowledge, an Essay Concerning the Social Dimension of Science*. Cambridge: Cambridge University Press.

CHAPTER THREE

Abstracting and indexing services
R.T. BOTTLE AND J.F.B. ROWLAND

In this chapter we are particularly concerned with finding our way through the vast mass of primary material described in Chapter 2. It is clearly impossible for any person to read or even skim the titles of more than a fraction of the newly published periodical literature which may contain something of interest, however specialized the field may be. *Chemical Abstracts*, published by the American Chemical Society (ACS), is the major tool for accessing this mass of literature, and is described in detail later in this chapter; it selects relevant chemical material from over 15 000 periodicals for abstracting in its effort to cover the world's chemical literature comprehensively. This chapter describes the main abstracting and indexing publications, and Chapter 5 will describe in detail the use of computer-based online information services, which has substantially replaced the use of the abstracts journals in printed form (except for the older, pre-1967, literature).

Abstracting and indexing publications, appearing quite soon after the initial appearance of the primary articles, provide access to the mass of literature; they are known as the secondary literature. Later, other authors distil the content, give the story some structure, and publish review articles and books, and data are compiled into handbooks (see Chapter 7); these are sometimes referred to as the tertiary literature and are discussed in Chapter 4 and in the quick reference section of Chapter 17.

Historically, a distinction has been made between an indicative abstract, which tells the reader what the paper is about, and an informative abstract, which summarizes what it says; the former can appear more quickly after publication of the original paper than the latter. This distinction perhaps has less relevance in the modern world of computer-based online retrieval, especially as the original full texts begin to be available online (see Chapters 2 and 5), and different databases covering the same literature are available in a user-friendly fashion from the same online host. It can be argued, indeed, in the electronic era that major intellectual effort put into abstracting is misplaced; the author's abstracts which most primary journals carry can be used and the secondary publications should concentrate on indexing and effective free-text information retrieval systems.

There is also some controversy about comprehensive coverage. *Chemical Abstracts* has always attempted to include all the world's chemical literature; this policy results in published work of doubtful quality — from unrefereed journals, perhaps — appearing alongside that of unquestioned status, but does mean that a proficient searcher can be quite confident that an exhaustive search of *Chemical Abstracts* will not miss anything relevant simply because it is not there in the database. The *Science Citation Index*, on the other hand, bases its coverage on the concept of the core literature of science; its virtues are that work listed in it can be assumed to be of a reasonable quality, since only major journals are covered, and also that it is interdisciplinary within science. The well-known Bradford–Zipf law (Brookes, 1977) applies; even though *Chemical Abstracts* staff scan over 15 000 journals regularly, 2000 provide 90 per cent of the entries (Anon., 1967).

Composition and layout of an abstract

An abstract is a summary of the publication or article accompanied by an adequate bibliographic description to enable the article to be traced. In practice, with minor variations between abstracting publications, the abstract contains the following information.

1. The title in the original language and/or the language of the abstracts publication; uninformative words (such as 'studies in') may be omitted, or if the entire title is insufficiently informative it may be supplemented with keywords.

2. The names of the authors; their addresses may be given, which is valuable if the primary publication concerned is an inaccessible one.

3. The bibliographical reference of the original article, in a standard and complete form.

4. The body of the abstract, which is a summary of the content of the paper. Where the original is itself a review and thus hard to summarize usefully, the form 'a review with 275 references' may be substituted.

There are variations for non-periodical original literature; for example, for a patent (see Chapter 11) the assignees as well as the authors may be given, and the abstract may contain the claims and one or more examples.

Early abstracting services

The British *Journal of the Chemical Society* and *Journal of the Society of Chemical Industry* carried abstracts of papers in other journals, as well as their own full primary papers, from as early as 1871 and 1882, respectively. In 1926 their abstracts sections were combined to form *British Abstracts*; this journal aimed to be comprehensive in its coverage of the world's chemical literature, but its indexing was never as thorough as that of *Chemical Abstracts*. However, with the great growth of the scientific literature after the Second World War it quickly became obvious that it was not economically possible for the UK to maintain a comprehensive chemical abstracting system independent of the American Chemical Society's *Chemical Abstracts (CA)*, and *British Abstracts* ceased publication in 1953. However, the Society of Analytical Chemistry (SAC) was not satisfied with the coverage of analytical chemistry in *CA*, and decided to continue the analytical section of *British Abstracts* under the name of *Analytical Abstracts*. The SAC eventually became one of the constituent societies of the Royal Society of Chemistry (RSC) and the RSC continues to publish *Analytical Abstracts* in both printed and machine-readable forms.

The corresponding German publication was originally called *Pharmaceutisches Central-Blatt* when it was founded in 1830; it subsequently became *Chemisch-pharmaceutisches Zentralblatt* and finally *Chemisches Zentralblatt (CZ)*. Before the Second World War, many chemists regarded *CZ* as superior in coverage to *CA*; its abstracts were longer and more informative, and though they were in German this was not a major difficulty as most professional chemists learned scientific German at that time. However, the Second World War and its aftermath put great difficulties in its way, and combined with the general swing to English-language publication in science (see Chapter 2), this led to the abandonment of *CZ* in 1970. Because of coverage differences, it is still useful to search both *CZ* and *CA* for pre-1930 literature, unless the information required can be found in *Beilstein* or *Gmelin*.

Chemical Abstracts

CA is now the sole abstracting publication which aims to give comprehensive coverage of all chemical, biochemical and chemical engineering publications in the world. Unlike the writing of primary papers, abstracting is generally done for payment, and the vast majority of papers are now abstracted and indexed by *CA*'s in-house full-time staff; indeed, the staff of Chemical Abstracts Service (CAS), over 1000, outnumber the remainder of the ACS's staff by a factor of three. It is these heavy staff costs which have rendered it uneconomic for any other organization to compete directly with CAS. *CA* is intended as a tool for retrospective searching rather than current awareness; its thorough abstracting and indexing procedures lead to some months' delay after the primary publication before an item appears in *CA*.

The ACS published *Review of American Chemical Research* from 1897, but this early version had only limited coverage. It was in 1907 that the ACS decided to embark on the publication of a comprehensive abstracting journal. It originally appeared twice a month, with one volume per year; since 1962, it has appeared weekly with two volumes per year. Each volume is provided with copious indexes (listed below), and these are collected into decennial (to 1956) or more recently quinquennial indexes, known as the Collective Indexes.

The first indexes were Author, Patent Number, Molecular Formula and Subject Indexes. There was a single Collective Molecular Formula Index for the period 1920–46 and a Collective Patent Number Index for 1907–36. Since 1972, the subject index has been divided into a Chemical Substance Index and a General Subject Index. In 1957, the very useful Index of Ring Systems was introduced, and this was followed in 1963 by a Patent Concordance which brought together families of patents issued by different countries (see Chapter 11). For five years (1967–71) there were Heteroatom-in-Context (HAIC) indexes, and from 1969 there was a Registry Number Index; this has now become the Registry Handbook, Number Section. As the literature has grown, these indexes have become very unwieldy in size; indeed, it is alleged that the latest Collective Index (1981–86) is the largest single publication ever produced, running to well over 80 large books and costing a sterling sum well into five figures. There is now serious doubt whether such enormous printed publications can continue; more than half of CAS's revenue now comes from electronic use of its databases (see Chapter 5), and optical media such as CD-ROM must now be seen as obvious candidates for the archival publication of such massive compilations.

When performing a manual search of *CA*, one searches the Collective Indexes (decennial and quinquennial) first for the older material, and then the volume indexes to cover the period since the most recent Collective. Over the period of a Collective Index, there is always some development of terminology; *CA*'s controlled indexing language is not static, but at the end of a Collective Index period, the language will be standardized over the whole index. Any errors will also be corrected. The Collective Index is therefore more than just a compilation of the ten volume indexes.

It should also be mentioned that *CA* is also available on microfilm. This is of particular benefit for libraries that are short of shelf space, as the entire collection of abstracts issues and Collective Indexes published since 1907 is contained on less than 500 cassettes of microfilm. Each cassette contains six to eight weekly issues, so the microfilm collection is never quite as up to date as the printed version.

A further printed service from CAS is *CA Selects*, an extensive series of specialized bulletins printed in the format of *CA* but selected from the database by a computer search. Each bulletin is published fortnightly. This series originated in the *Macroprofile* service from UKCIS (see p. 38), which was taken over by CAS themselves, enriched with the abstract text, typeset and marketed internationally.

Another useful publication from CAS that has already been mentioned in Chapter 2 is the *Source Index (CASSI)*. This lists all the periodicals which are regularly scanned by CAS for inclusion of material in *CA*. It is available in both printed and machine-readable forms, with quarterly updates. In addition to providing a valuable listing of chemical periodicals, it also provides information on frequency, date of first appearance, title changes and so on, and also gives the recommended IUPAC abbreviations for the titles of journals and holdings of the primary journals in major North American and a few European libraries. In chemistry it can probably be regarded as more complete and authoritative than the *World List* and *BUCOP* (p. 21), though these were of course interdisciplinary.

In view of the enormous quantity of literature to be covered, it was to be expected that CAS would be a pioneer in the use of computer technology for handling its data, and this was indeed the case. In the 1960s, CAS received a total of about $15 million from the Federal Government of the USA as development funding to facilitate the computerization of its operations. Initially, this computerization was primarily concerned with the immense sorting operations involved in producing the massive indexes, but it has subsequently extended to every aspect of the operation. In the late 1960s, the computer-readable version of the database began to be made publicly available, and during the 1970s this became the basis of the online information services described in Chapter 5. As computer typesetting developed, CAS also began to do its own typesetting from its in-house database, contracting out only the actual printing and binding. A major aspect of the computerization was the creation of the Registry System (see Chapters 5 and 6). This is a file of the compounds mentioned in the literature, now amounting to about ten million chemical substances. Originally the Registry was seen as an aid to correct indexing of the compounds in-house at CAS, as it provides for unique and unambiguous identification of each compound and the assignment of the correct systematic name. Each different compound is also assigned a unique number, the CAS Registry Number. More recent advances in computer technology have now made it possible to provide a public search service based on this file *(CAS ONLINE)* which enables customers to carry out chemical substructure searches directly on this database.

Arrangement of Individual Issues of *Chemical Abstracts*

Since 1962, the content of *CA* has been subdivided into 80 sections; each abstract is allocated to one of these sections, though cross-references are included in cases where more than one section might be appropriate. The weekly issues are of two types, published alternately: organic chemistry and biochemistry (sections 1–34) one week and physical and applied chemistry (sections 35–80) the other. Each weekly issue contains Keyword, Numerical Patent, Patent Concordance and Author Indexes. Currently sections 1–20 cover biochemistry, 21–34 organic chemistry, 35–46 macromolecular chemistry, 47–64 applied chemistry and chemical engineering, and 65–80 physical and analytical chemistry. Each of these five groupings can be purchased as a separate publication, with the Keyword Index to the complete weekly issue bound in; these subscriptions are relatively inexpensive but do not include any of the sophisticated volume indexes.

The abstracts in each volume (half-year) of *CA* are numbered consecutively; the number also has a computer-generated check letter on the end. This system has applied since the introduction of computer production of the journal in 1967. Prior to that a cumbersome system was used: columns, not pages, were numbered, and columns were lettered downwards (a) to (i) so the number 2149c, say, implied that the abstract started about one-third of the way down column 2149. Before 1947, a superscript number was used instead of the letter, e.g. 2149^3. Unlike the current system, these older systems did not give each abstract a unique identifier.

Within every one of the 80 sections, the abstracts are arranged in the following order.

1. Papers, including reviews, grouped into subclassifications; these subclassifications are not printed, but they are present and searchable in the machine-readable file.

2. Books, monographs, etc., listed by title, author and publisher; standards and theses appear here.

3. Patents.

4. Cross-references by title and abstract number to other sections where relevant abstracts appear.

Methods of using printed *Chemical Abstracts* and its indexes

Chemical Abstracts is such a large publication and its indexing so thorough and complex that one cannot use it effectively without some training and preparation. For current-awareness scanning, the arrangement into 80 sections and the weekly Keyword, Author and Patent Indexes are reasonably simple to understand, but for retrospective searching one has to use the Collective and Volume Indexes which are more structured and use controlled language. Each Collective Index is accompanied by an Index Guide, and for the inexperienced user of *CA* a little time spent with this is valuable (see Chapter 17). It is also useful to read the introduction to the particular index that one is thinking of using. It is always important to remember that the indexer worked from the whole paper, not just the title and abstract, and so the paper may well be retrievable by search terms that do not appear in the printed abstract.

In general, it is usual in performing a retrospective search to start with the most recent volumes and work backwards. Unfortunately, this entails using the indexes in the weekly issues for the very newest period, and then the Volume Indexes for the period before that; as the Collective Indexes take about three years to appear after the end of the five-year period to which they refer, up to 16 individual Volume Indexes may have to be consulted. Only for the earlier period can the more convenient Collective Indexes be used. For this reason, it may be more appropriate to use an online search of the machine-readable database for the recent period, and turn to the Collective Indexes for the period before the machine-readable database became available (before 1967).

It is also important to select the most appropriate index. This will depend on the nature of the query one is trying to answer. Author indexes are probably easiest to use, though even here one has to allow for problems with prefixes such as 'de' and 'von', 'Mc' and Mac', replacement of umlauts with 'ae', 'oe' and 'ue', difficulties in identifying the surname in Chinese names, and so on, all of which can affect the alphabetical position of an author.

A lot of chemical searches will be looking for chemical compounds. The easiest way of doing this is to use the *CAS ONLINE* service (see Chapter 5), but this is not accessible to all and in particular may well not be available to students. The Chemical Substance Indexes use systematic nomenclature, using 'benzeneamine' for 'aniline' for example, and this nomenclature has changed somewhat with each Collective Index period, so the user of this index needs significant chemical nomenclatural expertise. May we again advise the less experienced user to spend some time reading the introductory pages of these indexes before starting to search them. (Other guides to nomenclature are discussed briefly in Chapter 4.)

An easier alternative in many ways is to use the Molecular Formula indexes. Each compound appears here under its empirical formula, arranged according to the Hill (1900) system. The Hill system arranges symbols alphabetically, except that in carbon compounds C always comes first followed immediately by H. Thus organic compounds are arranged C, H, with other elements following alphabetically and with the lower number of atoms preceding the higher, i.e. C before C_2, before C_3, etc. then H before H_2, before H_3, etc. The arrangement of the formulae is in the order carbon + one different element, carbon + two different elements, carbon + three different elements, etc. and then by number of atoms, i.e. C_1 before C_2 before C_3. Isomers are listed alphabetically by their systematic name, so a search in the Formula Index may be the easiest way of discovering that name, which can then be looked up in the Chemical Substances Indexes.

Formula Indexes are useful for fairly simple structure searches, but not for substructure searches, where one would have to search not only for the molecular formula of interest but for every formula containing those atoms and more. In these cases, if it is not possible for financial or other reasons to undertake a *CAS ONLINE* computer search, one will have to get to grips with nomenclature.

If the search is for a topic, rather than a substance, the General Subject Indexes are appropriate; these use a controlled language, so use of the Index Guide to find the appropriate thesaurus terms is necessary. Difficulties tend to arise in new fields, or in interdisciplinary areas, where terminology may be rather fluid.

Abstracting and indexing services of the Royal Society of Chemistry

The earlier involvement of the Chemical Society (CS) with the production of *British Abstracts* and the subsequent development of *Analytical Abstracts* have already been mentioned. In the mid-1960s, the CS decided to become involved with the then-novel activity of computer-based information retrieval in chemistry, and in 1966 it set up the Chemical Society Research Unit in Information Dissemination and Retrieval based at the University of Nottingham. This unit developed software for searching the first computer-readable database products from CAS, *Chemical Titles* and *CA Condensates*, and provided experimental batch current-awareness services to UK chemists. This batch service became commercial in 1969, when the unit became known as the United Kingdom Chemical Information Service (UKCIS), and continues to this day. (A similar service based on the machine-readable version of *Biological Abstracts* and *Bioresearch Index* is also operated.) UKCIS also became the sales agent for the UK for the printed publications of CAS.

In addition to a batch current-awareness service based on CAS databases and providing a selective search service for the individual customer, UKCIS also provided bulletins based on broader search profiles and marketed to multiple customers. These bulletins, called *Macroprofiles*, were replaced by *CA Selects* produced by CAS themselves.

At about the same time, the agreement between the ACS and the CS was extended to include some input to the databases (abstracting and indexing) by Nottingham-based CS staff as well. In the mid-1980s, the *CA* input activity at Nottingham was run down and eventually ceased, but advantage was taken of the staff expertise gained to launch a range of specialized abstracting and indexing publications. By this time the merger of several societies that created the Royal Society of Chemistry (RSC) had taken place, and the publishing and information activities of the RSC are now known collectively as RSC Information Services. In 1989 the Nottingham unit relocated to Cambridge where it is now housed in the same building as the traditional journal and book publishing activities of the RSC, themselves previously in London.

The RSC's own abstracting and indexing publications now include the following periodical titles, some of which have been taken over from other organizations and others founded from scratch. There is no intention of competing with CAS, with whom the RSC has a good working relationship; the RSC publications are intended to supply niche markets, and are mostly quite small.

Chemical Business NewsBase
Analytical Abstracts
Mass Spectrometry Bulletin
Laboratory Hazards Bulletin
Chemical Hazards in Industry
Chemical Engineering Abstracts
Theoretical Chemical Engineering Abstracts
Current Biotechnology Abstracts
Methods in Organic Synthesis
Natural Products Updates

Of these, the first-named is not published in printed form and will be considered in Chapter 12, and the two last-named exist at present only in printed form. *Methods in Organic Synthesis* and *Natural Products Updates* (like *Current Abstracts in Chemistry and Index Chemicus*, see p. 41) emphasize the use of chemical structure diagrams in their abstracts; development work aimed at putting them online complete with the diagrams is in progress. The remaining titles exist both as printed abstracts and indexes mostly appearing monthly and as cumulating online databases available from various online vendors (see Chapter 5).

Analytical Abstracts is the largest of the titles and aims at comprehensive coverage of the analytical chemistry area; it has been published for 35 years and the machine-readable file extends back ten years. Unlike *CA*, it aims to provide sufficient detail of laboratory method in its abstracts that the skilled analytical chemist could use the method without necessarily seeing the original paper. It survived the demise of the rest of *British Abstracts* because analytical chemists were dissatisfied with *CA*'s coverage of analytical chemistry.

An agreement has been signed with DECHEMA of Germany which resulted in *Chemical Engineering Abstracts, Theoretical Chemical Engineering Abstracts* and *Current Biotechnology Abstracts* being merged with corresponding DECHEMA publications in 1991.

RSC Information Services also produces some non-periodical secondary publications, and these also are available in printed as well as machine-readable form. They are:

The Eight-Peak Index of Mass Spectra
The Agrochemicals Handbook
European Directory of Agrochemical Chemicals
World Pesticide Index
Nutrition Databank (McCance and Widdowson)
Hazards in the Chemical Laboratory (Bretherick)

In each of these cases the machine-readable databank is continuously updated while the printed book appears in updated editions at intervals. The last-named book, which is a product of the books department rather than of the former UKCIS, is not yet available in online form but is likely to become available in this form in the future.

Other chemical abstracting services

While no service other than *CA* attempts to provide comprehensive coverage of chemistry, there are several that may be useful to supplement a search of *CA*, especially in certain non-English language areas. Services of adjoining disciplines may also usefully be consulted if the field of search is interdisciplinary or near to the disciplinary boundary. The interdisciplinary *Science Citation Index* also needs to be mentioned in this context.

Referativnyi Zhurnal, Khimiya

Although *CA* makes considerable effort to cover Russian literature, until recently political difficulties had hampered its efforts, and for complete coverage of Russian, mainland Chinese and Eastern European chemistry one also has to consult the national abstracting publication of the former USSR, *Referativnyi Zhurnal*, published since 1953 and in Russian. This is published in sections for the different scientific disciplines, and in addition to *Khimiya* the biological and metallurgical sections may be useful in a chemical search. Each issue has a subject and author index; non-Cyrillic authors' names are of course transliterated into Cyrillic, as well as being present in Roman script if originally so. The annual indexes are author, subject, patent, molecular formula, cyclic compounds and elements (in this last, elements are listed alphabetically and compounds are listed below each element in ascending order of carbon-atom content). British Library publishes a *Guide to the Referativnyi Zhurnal* in English, which lists the sections and gives guidance on transliteration from Cyrillic script.

Bulletin Signalétique

This publication fulfils a similar function for the French language to that of *Referativnyi Zhurnal* for the Russian. However, the political and linguistic barriers being so much smaller between English and French, the *Bulletin Signalétique* has declined in importance in recent years. It is published by the Centre National de la Recherche Scientifique, appears ten times per year, and is divided into sections: Section VII covers chemistry, but other sections relevant to chemistry are: Biochemistry and Biophysics; Atoms, Molecules, Fluids and Plasmas; and Crystallography. The chemistry section is subdivided into: General and Physical Chemistry; Analytical Chemistry; Organic Chemistry, Theory; Organic Chemistry, Preparation and Properties; Molecular Evolution and Origin of Life. The abstracts are indicative rather than informative, and each issue contains author and subject indexes; the language, of course, is French.

Chemischer Informationsdienst

This publication came into existence after the demise of *Chemisches Zentralblatt* in 1969, using some of *CZ*'s former staff. It is intended as a German-language alerting service, rather than an archival retrospective searching tool, and is mainly aimed at the chemical and pharmaceutical industries in German-speaking countries. It combined the *Schnellreferate* (quick reference) section of the old *CZ* (which had been published by the Gesellschaft Deutscher Chemiker) with the Bayer company's *Fortschrittberichte*. These two organizations produce it and it is published by VCH (Verlag Chemie). *Chemischer Informationsdienst* is published weekly, each issue containing 600–700 abstracts; about 350 periodicals are scanned for relevant items. It is divided into five sections: general; physical-inorganic and inorganic chemistry; physical-organic and organic chemistry; applied chemistry; and analytical chemistry. The abstracts concentrate on graphical depiction of reactions, and effort is concentrated on fast publication.

Current Abstracts of Chemistry and Index Chemicus

This journal, published by the Institute for Scientific Information (ISI) in Philadelphia, concentrates on papers that report new compounds, intermediates or methods, selects from a limited list (about 100) of organic chemistry journals that are alleged to contain 90 per cent of significant new compounds, and uses a graphical style of abstract. It claims to note

3500 new compounds per issue.

Each abstract in *CAC&IC* (registered trade mark) consists of the paper's title, bibliographical citation, authors' names, addresses and corporate affiliation, the authors' abstract taken from the original paper, and a number of graphical features: reaction schemes showing the main original steps described in the paper; an 'instrumental data alert' showing which main analytical techniques (e.g. mass spectrometry, nuclear magnetic resonance) were used to characterize the new compounds — the relevant positions on a printed 'wheel' being shaded in; an indication of whether any innovative synthetic method was used; highlighting of the new compounds; indications of any isotopically labelled compounds made and of the biological activity of the new compounds; and an indication of hypothetical intermediates not isolated. All of these items of information appear in a standardized format so that the regular user knows exactly where in the layout of the abstract to find each item.

The indexes provided in each issue of *CAC&IC* are: molecular formulae; authors; keywords; corporate authors; and an index to the 'instrumental data alerts'.

For about 20 years, until 1987, *CAC&IC* was accompanied by a machine-readable database called the *Index Chemicus Registry System (ICRS)* which included the structures of the new compounds in Wiswesser Line Notation (WLN; see Chapter 5), and a printed version of this database was published under the title of *Chemical Substructure Index*. With the greatly increased power of computers now, and the widespread availability of the *CAS ONLINE* service based on connection-table searching, WLN has become an obsolete system and *ICRS* was unable to compete with *CAS ONLINE*, hence its discontinuance. However, the fundamental concept that organic chemists think in terms of structural formulae and reaction schemes is sound; no direct printed equivalent of the CAS Registry System exists, so for the user without access to *CAS ONLINE*, or who wants a smaller file containing only organic chemistry, only from the core journals, and presented in graphical form, *CAC&IC* remains useful. A similar rationale lies behind the RSC's *Methods in Organic Synthesis*, which, though smaller, can be seen as a competitor of *CAC&IC*.

Another companion publication still appearing is *Current Chemical Reactions* (*CCR*, 1979–) which reports novel reactions; the 'new reaction' alert in *CAC&IC* cross-refers to *CCR*. Although *ICRS* has ceased, machine-readable files of the bibliographical material in *CAC&IC* and *CCR* are still available.

Science Citation Index

The Institute for Scientific Information also publishes the *Science Citation Index* (*SCI*) in both printed and machine-readable forms (including a compact disk edition); indeed, it is arguably that organization's best-known and most innovative product. Like ISI's other publications, it makes the assumption that the best science is published in a limited number of periodical titles. This contention is undoubtedly largely true, and Bradford's Law of Scattering (Brookes, 1977) applies to scientific literature; however, the criticism has been made that 'core literature' as defined by ISI tends to discriminate against non-English language literature and, within the English-language material, against non-US titles. CAS avoids such criticism, of course, by seeking to cover everything chemical regardless of language, country of origin, or indeed quality. ISI would contend that CAS's approach is not only impractical on economic grounds for a for-profit organization, but also undesirable as it fills *CA*'s columns with some low-quality material.

SCI covers about 3500 (since 1979) core periodicals covering the whole of science; it includes all of the contents of each journal selected. The most distinctive feature, however, is not its interdisciplinary character, but the fact that it is designed for citation searching. The rationale is that the reference list appended by an author at the end of each paper can be said to define the information content of the paper, at least as accurately as indexing or thesaurus terms allocated to the paper later by someone other than the author. When the author compiles the reference list, he or she is in a sense indexing the paper. Thus the cited references can be used as search terms, just as indexing terms can. So if one knows of a relevant reference from several years ago, one can search for newer papers that have cited it. This is rather easier to do in practice by means of an online search than with the printed indexes; the procedure for use of the printed indexes will, however, now be described. *SCI* has appeared since 1961; it became quarterly in 1964 and bi-monthly in 1979. The final issue of the calendar year is an annual cumulation and therefore replaces the earlier issues of the year. Five-year cumulations are also available. Two decennial issues have retrospectively covered about 500 journals from 1945 to 1964. Whilst they may be interesting for scientometric studies, the period is much better covered by *CA, Beilstein*, etc.

SCI in its printed form comes in three parts, the Citation Index, the Source Index and the Permuterm Index. The last-named, which can be purchased separately, is simply an index of permuted pairs of keywords from the titles of papers, and is thus a very crude subject index; its value (if any) is to help to identify authors working in the field of interest if one does not know of any to use as starting point in the citation search.

The Source Index lists the papers published in the period of the index, indexed by authors' names and giving full bibliographical citations so that the original papers can be retrieved. It enables the searcher to ascertain whether an author known to be active in a particular field has published anything during the period in question. It also has a corporate author index so that papers can be traced if a company or university is known to be active in the field, even if no individual author's name is known. However, no effort is made to standardize names and addresses of organizations, so one must look up, for example, both Birmingham University and the University of Birmingham.

The Citation Index is the heart of the system. It is also an author index, but it lists cited authors, not citing authors. Thus, if a paper by A.N. Author written in 1980 (or even in 1880!) is cited in 1990 by A.N. Other, the name of A.N. Author and an abbreviated form of his 1980 reference will appear in the 1990 Citation Index; under this heading in the alphabetical list will appear (amongst others, perhaps) the name of A.N. Other and his or her 1990 reference. Thus, if one knows that the 1980 paper is relevant to one's query, one can retrieve the 1990 paper that cited it (providing it was published in one of the 3500 journals scanned by *SCI*). The traditional method of using references for literature searching has always been to take a new paper and use its reference list to work *backwards* to older relevant papers; with the *SCI*, one reverses the process and works *forwards* in time from a known older paper to later papers that have cited it. It must be noted, however, that if one's known relevant paper was published in 1980 (say), one has to look up that paper in the Citation Index of every subsequent year down to the present, if an exhaustive search is to be performed. Also, having located a newer relevant paper in the Citation Index, one has to look it up in the Source Index of the appropriate year in order to find its full bibliographical citation, necessary if one wants to obtain a copy of the original paper. Thus the process, though effective, is somewhat long-winded, and its computer-based equivalent is easier and quicker. And of course if A.N. Other, inadvertently or from choice, fails to cite the relevant paper by A.N. Author (1980) or any other of one's starting references, then his or her 1990 paper will not be retrieved.

The *SCI* is particularly useful in interdisciplinary areas, where one is unsure which specialized abstracts journal to use, and in areas where terminology is imprecise or rapidly changing, where the use of citations rather than index terms may be valuable. It has also been extensively used by its publisher, Eugene Garfield, for analysing the structure of science and the scientific community through its networks of cross-citation; one can demonstrate clusters of scientists who cite each other a lot but outsiders little, for example, or measure the strength of linkage between different subjects by their citation connectivities. Somewhat more controversially, it is alleged that one can measure the quality of a journal by the number of citations it receives from, and gives to, other journals, as an 'objective' measure of journal pecking orders (see Chapter 2). A method of using the *SCI* to obtain a list of the most relevant journals for a particular scientist was described by Bottle (1969) in the second edition of this book. The scientist is asked for a list of 10–12 key references, and these are then used to search the *SCI*; a ranking list is then produced of the journals yielding the relevant hits retrieved.

The journal pecking orders produced by this type of analysis for different fields of science can be very useful for a new worker in the field, or for an information worker performing a search for a client outside his or her original field of expertise. Acquisitions librarians can also use them for assessing the value of journals when deciding on new subscriptions or (more commonly nowadays) on journals to cancel.

It is similarly alleged that the quality of a scientist's work, or even of that of an institution, can be measured not just by the weight of his or her own reprints, but more accurately by the quantity of citations received from other workers in their papers. However, the most heavily cited papers tend to be those which describe standard analytical or other techniques.

SCI has also demonstrated that most citations are to quite new papers; over 50 per cent are to papers published in the preceding six years. This is likely to be rather less true of chemical papers than of those in some of the other fields covered by the *SCI*; chemistry is unusual among sciences in that quite old work can often be quite relevant to new research.

Services of Derwent Publications Limited

Derwent Publications Limited, a London company now part of the Thomson Organization, has for a considerable period been involved in producing specialized information services for the chemical and pharmaceutical industries. They have a heavy emphasis on patent information services (see Chapter 11), but also provide a journal abstracting service called *Ringdoc* (see Chapter 14). The name arises from its origins as a product of the Pharma Dokumentationsring, a co-operative operation of a number of German and Swiss companies; Derwent eventually took over the production of the abstracts journal from the Ring. *Ringdoc* staff scan over 600 journals to abstract over 60 000 papers per year, but like ISI they have a philosophy of selection for quality rather than comprehensiveness. As well as the complete weekly journal, specialized subset journals are also printed. Cumulative author and subject indexes are provided. In addition, terms are highlighted in the text of the abstracts themselves to facilitate scanning, by the use of a thesaurus of terms for what Derwent calls 'codeless scanning'. The chemical substances were coded by the Ring Code (a type of molecular fragment or functional group code) for structure searching using a punched-card method, but with the advent of computer-based searching a connection-table system is now used (see Chapter 6). The machine-readable version of the file is now available on the online networks, but Derwent has always operated on the basis of serving a relatively small number of large companies, and its charging system therefore involves a large annual base subscription with relatively low charges for online service use or for the purchase of individual abstracts journals.

Abstracts journals in adjoining subject areas

These are described in detail in companion volumes in this series, and will be described only briefly here.

Biological Abstracts (fortnightly) and the accompanying monthly *Bioresearch Index* (covering non-serial publications) are published by BIOSIS, an US not-for-profit organization, and cover mainly biology, but border areas such as physiological chemistry and pharmacology are included. It has half-yearly indexes of authors, biological taxa, and keywords taken from the title and 'enrichment terms', and two other more specialized indexes: the Biosystematic Index which groups abstract numbers by broad categories, and the Cross Index which lists them under 650 fixed headings. The machine-readable version is carried by several online hosts, and a batch current-awareness service based upon it is operated in the UK by RSC Information Services alongside their corresponding *CA* service.

Index Medicus, an index journal without abstracts, has been published by the National Library of Medicine (NLM) of the USA since 1879; its machine-readable equivalent, MEDLINE, has been available since 1963. The printed version includes less information than the online version. The printed *Index* uses a controlled thesaurus of headings, *MeSH (Medical Subject Headings)*, and each item of literature is listed three times under the three most appropriate *MeSH* headings, whereas in MEDLINE as many index terms as necessary are assigned to each item. The NLM also produces three more specialized online databases, TOXLINE, CHEMLINE and CANCERLINE, but as these have no corresponding printed publications they are discussed in Chapter 5.

Engineering Index is another large American inter-disciplinary abstracting service; it has appeared monthly since 1884, and covers 1400 technology journals, mainly in English, and may be useful for chemical engineering papers, especially in earlier years when *CA* did not cover this area so thoroughly as it does today. *Science Abstracts*, a UK publication produced by the Institute of Electrical Engineers, is the printed version of their well-known INSPEC database; despite its very general title, the only area of pure science it covers is physics *(Physics Abstracts)* which may sometimes include areas of chemical physics inadequately covered by the chemical abstracting publications.

Current Contents and *Chemical Titles*

Hardly abstracts or indexing journals in the proper sense, these two periodicals are only title-lists but their value lies in their rapid publication after the appearance of the primary literature to which they refer. *Current Contents* is published by ISI in several editions covering different subject areas, including *Current Contents, Physical, Chemical and Earth Sciences*, and simply reproduces photographically the contents pages of the primary journals. It also has a simple keyword index in each issue, based of course on keywords from the title only, plus author and address indexes. *Current Contents, Life Sciences* also contains relevant material in the biochemistry area. In May 1991, *Current Contents on Diskette with Abstracts* was introduced. Though expensive, it may eventually make *Current Contents* of some use for retrospective searching.

Chemical Titles is a CAS publication, and is a computer-produced title list; indeed, it was CAS's first computer-produced publication, dating from 1961. It is based on the contents lists of the core journals of chemistry, which yield about half the contents of *CA* itself. As it does not have to wait for the lengthy abstracting and indexing process, it is much more up to date than *CA*. It contains a keyword-in-context index to the titles, an author index, and a bibliographic list of the papers included. It is published fortnightly, and a machine-readable file is also marketed.

References

Anon. (1967), *CAS Today: 60th Anniversary Edition*. Columbus, Ohio: Chemical Abstracts Service.
Bottle, R.T. (1969), 'Abstracting and information retrieval services,' Chapter 5 in *Use of Chemical Literature*, 2nd edn, ed. R.T. Bottle. London: Butterworth, pp. 71–72.
Brookes, B.C. (1977), *Journal of Documentation*, **33**, 180.
Hill, E.A. (1900), *Journal of the American Chemical Society*, **22**, 478.

CHAPTER FOUR

Books, reviews, dictionaries and encyclopaedias

J.M. SWEENEY

Reference works and summaries of the literature such as encyclopaedias, dictionaries and reviews are valuable sources of background information. They are required reading when it is necessary to obtain familiarity with a new subject or even one on the periphery of one's main interests. Equally, many works of this kind are essential when the answer to a simple question is needed, e.g. the definition of a chemical term. For this sort of enquiry, or indeed for general (i.e. non-specialist) background reading, encyclopaedias and dictionaries are the best publications to use and some at least will be available even in small public or industrial libraries. At a rather more advanced level are books or monographs, the latter title being usually reserved for those books of a particularly specialized nature written by an expert in the field. Allied to monographs are reviews, which can be for the non-specialist but which increasingly are of a highly specialized nature and which critically evaluate the literature rather than merely summarize it. Reviews will, of course, give comprehensive bibliographies but the latter can also appear on their own and can be a useful alternative to a literature search.

Other, and sometimes neglected, forms of literature can provide helpful state-of-the-art surveys. Theses, for example, should provide a survey of previous work on their particular subject and are described in Chapter 2. Conference proceedings can also be useful for background information as many conferences summarize the latest position in their field.

Apart from theses, all the subjects mentioned above, i.e. encyclopaedias, dictionaries, monographs, reviews, bibliographies and conference proceedings, are described in this chapter. In order to follow up the information they give it is often essential to contact the organizations and people currently in the forefront of research and the reader is referred to Chapter 17 for details of how to keep up to date. There are also short sections on nomenclature and the history of chemistry at the end of this chapter.

Encyclopaedias

Encyclopaedias are the best source of background information on an unfamiliar subject but care should be taken to check their currency. For example, in the *Encyclopaedia Britannica* article on the structure of molecules in the 15th edition (1986) there are no references to works later than the early 1970s. Similarly, in the article on the chemical elements the latest reference is to 1981 work. It is therefore reasonable to assume that this article was last revised about 1982 or 1983. Those encyclopaedias useful to the chemist fall into three groups: general ones, scientific and technical encyclopaedias and those devoted specifically to chemistry. Each group will be considered in order.

General encyclopaedias are dominated by the *Encyclopaedia Britannica* (15th edn, 1986, reprinted 1990) which, apart from the excellence of its articles such as those mentioned above, is quite frequently the only encyclopaedia available in small reference libraries. It is now split into four parts: the *Macropaedia* (17 vols) carrying in-depth articles, the *Micropaedia* (12 vols) with short entries, the one-volume *Propaedia* or general guide and the *Index* (2 vols). The *Micropaedia* is a handy quick-enquiry source but lacks references (except to the *Macropaedia*). The *Britannica* also produces a *Science and the Future Year Book* which has a mixture of review articles (e.g. including pesticides and zeolites in the 1987 yearbook) and shorter entries recording the latest scientific advances. Many other general encyclopaedias exist but it is unlikely that any will be found as useful as the *Britannica*.

Among scientific and technical encyclopaedias the most important is the McGraw-Hill *Encyclopedia of Science and Technology* (20 vols, 6th edn, 1987). Like *Thorpe*, but unlike *Kirk-Othmer* or *Ullmann* (see below) the articles are mainly short entries rather than comprehensive reviews. It is therefore particularly useful for quick reference purposes, e.g. for properties of some individual lanthanide elements which do not have separate entries in *Kirk-Othmer* or *Ullmann*. References are given after some entries. As with the *Britannica*, there are yearbooks which have a similar format. A recently published scientific and technical encyclopaedia is the *Encyclopedia of Physical Science and Technology* (15 vols, Academic, 1987) which includes chemistry in its coverage.

Materials and Technology (8 vols, Longmans 1968–75) covers a rather narrower field but in great depth. It is based on the sixth edition of *Warenkennis en Technologie* (J.W. van Oss, ed., De Bussey, Amsterdam). The arrangement is not alphabetical but instead each volume is devoted to particular subjects as follows:
   1. Air, Water, Inorganic Chemicals and Nucleonics, 1968
   2. Non-metallic Minerals and Rocks, 1971
   3. Metals and Ores, 1970
   4. Petroleum and Organic Chemicals, 1972

5. Natural Organic Materials and Related Synthetic Products, 1972
6. Wood, Paper, Textiles, Plastics and Photographic Materials, 1973
7. Vegetable Food Products and Luxuries, 1975
8. Edible Oils and Fats and Animal Food Products; General Index, 1975

More up-to-date is a recent treatise from Pergamon, the *Encyclopedia of Materials Science and Engineering* (8 vols, 1986) which covers all major classes of materials including composites. Supplementary volumes are being produced.

Among shorter encyclopaedias of science and technology are the one-volume *Concise Encyclopedia of Science and Technology* (2nd edn, McGraw-Hill, 1989), many of whose articles are shorter versions of those in the main encyclopaedia mentioned above, and the two-volume *Van Nostrand's Scientific Encyclopedia* (6th edn, 1983). This carries references although few are to articles published after 1980.

Chemistry and chemical technology proper are well served with encyclopaedias. The *Encyclopedia of Chemical Technology* (Wiley) is the best-known and is usually referred to as *Kirk-Othmer* after the editors of the first edition. The third edition appeared from 1978 to 1984 and has 26 volumes including an index volume and a supplementary volume. The title is somewhat unfortunate as much more chemistry is covered than it would lead one to expect. Articles are substantial, often over 20 pages in length, and have comprehensive bibliographies which in many cases include patents. Many references date from only one year before the articles which cite them. *Kirk-Othmer* is a useful source of physical property data although care must be exercized in using these as they are not always critically evaluated. The index gives CAS Registry Numbers against entries for individual compounds and is one of the best sources in which to find these numbers quickly.

The fourth edition of *Kirk-Othmer* has recently commenced publication. However, *Kirk-Othmer* is also available online (via Dialog, etc.) and the online version is constantly updated. Searches can be made on several parts of the entries besides the text, for example tables, figure titles, references and authors. The database includes the *Encyclopedia of Polymer Science and Engineering*. See Chapter 5 for more details.

More up to date than the latest complete printed version of *Kirk-Othmer* is the fifth edition of *Ullmann's Encyclopedia of Industrial Chemistry* (1985–, VCH). *Ullmann* was first published in 1914 and the latest edition is the first to appear in English rather than German. As is the case with *Kirk-Othmer* the name fails to indicate the considerable coverage of pure as well as applied chemistry. The encyclopaedia is in two parts: part A, in alphabetical sequence which will comprise 28 volumes and part B, an eight-volume section devoted to fundamental techniques as follows:

B1: Chemical Engineering Fundamentals

52 Books, reviews, dictionaries and encyclopaedias

B2 and B3: Unit Operations
B4: Chemical Reaction Engineering and Materials Science
B5 and B6: Analytical Methods
B7 and B8: Environmental Protection and Plant Safety

Publication of the B series has now commenced. The A series commenced publication in 1985 and the entire fifth edition should be complete by about 1995–96.

Articles are similar in length and coverage to *Kirk-Othmer* and references are likewise given up to one year before publication date. It is possible that *Ullmann* will eventually be available online although not until most of the fifth edition has appeared. The fourth, German language, edition of *Ullmann* (1972–1984, 25 vols) consists of six volumes on fundamental techniques with the remaining volumes being an alphabetical sequence of articles.

The *Encyclopedia of Chemical Processing and Design* (ed. J.J. McKetta, Dekker, 1976–) is biased towards chemical engineering. Volumes are appearing at a very slow rate and by 1991 had reached letter P in volume 38. At such a rate it is unlikely to be completed in the near future. An older but still very useful treatise is *Thorpe's Dictionary of Applied Chemistry* (12 vols, 4th edn, Longmans, Green, 1937–56). This, despite its title, is an encyclopaedia not a dictionary, with articles ranging in length from a paragraph to several pages. Pure as well as applied chemistry is covered. *Thorpe* is available in many small libraries which may lack other encyclopaedias mentioned in this chapter.

A German six-volume encyclopaedia is *Römpps Chemie Lexikon* (revised by O.A. Neumüller, 9th edn, Franckh'sche Verlag, Stuttgart, 1988–9). It has the useful feature of giving references to other chemistry encyclopaedias such as *Beilstein*, *Gmelin*, *Kirk-Othmer* and *Ullmann*.

The advent of CD-ROM has meant that an additional storage medium for works such as the large encyclopaedias is now possible. The *McGraw-Hill* and *Kirk-Othmer* encyclopaedias are already available on CD-ROM and similar titles will doubtless follow.

Besides multi-volume encyclopaedias there are a number of single-volume works. The *Kirk-Othmer Concise Encyclopedia of Chemical Technology* (Wiley, 1986) is simply a condensed version of the third edition of *Kirk-Othmer* proper and contains shorter versions of the same articles with some references, data, etc. Another single-volume chemical encyclopaedia is Van Nostrand Reinhold's *Encyclopedia of Chemistry* (D.M. Considine, ed.-in-chief, 4th edn, 1984). The *Cambridge Illustrated Thesaurus of Chemistry* (A. Godman and R. Denny, CUP, 1985) groups its definitions under topics and is useful if one's chemistry is a little rusty.

Dictionaries

Dictionaries (i.e. those publications giving definitions of chemical terms and/or lists of chemical compounds) can be divided into those covering science and technology as a whole and those covering chemistry in particular. Those dictionaries covering particular classes of compound, e.g. Heilbron's *Dictionary of Organic Compounds*, are described in the appropriate chapters of this book.

A comprehensive general dictionary is the *McGraw-Hill Dictionary of Scientific and Technical Terms* (4th edn, 1989) which includes a large number of chemical definitions. Over 9000 such terms have been selected and published separately as the *McGraw-Hill Dictionary of Chemistry* (1984). The *McGraw-Hill Dictionary of Science and Engineering* (1984) is another shorter version of the main dictionary, still covering all scientific and technical disciplines but with fewer terms. *Chambers' Science and Technology Dictionary* (rev. edn, 1988) is a new version of one of the most useful technical dictionaries.

There are many chemical dictionaries proper of which several are lists of chemical compounds, usually with some property data (often qualitative) and frequently with further information such as method of preparation, uses, etc. The *Condensed Chemical Dictionary* (rev. G.G. Hawley, 11th edn, Van Nostrand Reinhold, 1987) lists properties (including hazards), technical grades, suitable containers and uses for many thousands of inorganic and organic chemicals. It also very usefully includes trade names (mainly US) together with the addresses of the manufacturers. The *Concise Chemical and Technical Dictionary* (H. Bennett, ed., rev. edn, Edward Arnold, 1986) lists more compounds but the information under each is confined to properties. Although it includes trade names it gives no details of their manufacturers. Another quite recently revised work is *Grant & Hackh's Chemical Dictionary* (5th edn, rev. R. & C. Grant, McGraw-Hill, 1987). Although this contains over 55 000 generic and trade names the information under individual entries is limited. A dictionary with a rather different bias is *Gardner's Chemical Synonyms and Trade Names* (J. Pearce, ed., 9th edn, Gower, 1987). This includes both UK and US trade names together with addresses of the manufacturers. Finally, mention must be made of the extremely useful *Merck Index: An Encyclopedia of Chemicals, Drugs and Biologicals* (12th edn, 1989) which lists over 10 000 compounds with a bias to, but not exclusively dealing with, pharmaceuticals. Information for each compound is detailed and generally includes preparation, properties and toxic effects. Relevant patents are often mentioned. There is also a section on named organic reactions and various tables of data. A cross-reference name index includes generic and trade names. The *Merck Index* is now available online on Dialog, CIS, Questel, etc.

Turning to dictionaries of chemical terms one finds that these will of course contain entries for chemicals but in less detail than the dictionaries mentioned above. The *McGraw-Hill Dictionary of Chemistry* has already been mentioned. *Miall's Dictionary of Chemistry* (D.W.A. Sharp, ed., 5th edn, Longmans, 1981) contains biographic details among other entries. Other dictionaries include the *Dictionary of Chemistry* (J. Dainith, ed., Harper & Row, 1981), *Macmillan Dictionary of Chemistry* (D. Hibbert and A.M. James, 1987), the *Concise Dictionary of Chemistry* (OUP, 1985) and the *Glossary of Chemical Terms* (C.A. Hampel and G.G. Hawley, 2nd edn, Van Nostrand Reinhold, 1982). For those interested in the derivations of chemical terms there is the *Concise Etymological Dictionary of Chemistry* (S.C. Bevan, S.J. Gregg and A. Rosseinsky, Applied Science, 1976). Somewhat different from the above dictionaries is the *Compendium of Chemical Terminology* (V. Gold et al., Blackwells, 1987). This gives IUPAC recommended definitions with references to IUPAC source documents (the majority of these being articles in *Pure and Applied Chemistry*). A *Dictionary of Named Effects and Laws in Chemistry, Physics and Mathematics* (D.W.G. Ballentyne and D.R. Lovett, 4th edn, Chapman & Hall, 1980) is useful for definitions of organic named reactions and of laws and equations in physical chemistry.

Pharmacopoeias offer sources of information on properties, official testing methods, analyses, etc. of pharmaceuticals. They are produced by various countries, the UK ones being the *Pharmaceutical Codex*, formerly the *British Pharmaceutical Codex* (11th edn, Pharmaceutical Press, 1979) and the *British Pharmacopoeia* (2 vols, HMSO, 1988). The loose-leaf *European Pharmacopoeia* (2nd edn, Maisonneuve, 1980–) is constantly updated. In the US the standard works are the *United States Pharmacopoeia/National Formulary* which are published in a single volume (22nd revision and 17th edn, respectively, American Pharmaceutical Association, 1990). Both the UK and US pharmacopoeias are regularly updated with addenda and supplements respectively. *Martindale's Extra Pharmacopoeia* (29th edn, Pharmaceutical Press, 1989) is similarly useful, particularly for information on proprietary products, as is the *British National Formulary* (British Medical Association, 1983). *Martindale* is now quite widely used in its online version. See Chapter 14 for other pharmacopoeias, excipient handbooks, etc.

A good guide to sources of natural products is W. Karrer's *Konstitution und Vorkommen der organischen Pflanzenstoffe* (2nd edn, Birkhauser, 1976–85). It has a compound, formula and plant index but specifically excludes alkaloids.

Much of the industrial applications of chemicals is concerned with formulation problems. Apart from the advice available from manufacturers and their trade literature, an invaluable collection of recipes for most conceivable products from adhesives to varnishes, bubble gum to liqueurs, as well as the nostrums covered by the pharmacopoeias, etc. is contained in H. Bennett's *Chemical Formulary* (20 vols, Chemical Publishing, 1933–74), for which a cumulated index of Vols 1–10 is available. One must, of course, check current legislation to see if they are permitted, as 'Bennett' contains some recipes which are not.

A vast amount of information is provided in the *Colour Index* compiled by the Society of Dyers and Colourists (7 vols, 3rd edn, 2nd rev., 1982). Each dye or pigment is assigned a Colour Index number (according to its chemical type) which is often quoted to avoid ambiguity. In volumes 1–3 dyes are arranged according to use and information on applications, fastness, general data and trade names is given. Volume 4 gives structural formulae (where available), solubility information and literature references to preparations. Information on intermediates, reducing agents, brighteners, etc. is also given, and there is a trade name index and list of manufacturers in volume 5. Volumes 6 and 7 are updates of volumes 1–4. A separate volume on pigments and solvent dyes compiled from information in volumes 2–7 was published in 1982.

Monographs

A good monograph, written by an expert in the field, can be a valuable guide to the development and present state of the subject that it covers. Many monographs are issued as members of series which are published both by learned societies and by commercial publishers. The American Chemical Society produces two important series, the *Advances in Chemistry* series and the *ACS Monographs* series. Quite frequently individual monographs in these series appear in subsequent editions, for example *Chemical Carcinogens* (2nd edn, C.E. Searle, ed., 1984: ACS Monograph 182). Other titles in these series are *The Electrochemistry of Biomass and Derived Materials* (H. Li Chum and M.M. Baizer, 1985: ACS Monograph 183), *Rubber-Toughened Plastics* (C.K. Riew, ed., 1989, *Advances in Chemistry* 222) and *Silicon-based Polymer Science: A Comprehensive Resource* (J.M. Zeigler and F.M. Gordon Feardon, eds, 1990: *Advances in Chemistry* 224). Also from the American Chemical Society is the *ACS Symposium* series which is complemented in the UK by the Royal Society of Chemistry's *Special Publications* series, each issue of which consists of proceedings of a conference organized by the Society. Included in this series are the BOC Priestley Conferences which are devoted to aspects of oxygen chemistry and technology in commemoration of Priestley's discovery of the gas in 1774. The third, held in 1983, the 250th anniversary of Priestley's birth, was on *Oxygen and the Conversion of Future Feedstocks* (Special Publication no. 48). The proceedings of the fourth conference, *Membranes in Gas Separation and Enrichment*, were published in 1987

(Special Publication no. 62). The RSC also produces the *RSC Paperbacks* series in which important topics are dealt with at a less advanced level.

Monograph series are also produced by a number of commercial publishers of which a typical series is *Lecture Notes in Chemistry* from Springer. This has a bias towards theoretical and physical chemistry, including such titles as *Synthon Model of Organic Chemistry and Synthesis Design* (J. Koca *et al.*, 1989, no. 51) and *Many-body Methods in Quantum Chemistry* (U. Kaldor, ed., 1989, no. 52). Many other series are produced by Elsevier, Pergamon, Butterworths, McGraw-Hill, Wiley, Academic Press and others. These are often devoted to particular areas of chemistry such as Elsevier's *Studies in Organic Chemistry* and *Journal of Chromatography Library*.

As is the case with encyclopaedias, monographs can become out of date rather quickly. While this is not too great a problem in areas where there is little development, it can be serious in rapidly expanding fields such as organometallic chemistry or composite technology. In such cases it is unwise to rely on any treatise more than five years old as in that time there may have been advances in the underlying theory, the experimental techniques used or even drastic revisions of the data relevant to the subject. One can only endorse the opinion given in the third edition of this book, namely that journals such as *Chemistry in Britain* and *Journal of Chemical Education* could carry annotated lists of selected current monographs by experts in the field surveyed. Those areas of chemistry or chemical technology in which rapid advances are being made could be surveyed every few years, others less frequently.

Reviews of monographs are of course an important feature of *Chemistry in Britain, Journal of Chemical Education, Angewandte Chemie, Nature, Chemistry and Industry* and many specialist journals. Unfortunately, these reviews often appear a considerable time after the monographs have been published although the same journals often carry advertisements from publishers giving details of their forthcoming publications. Reviews could be traced through *Technical Book Review Index* which appears monthly and which contains abstracts of reviews from US and UK journals, but it ceased publication in 1987. An author index for the year appeared in each December issue. Book reviews can also be traced through the *Science Citation Index* (see Chapter 3) if the author is known.

Besides the advertisements in the journals mentioned above books and monographs on suitable subjects can be traced through *Subject Guide to Books in Print, British National Bibliography* or any of the other standard guides mentioned in Chapter 1. It should also not be forgotten that books are noted (although generally not abstracted) towards the end of the subject sections in each issue of *Chemical Abstracts*. In the *CA* indexes, abstract numbers which refer to books are distinguished by a B prefix, and they can also be searched separately online. Subject catalogues of large libraries are also worth perusing, and in the case of the Science Reference and Information Service can be searched online by subject through *BLAISELINE*.

Finally, it should be noted that each September issue of the *Journal of Chemical Education* carries a list of new books, mainly US ones.

Reviews

While monographs can be useful in giving a comprehensive survey of the area in question they can, as mentioned above, suffer from a lack of currency. Review articles, compiled by an authority in the subject who moreover will have a clear picture of research on that topic currently in progress around the world, can overcome this problem to some extent simply by being a more flexible format. They can be published more quickly than monographs so tend to be less out of date. Indeed, the existence of a good recent review is an excellent aid when making an exhaustive literature search on a topic. Even if a review (or indeed a monograph) is some years out of date later work can be traced from the references given via the *Science Citation Index* (see Chapter 3).

Many journals which publish articles on original work also carry reviews from time to time. Some, such as *Angewandte Chemie*, *Experientia* and the *Journal of Materials Science*, have one or more review articles in each issue. However, there are a number of journals devoted solely to reviews. The best of these journals contain reviews that present a critical and comprehensive evaluation by an acknowledged expert in the field together with a good bibliography. Even good reviews, however, often fail to include patent literature, a particularly serious omission in any field characterized by rapid technical advances. Grey literature (reports, conference proceedings, etc.) is also treated less thoroughly than it should be. Another omission tends to be work in non-English languages such as Russian, Japanese and Chinese.

Reviews are also becoming increasingly specialized. This is inevitable with the rapid growth of chemistry which results in the literature of even a fairly narrow area (such as the chemistry of individual elements) being too large for an overall comprehensive review. Monographs can be useful for these broader subjects and it is worth remembering that the articles on more general topics in encyclopaedias such as *Kirk-Othmer* and *Ullmann* can be regarded as review-type treatments and often have excellent bibliographies.

Many review series are produced by learned societies and the Royal Society of Chemistry publishes one which retains the name of one of its predecessors, the *Chemical Society Reviews* (1972–). This quarterly publication is the successor to the *Quarterly Reviews of the Chemical Society* and the *RIC Reviews*. Each issue carries about four reviews,

58 Books, reviews, dictionaries and encyclopaedias

some of them over 40 pages long. The equivalent publication from the American Chemical Society is *Chemical Reviews*, published every two months. The type of reviews is similar to those in the UK publication although there are generally more per issue. *Chemical Reviews* also contains the contents pages of another ACS journal, *Accounts of Chemical Research*, which publishes shorter reviews of research areas currently under investigation. It also publishes some reviews of areas where the author(s) attempts to clarify subjects that have grown too fast and have become muddled.

Commercial publishers are also sources of many review series. Typical is *Survey of Progress in Chemistry* (Academic) which contains about four lengthy reviews per issue, i.e. it does not attempt to cover the whole of chemistry as do the *RSC Annual Reports* (see below). However, no volumes have been published since 1983. Another series is *Topics in Current Chemistry* from Springer. There are usually several volumes per year in this series. Two recent volumes are *Transition Metal Coordination Chemistry* (no. 160, 1992) and *Macrocycles* (no. 161, 1992).

There are some useful reviews from the former Soviet Union which often cover sources neglected by review articles published in the West. *Russian Chemical Reviews* is a cover-to-cover translation of *Uspekhi Khimii*. It is published monthly by the British Library Document Supply Centre with the cooperation of the Royal Society of Chemistry. Publication is usually about six months after the appearance of the original Russian work. The reviews tend to be shorter than those in *Chemical Society Reviews* or *Chemical Reviews*. Also from Russia is *Soviet Scientific Reviews* of which section B is entitled *Chemistry Reviews*. Volumes appear approximately once a year. Recent (1991) issues have included reviews on paired polymers and on synthetic aspects of the fluorination of organic compounds. The reviews are published under the auspices of the Russian Academy of Sciences and distributed by Harwood Academic Publishers.

Rather different from the above review periodicals are the *Annual Reports of the Progress of Chemistry* published by the Royal Society of Chemistry and its predecessor since 1904. The intention is to provide authoritative reviews of the whole of chemistry for the year in question. There are now three sections: *A, Inorganic Chemistry; B, Organic Chemistry;* and *C, Physical Chemistry*. Parts A and B have chapters corresponding in most cases to chemical groups whereas part C reviews particular topics (e.g. nuclear magnetic resonance spectroscopy) every few years. The volumes are usually published about one year after the year that has been reviewed. The bias is towards pure chemistry, and although large numbers of references are given, few are to patents, conferences or other non-journal literature. Furthermore, the reports are written for specialists and quite detailed subject knowledge is assumed.

Books, reviews, dictionaries and encyclopaedias 59

There is a collective index to volumes 1–46 (1904–49) and each volume has an author index. The increased amount of literature to be reported on has meant that the task of the contributors of the *Annual Reports* has become steadily more difficult and the coverage more selective. In order to overcome this difficulty the then Chemical Society introduced its *Specialist Periodical Reports* in 1967. These cover particular topics and appear either annually or biennially. Over 30 topics were dealt with including *Organophosphorus Chemistry, Spectroscopic Properties of Inorganic and Organometallic Compounds* and *Mass Spectrometry*. The topics can be reviewed in greater detail than is possible with the *Annual Reports*. Several titles in this series have now ceased publication, and others are erratic in appearing.

Applied chemistry was covered by the *Reports on the Progress of Applied Chemistry* (1916–76, Society of Chemical Industry) but these have now ceased publication. John Wiley do, however, publish for the Society of Chemical Industry the series *Critical Reports on Applied Chemistry*. About 30 titles have appeared so far, recent ones being *Chemotherapy of Tropical Diseases* (1987, vol. 21), *Thermotropic Liquid Crystals* (1987, vol. 22) and *Chemistry with Ultrasound* (1990, vol. 28). Most volumes are about 100–150 pages in length. Good reviews on applied chemistry can also be found in periodicals such as *Processing, Process Engineering* and *Manufacturing Chemist*, as well as in periodicals devoted to particular industries.

The *International Review of Science* (Butterworths, 1976) was a successor to the *MTP International Review of Science*. It provided a comprehensive review of selected topics in chemistry, although the total number of topics covered was about 100, most of them fairly broad fields (e.g. boron hydrides). This series has now been discontinued. There are however many series having titles such as *Advances in...* or *Progress in...* which will provide many excellent reviews. Many of these titles are included in the *List of Annual Reviews of Progress in Science and Technology* (2nd edn, Unesco, 1969) which, although now out of date, gives some idea of the range of titles available. One publisher which perhaps deserves special mention is CRC Press: it has produced a series of quarterly review titles such as *CRC Critical Reviews in Analytical Chemistry* (1970–) and *CRC Critical Reviews in Solid State and Materials Sciences* (1971–).

The best way to trace a review on a particular subject is through *Chemical Abstracts*. Reviews in *CA* can also be searched online. The Institute for Scientific Information publishes the *Index to Scientific Reviews*. This multidisciplinary work retrieves all items in 225 review serials but also (a) all items in the *Science Citation Index* database with words such as 'Advances', 'Progress', etc. in their titles, (b) all articles with 50 or more references and (c) all articles coded R for review or bibliography. It appears twice a year and has yet to be cumulated. The print size, although larger than in the *Science Citation Index*, is still rather small. *CA Reviews Index*, an index to all the reviews in *Chemical Abstracts*, unfortunately ceased publication in 1982, having started in 1975.

Some areas of chemistry have had their own review guides. Organic chemistry for example was served by the *Index of Reviews in Organic Chemistry*, which was started at ICI by D.A. Lewis and which was taken over by the Royal Society of Chemistry. The last issue covered the year 1985. A similar guide, *Index of Reviews in Analytical Chemistry*, also from the RSC, managed only two issues (1982–3) before it ceased publication.

Bibliographies

A good recent bibliography on a key subject is a valuable aid which will save a great deal of time searching the literature. Reviews, encyclopaedia articles, etc. will of course include bibliographies but many other bibliographies appear either on their own or as parts of series. Prominent among the latter are those on physical property data published by the National Institute of Standards and Technology as part of their *National Standard Reference Data Series;* see Chapter 7.

The remarks made above about encyclopaedias, reviews, etc. on such topics as currency and coverage of non-journal literature apply equally to bibliographies. Two publications are of particular importance in tracing down bibliographies. *A World Bibliography of Bibliographies* (T. Besterman, 4th edn, Societas Bibliographica, Lausanne, 1966) is the most comprehensive guide to bibliographies and covers all subjects. The arrangement is alphabetical by subject. The section on chemistry contains few references to post Second World War material and is best used for historical research. There is a supplement covering material published in 1964–74 (A.F. Toomey, Rowman and Littlefield) which reproduces Library of Congress printed catalogue cards. Of greater use for recent material is the *Bibliographic Index* (H.W. Wilson), published three times a year, the last being an annual cumulation. About 2600 periodicals are scanned and all bibliographies with 50 or more citations are included. Most of the chemistry references are to the latest volumes of the *Advances in...* type of reviews. The Wilson company's databases are also available online (*WILSONLINE ORBIT*) and on CD-ROM (*WILSONDISC*).

Conferences, symposia and meetings

Meetings of all sorts provide chemists with the opportunity of hearing directly from their peers about the latest advances in their field. Conferences vary of course in quality and sometimes one suspects that certain papers are given merely to give the speaker an opportunity to claim expenses so that he/she can attend the rest of the conference. However, many papers given at conferences are by distinguished experts who present either new work or a state-of-the-art review. Unfortunately, most people who could benefit from a conference will be unable to attend in person. It is therefore important to have the proceedings of the conference published, including not only the lectures but the subsequent discussions which often provide valuable ancillary insights and information. However, conference proceedings are notoriously difficult to track down even if they are published. One particular problem concerns prestigious international conferences whose content is especially valuable. These are generally held as series with conferences taking place at intervals of one to about four or five years. Unfortunately each conference is usually held in a different country and the proceedings are entrusted to different publishers. The acquisition of a full set of proceedings by a library is consequently difficult. In general, proceedings may be published a considerable time after the conference took place. Sometimes preprints of papers are available.

The American Chemical Society holds half-yearly meetings which are attended by many thousands of people, there being a number of parallel sessions. Abstracts of the papers are available and contain author and keyword subject indexes. The schedule of papers is published in *Chemical and Engineering News* prior to the meeting. Some of the papers will eventually be published in full in the appropriate ACS journals or in other ACS publications. Many of these symposia appear in the *ACS Symposium Series*.

Perhaps because of the difficulties involved in their identification and acquisition, conference proceedings fall into the class of grey literature which tends to be ignored by authors of research papers, reviews, etc. There are however a number of guides to conference proceedings of which the most important is the British Library Document Supply Centre's *Index of Conference Proceedings*. BLDSC has collected conference proceedings since its formation in 1964 and now has over 200 000 proceedings. The *Index*, in subject keyword format, appears monthly with annual cumulations. An 18-year microfiche cumulation covers the years 1964–1981. It includes conferences appearing individually, those appearing in serials and serials devoted wholly to conference proceedings. All those listed are available on loan to organizations in the UK. The entire file can be searched online via *BLAISELINE* and there is also a CD-ROM version.

The Institute for Scientific Information publishes the *Index to Scientific and Technical Proceedings* (1978–) which gives lists of individual papers given at the conferences whose proceedings it covers. Both proceedings published as books and those in journals are listed. *ISTP* is published monthly with annual cumulations and includes subject, author/editor, sponsor, meeting location and corporate indexes. The *Conference Papers Index* (bimonthly, Cambridge Scientific Abstracts, 1973–) also gives details of individual papers including those not subsequently published. Unlike *ISTP* it is arranged in subject order so that it is easier to browse. The subject and author indexes are cumulated annually and the service is available online (Dialog, ESA-IRS). The *Directory of Published Proceedings* from Interdok (monthly) appears in three parts of which the series *SEMT* (Science, Engineering, Medicine and Technology) (1965–) is of most interest to chemists. There are editor, location, subject, sponsor and acronym indexes in each issue but no details of individual papers. Cumulative indexes appear in some of the earlier volumes. *Proceedings in Print* (bimonthly, Proceedings in Print Inc., 1964–) also lacks details of individual papers. The index is a corporate one covering subjects, sponsors, corporate authors and editors.

In addition to finding out whether conference proceedings have been published it is often necessary to have details of forthcoming conferences. The best way to find out about these is to consult appropriate trade or news-type pure chemistry journals. *Chemistry in Britain* is a particularly useful source of this type of information. *Chemistry and Industry* publishes supplements to its January and June issues which contain six-monthly calendars of meetings. These journals and similar ones often carry short reports of conferences which can be helpful if no proceedings have been published. It should not be forgotten that forthcoming exhibitions can also be traced in this way, including those which are not attached to a particular conference.

There are a number of publications that are devoted to news of forthcoming conferences. Aslib has published since the early 1970s the quarterly *Forthcoming International Scientific and Technical Conferences*. The main issue is in February followed by a supplement in May and cumulative supplements in August and November. Conferences are listed in date order with subject, location and organization indexes. More restricted in subject scope is the *World Calendar* (1975–) published by Materials Information which lists forthcoming meetings in metallurgy and materials science. However, these subjects are interpreted quite widely. There are subject, sponsor and location indexes in each quarterly issue. *Exhibitions and Conferences* (York Publishing, 1984–), although comprehensive, has no indexes. As with the Aslib publication, the first issue each year is the main one.

Details of individual papers in conferences can sometimes be found in abstracting journals, especially *Chemical Abstracts*. Coverage of conference papers, however, is still patchy and many secondary sources do not cover them at all.

Finally it is worth mentioning Festschriften, volumes commemorating the work of distinguished living chemists. These often contain contributions from other prominent chemists working in the same field and thus serve to provide a state-of-the-art survey. For example, Sir Nevill Mott's eightieth birthday was the occasion of a Festschrift in *Philosophical Magazine* (Part B, 1985, volume 52, no. 3) that contained contributions on many areas of solid-state physics and chemistry.

Nomenclature

The great variety and complexity of chemical compounds makes their nomenclature a particularly difficult problem. Besides trivial and proprietary names each compound should have a unique systematic name which takes into account its particular structure, including if necessary its isomers. Standardization of names is particularly important in information retrieval when it can sometimes be essential to know that all references on a substance have been obtained. Hence there should be a particular set of rules by which compounds can be named in a unique manner. Such rules were drawn up by the International Union of Pure and Applied Chemistry (IUPAC) in 1957. The latest version in inorganic chemistry is *IUPAC: Nomenclature of Inorganic Chemistry: Recommendations* (Blackwell, 1990), the 'Red Book'. For the 'Blue Book' which deals with organic chemistry the latest edition is *IUPAC: Nomenclature of Organic Chemistry* (Pergamon, 1979). Unfortunately the IUPAC rules do not lead to unique names for many compounds so that their effectiveness for information retrieval is limited.

Chemical Abstracts Service have overcome this problem by devising their own system of nomenclature which is based on that of IUPAC. However, the *CA* system does assign a unique name to every chemical compound and so ensures consistency in the indexes. The *CA* nomenclature is inevitably based on forms which are easily indexed and many compounds are treated as derivatives of simpler ones. Esters and salts, for example, appear as derivatives of the parent acids. Trivial names are kept to an absolute minimum so that even a common name such as aniline is replaced by benzeneamine.

The latest guide to *CA* nomenclature can be found in appendix 4 of the 1987–1991 *CA Index Guide*. It should be noted, however, that as *CA* nomenclature is constantly revised, a particular compound can have different names in different *CA* Collective Indexes. The *Index Guide* for the appropriate Collective Index should always be consulted. As this can be time-consuming it is often advisable to search for a compound online by using its unique Registry Number which once assigned is not altered.

CA names are used in both the *European Inventory of Existing Commercial Chemical Substances (EINECS)* and the equivalent *Toxic Substances Control Act Chemical Substances Inventory*. These list substances marketed in the EC and USA, respectively.

Compounds used for particular purposes are often given generic names or other descriptors unique to that particular usage. Dyes and pharmaceuticals are good examples. The *Colour Index* has already been mentioned. Pharmaceutical generic names are listed in the *USAN and the USP Dictionary of Drug Names* (US Pharmacopeial Commission, 1984). The *Merck Index* is also a useful source of such information. Recommended names for chemicals used in industry are listed in *British Standard 2474: 1983*.

The standard guide to chemical nomenclature is *Introduction to Chemical Nomenclature* by R.S. Cahn and O.C. Dermer (5th edn, Butterworths, 1979) although it is now getting rather out of date. *Chemical Nomenclature Usage* by R. Lees and A.F. Smith (Ellis Horwood, 1983) is a record of a symposium on chemical nomenclature held in 1982. It is a very useful and readable work which does not describe chemical nomenclature itself but rather its use in various fields of activity such as education and industry.

Line-formula chemical notations were at one time widely used as an alternative to nomenclature but have now been largely superseded by topological systems. The current situation is discussed in Chapter 6. By far the most widely used line-formula system was the one devised by W.J. Wiswesser which is described in detail in *The Wiswesser Line-formula Chemical Notation (WLN)*, edited by E.G. Smith and P.A. Baker (3rd edn, Information Management Inc., 1975), or quite briefly on pp. 95–102 of the third edition of this book.

History of chemistry

Historical chemistry is well served by information sources only a few of which can be mentioned here. Standard treatises are dominated by J.R. Partington's *History of Chemistry* (Macmillan) which appeared in four volumes between 1961 and 1970. Unfortunately only part of Volume 1 was completed. A.J. Ihde's *The Development of Modern Chemistry* (Harper & Row, 1964) is another useful text. Joseph Needham's *Science and Civilization in China* treats chemical science and technology in Volume 5 (CUP, 1985–) and contains much non-Chinese material.

Information about distinguished chemists of the past can be obtained from the *Dictionary of Scientific Biography* or from Poggendorf's *Biographisch-Literarisches Handwörterbuch zur Geschichte der exacten Naturwissenschaften*. Chapter 17 of the previous edition of the present work also gives a list of biographies of many prominent chemists.

Although many journals such as *Chemistry in Britain, Education in Chemistry* and *Journal of Chemical Education* contain occasional articles on historical chemistry, the only one devoted to the subject is *Ambix* which commenced publication in 1937 and appears three times a year. Other journals devoted to the history of science generally include *Isis* (1912–) and the *British Journal for the History of Science* (1962–).

Studies of the growth of science which include its effects on scientific, including chemical, information are not common but the books by D.J. de Solla Price such as *Little Science, Big Science and Beyond* (Columbia University Press, 1987) and *Science since Babylon* (enlarged edition, Yale University Press, 1975) deserve particular mention.

A much fuller treatment of historical sources than can be given here is contained in another book in this series, *Information Sources in the History of Science and Medicine*, edited by P. Corsi and P. Weindling (Butterworths, 1983). Chapter 16 of that work, by W.H. Brock, is on the history of chemistry. The Royal Society of Chemistry has published *Recent Developments in the History of Chemistry* (C.A. Russell, ed., 1985) which also provides an excellent review of recent work.

CHAPTER FIVE

Online searching for chemical information

G.G. VANDER STOUW

The third edition of *Use of Chemical Literature*, published more than a dozen years ago, devoted about two pages to the subject of online searching. While it was pointed out that 'the scientific community has to become geared to the use of computerized retrieval systems', it was not generally thought at that time that online searching played any more than a very peripheral role among chemical information sources. In the years since that edition was prepared, a revolution has occurred in the way in which chemical information is accessed and used. The number of online files has increased at a nearly explosive rate; if that growth is finally slowing, it is only because most major sources are now available online. Along with the growth in computer-readable sources, there has come the widespread availability and use of personal and mini (mid-sized) computers, with the result that many chemists now are computer-literate and have computer resources available to them in their workplaces. Similarly, it is now expected that the library or information centre that serves the chemist will have terminals or personal computers as well as personnel trained to carry out online searches or to advise the chemist in doing his or her own searching.

The period between the third and fourth editions of this book began and ended with very significant events in the development of online information resources for the chemist. The first was the introduction in late 1980 of two services which provided online access to chemical substances contained in the Chemical Abstracts Service (CAS) Registry System. The capability to search a file of chemical structures for specific structures or families of related structures is a powerful tool uniquely available to users of chemical information. The second major event is the introduction of online access to chemical structures and data from the *Beilstein* handbooks (see Chapter 9). Now that *Beilstein* is available, virtually all important databases in chemistry are searchable online; since the 1980s we have seen the development of files based on almost all important databases, including bibliographic files, chemical structure

68 Online searching for chemical information

and reaction files, and files of numeric data. There has also been developed online access to important primary sources, including the full text of journals from the American Chemical Society and the Royal Society of Chemistry, and of tertiary sources such as the *Merck Index* and the *Kirk-Othmer Encyclopedia of Chemical Technology*.

The intention of this chapter is to describe available sources of online chemical information and to indicate some of the considerations involved in selecting and using an online source. While this discussion includes material on well-known databases, there are also many lesser-known files which may be important for particular searches, and it is hoped that the chapter will provide some awareness of these and what they have to offer. The reader must be aware, however, that the online scene is an evolving one, and this chapter will be out of date within a few years. The prospective online searcher is therefore urged to seek current information about databases and files; and the discussion of information sources later in this chapter, as well as the knowledge of professional information specialists, will help in the acquiring of this information.

Although online searching is a widespread and powerful tool for acquiring chemical information, it is not yet nearly as simple to use as may someday be the case. For that reason, the majority of online searches are probably done today by the trained information specialist. Nevertheless, the use of search systems by the chemist or other 'end user' of the desired information is growing, particularly in those organizations where end-user training is encouraged and the necessary equipment is made available. There is a considerable amount of knowledge that the searcher must employ both in selecting a file or files to be searched and in carrying out the search. These areas include basic search logic, relationship between information files and vendors, search languages and protocols, contents of specific files, costs, and where to get more and more up-to-date information. These areas are discussed in more detail in the following paragraphs.

Basic search logic
The searcher must understand the logical principles involved in constructing a search query, including the application of the Boolean logical operators AND, OR and NOT, as well as concepts such as term truncation. Virtually all online search systems apply these in one way or another.

Vendors
Most online searching today is done on the computers of organizations who have acquired computer-readable databases from a number of sources, and have made them searchable using common protocols and command language; usually they are accessed via long-distance telecommunications networks. The organizations who provide the

search capabilities are called 'vendors' or 'hosts'. Generally, the user will have established an account with the vendor and will be billed for each search. In some cases an annual subscription fee is also required, though this is the exception. Up-to-date information on vendors, and the files available from them, can be obtained from the most recent editions of the *Directory of Online Databases* published by Gale Research. Appendix 1 shows a list of the publicly available databases cited in this chapter and of vendors who provide access to theses databases, or have announced plans to do so. Addresses of vendors mentioned in this chapter, and of suppliers of software for in-house use discussed here, are given in Appendix 2.

Search languages and protocols
The performance of an online search takes place in four basic steps:

1. Establish a connection with the desired vendor's system, and then with the desired file within that system.

2. Construct the query according to the rules of the search language used by the vendor, and instruct the system to carry out the search.

3. Examine initial results, and refine the search if necessary.

4. Display the results online or instruct the system to print them offline for mailing to the user's location.

While the above operations apply equally to any online search, the ways in which the user carries them out are different on each vendor's system. Significant differences can be found in the protocols followed for logging on to a system and selecting a file, and in the search language to be followed in stating the query and giving the system instructions regarding search and display operations. Vendors frequently introduce new or improved features into their systems in order to meet the needs expressed by users or to respond to improvements introduced by competitors' systems. Specific and current information about languages and protocols will be found in the manuals published by the various vendors and in their training classes.

Recent years have seen several attempts to help the user surmount the differences between specific search systems by the use of gateway services, through which a user connected to one search system can access files on another system without changing system or language. Information on gateways can be obtained from specific vendors or their publications; organizations who offer gateways, and their addresses, are listed in the directory of databases previously cited.

Files

A computer-readable database is generated by an organization, usually referred to as the database producer, which compiles the information from source materials. Often the information is made available in both printed and computer-readable forms. A given database may be acquired by one or several vendors who make it available as part of their online services. The specific implementations of that database by the vendors may differ in various ways: the information may be organized differently, a database covering a number of years of material may be split up into several small files, and so on. For example, the information which appears in *Chemical Abstracts* indexes appears in semi-annual printed indexes published by CAS. This information is also available online as part of the services of at least eight vendors. All of these implementations of the CAS data are different, and the user must have information regarding the specific system in order to use any one of them.

The prospective searcher is therefore faced with making two selections. The first is to determine which database or databases are likely to contain the desired references or data. The second choice is that of which of the files based on a specific database to search. The latter choice may simply be a matter of preference for the way in which one or another vendor has implemented a particular file. However, it may well be that the search will need to make use of more than one database, and the selection of vendor will then take into account which vendor has the most appropriate mix of files or the most convenient facilities for handling searching of multiple files and manipulation of the search results.

Costs

The user of printed information services usually does not have to think about the costs of the searching, except in terms of the labour costs of doing a search and the annual costs of subscriptions; the latter is usually the concern of the librarian. Online searches, on the other hand, incur costs immediately, and the searcher almost certainly will have to be conscious of this point as the search is being carried out. The primary means of measuring cost has, until recently, been 'connect time', i.e. a charge based on the length of time that the user has been connected to a file. In 1987, CAS announced a new licensing policy for its databases which included charging based largely on the number of search terms used in a search. Whether other vendors or database producers will follow this precedent is not yet clear, but it is likely that changing technology will encourage a general move away from 'connect hour' pricing.

Getting information

The above discussion may seem rather daunting to the inexperienced searcher who needs a variety of information which is continually subject to change. Any given source of printed information, including this text, will become out of date in short order. However, a great deal of information is available to the searcher who is willing to make an effort. Search service vendors offer manuals, newsletters and training classes, and the major vendors maintain staffs who are trained to answer telephone enquiries at 'help desks', sometimes through toll-free telephone numbers.

Another major source of information is through meetings and conferences. Within English-speaking countries, major online conferences are held annually in London and in major cities in the US, Canada and Australia. These provide not only exhibits and demonstrations by vendors, but also papers, seminars, etc., about new or changed services and files. These conferences are, of course, more likely to be attended by the librarian or information specialist than by the end-user scientist. The latter, however, can get a great deal of information from the exhibitions held at major scientific conferences, such as those of the American Chemical Society, where displays by information vendors are becoming increasingly prominent. These offer an opportunity to talk to representatives from the vendors, to get up-to-date literature, to see demonstrations, and so on.

Major journals of relevance to online searching for chemical information include the *Journal of Chemical Information and Computer Sciences*, published bimonthly by the American Chemical Society, *Online* and *Database*, both published by Online, Inc. and *Online Review*, published by Learned Information.

The files which contain information of interest for chemistry or related sciences can be divided into several major groups: bibliographic information, full texts, chemical structures and nomenclature, chemical reactions, and data about chemical substances. The following sections of this chapter will discuss in turn the major sources of online information in each of these areas.

Bibliographic files

A bibliographic database does not lead the searcher directly to data, but rather it provides citations to documents that can be expected to contain the desired information. The field of chemistry has had a long history of printed abstracting and indexing services which were designed to meet this goal of leading the user to references of interest (see Chapter 3). Thus, as computer technology developed, there were existing sources of bibliographic information which could form the basis of databases to be used in online files. In addition, new databases have come into existence which did not have a previous history as printed services. In general, searching a bibliographic file in chemistry is no different than searching a file from another discipline. The most important feature specific to chemistry is the use of substance identifiers, especially of CAS Registry Numbers, as search terms in searching for references to chemical substances. As discussed later in this chapter, the searcher for chemical information can use online structure or nomenclature files to identify Registry Numbers for substances of interest, and then transfer these numbers to the bibliographic file, possibly combining them with other search terms, to find references citing these substances.

Chemical Abstracts
Since 1907 the American Chemical Society has published *Chemical Abstracts (CA)*, which is described in greater detail in Chapter 3. Beginning in the late 1960s, CAS brought computer processes extensively into the manufacturing of *CA*, with the result that almost all of the data which appear in the printed abstract issues and semi-annual indexes became available in computer-readable form. Online files based on CAS data first became available in the mid-1970s, and have continued to evolve since that time. Today a number of different *CA*-based files are available.

The *CA File*®, searchable via the vendor STN International®, contains data for journal articles and patents covered by *CA* since 1967 (nearly 10 million records by 1991). Searchable data include the index entries from the semi-annual General Subject and Chemical Substances Indexes, the keywords from the biweekly Keyword Index, and the text of the abstracts published in *CA* issues, as well as such items as author names, type of document, document language, etc. The file can also be searched for specific chemical substances by the use of CAS Registry Numbers. CAS also offers *CA Previews*, a database containing the most recent document information before that information is added to the CA File. An additional CAS database available through STN, called *CAOLD*, contains Registry Numbers and references for abstracts published in *CA* in the period 1957–1966. References are added to CAOLD as substances from that time period are added to the Registry File. *CAOLD* contained some 650 000 references in 1987.

Several other online files based on *CA* are available through various vendors. All are generated from a CAS database called *CA Search*, which contains references, keywords, and index entries, including Registry Numbers, from *CA* since 1967. Thus all of these files have nearly the same content, but differ in the way they are searched because of differences between the search languages of the specific vendors involved. Among the files available are those from the following vendors (see Appendix 2 for vendor addresses): BRS, the Canada Institute for Scientific and Technical Information, Data-Star, Dialog, ESA-IRS, Life Science Network, Orbit and Questel.

Some vendors have broken the database into several files corresponding to the five-year Collective Index periods of *CA*, while others permit the entire period from 1967 forward to be searched in a single file.

Patent files

As indicated above, the *CA File* includes information from patents covered by *CA*. Several other files based on the patent literature are also available online. These include the following publicly available files; in addition, Derwent Publications Ltd prepares files, but these are available via the Orbit Search Service to subscribers only. See Chapter 11 for details of the printed version of these sources of patents information.

> *APIPAT*, produced by the American Petroleum Institute, covers some 175 000 patents, from 1964 forward, related to the petroleum industry. It is available via Orbit and STN.
>
> *CLAIMS*, a family of files produced by IFI/Plenum corporation, is available via Dialog, Life Science Network, Orbit and STN.
>
> *INPADOC*, produced by the International Patent Documentation Centre, covers patents from all areas, and is available via Dialog, Orbit and STN.
>
> WORLD PATENTS INDEX *(WPI)*, prepared by Derwent and available via Dialog, Orbit and Questel, contains more than 5 million patent citations.

Other bibliographic files

Several other databases important to chemistry are available online. These include the following.

> ANALYTICAL ABSTRACTS, published by the Royal Society of Chemistry, with more than 100 000 abstracts related to analytical chemistry, and available online through Data-Star, Dialog and Orbit.
>
> *BIOTECHNOLOGY*, produced by Derwent Publications Ltd, and available through Dialog and Orbit, contained more than 110 000 records in 1991.
>
> CHEMICAL ENGINEERING ABSTRACTS, produced jointly by Dechema and the Royal Society of Chemistry, contains more than 200 000 abstracts and is available via Data-Star, Dialog, ESA-IRS, Orbit and STN.
>
> CURRENT BIOTECHNOLOGY ABSTRACTS, from the Royal Society of Chemistry, contained more than 31 000 abstracts from 1983 forward to 1991, and is available online via Data-Star, Dialog, ESA-IRS and Orbit.
>
> *EMTOX* is a file corresponding to the Toxicology and Pharmacology and Toxicology sections of Excerpta Medica. It is produced by Elsevier Science Publishers and is available online via DIMDI.
>
> *JICST-E*, produced by the Japan Information Centre for Science and Technology, and available via STN, includes citations and

abstracts to literature published in Japan on chemistry and other sciences. The file includes more than 300 000 citations from 1985 to date.

TOXLINE is a database produced by the US National Library of Medicine and available via the National Library of Medicine, BRS, Data-Star, Life Science Network, Dialog and DIMDI. It contains abstracts and citations to literature in all areas of toxicology. See also Chapters 13 and 14 for details of other health & safety and biomedical databases.

Full-text files

Although the development of online searching has emphasized the use and availability of bibliographic files, there is no inherent reason why searching should be restricted to index terms and citations rather than to actual texts, if the latter are available in computer-readable form. As publishers of chemical journals have implemented computer-based typesetting and publication processes, full-text databases have increasingly become available. These can be searched using search terms and Boolean logic combinations, much as can bibliographic files, but there are differences in optimum search technique of which the searcher must become aware through experience. Some search languages offer capabilities that are particularly useful for searching of full-text files. These include, for example, the ability to specify the order in which search terms should occur in a document in order for matching to occur, or to specify limits on the number of words that can occur between two search terms in the document.

A number of chemical journals have become available as full-text files searchable through STN (see also Chapter 3). These include the following files:

CJACS — approximately 100 000 articles from 18 primary journals published by the American Chemical Society, covering 1982 forward.

CJOAC — articles from the *Journal of the Association of Official Analytical Chemists*.

CJRSC — articles from 10 journals published by the Royal Society of Chemistry, including over 13 000 articles from 1987 forward.

CJWILEY — articles from five polymer journals published by John Wiley including over 4000 records from 1987 forward.

CJVCH — articles from the international (English-language) edition of *Angewandte Chemie*.

Another full-text file of interest in the chemical industry is Drug Information Fulltext, which contains about 1200 monographs concerning about 50000 commercially available and experimental drugs. The database is produced by the American Society of Hospital Pharmacists and is available via BRS, CIS, Dialog and the Life Science Network.

Chemical structures and nomenclature

The science of chemistry is, by definition, about chemical substances and transformations between them. For many years, since long before the advent of computers, chemical information handling has made use of highly systematized and standardized methods for representing and naming chemical substances, especially through the use of structural diagrams. (See Chapter 6 for a further discussion of methods for encoding and naming substances.) Large paper-based files of chemical structures and names were already in existence for many years before computer-based searching became possible, and these have formed the basis for the large computer-readable structure and nomenclature files that are now available online.

Basic concepts of substance searching

In order to discuss available files and search systems for searching for chemical substances, it is necessary to introduce several concepts that may not be familiar to the reader.

Registry Number

Most chemical substance databases assign to each substance within the database an identification number, often called a Registry Number. Usually some type of computer processing is applied to substances entering the system to ensure that a given substance will have only one Registry Number. The Registry Number serves as a key for retrieving information about that substance from any databases within a system. (In the USA the Registry Number is often compared to a Social Security Number*, a number assigned by the federal government to an individual which serves as the address for all sorts of information about that individual in files maintained not only by the Social Security Administration, but also by banks, employers, etc.) For example, a system might include databases containing structures and/or names, which can be searched to determine Registry Numbers of interest, and also databases containing bibliographic information or numeric data, which can be searched using the Registry Numbers to find other types of information about the substances of interest.

The ability to identify Registry Numbers in one file, and then use those numbers to search other files, is a very powerful feature of many chemical information systems. Whether one can move between files in this way, and the convenience of doing so, may be important considerations in making a choice between competing vendors' search systems.

---

* One's National Insurance Number is increasingly serving a similar function in the UK (eds).

Identity search

One obvious type of query that one will want to search in a file of chemical substances is simply: Is this substance (for which the searcher will provide a name or structure) present in the file? If so, what is its Registry Number (or other identifier)? While this type of query would seem to require a simple yes-or-no answer, there is a question of just how specific the query is meant to be. The substance requested may be present in a file without any ambiguity, or there may be substances present which are identical to the substance requested except that they differ in stereochemistry, or they include isotopic labelling which was not specified, etc. The searcher must consider the precision required by a search, and must be aware of what capabilities the search system being used provides for specifying that precision.

Substructure search

Another type of question frequently put to a chemical structure file is one in which the searcher is looking for all structures on file that contain a given structural unit, possibly with some variations allowed. Consider, for example, a search for the structural unit

CO_2H (attached to benzene ring)

If identity search is specified, this search will be expected to lead only to retrieval of records for that specific structure (benzoic acid). However, if substructure search is specified, the search should retrieve additional structures such as

H_2N—⌬—CO_2H HO—⌬(CO_2H)—OH etc.

Substructure searching is a powerful concept in chemical information handling, since it allows the searcher to look in a single query not only for a specific substance of interest, but also for other substances that may be chemically similar to it, and therefore of potential interest for the search.

A structure search system typically provides the user with a means for drawing the structure or substructure of interest and then instructing the system to carry out the search. The retrieval will typically consist of structural diagrams of those substances which answer the query, along with their Registry Numbers and names.

Nomenclature searching

Information services that offer searching of structural files normally also offer searches of files of chemical names for the same substances. In addition, there are other services that offer only nomenclature search. The names in these files can usually be said to be of one of two types, used for two different purposes.

Trivial names or common names

These are names, often not carrying structural information, that are likely to be familiar to the searcher. Familiar examples include names such as 'caffeine' and 'vanillin', as well as trade names. Using these names as search terms is often a very convenient way to do an identity search; either the name is in the file or it is not.

Systematic names

These are names which are constructed from small nomenclature units which contain structural information and which are built up according to well-defined rules. A chemist can normally look at a name and from it determine the structure of the substance being named. Thus, for example, many chemists can probably draw a structure for '4-hydroxy-3-methoxybenzaldehyde' even if they do not remember what 'vanillin' stands for.

A file of systematic names provides an opportunity for a limited form of substructure search. For example, a searcher could ask for names including '3-methoxy' and 'benzaldehyde' and retrieve the name '4-hydroxy-3-methoxybenzaldehyde' shown in the preceding example. However, because of the complex and hierarchical nature of systematic nomenclature, many substances whose structures contained the structural unit defined by '3-methoxybenzaldehyde' would not have names containing those nomenclature fragments.

There are circumstances in which nomenclature searching is very useful. One is as a means of retrieving substances whose structures are not precisely known and therefore cannot be included in a structure file, at least in the case where the structure is partially known and those parts can be represented by systematic nomenclature terms. Nomenclature searching is also useful, obviously, if no structure file is present in the system being used.

Chemical substance databases

The number of chemical substance databases that are available publicly for online searching is quite small. Those from six database producers are discussed in the following paragraphs: Chemical Abstracts Service, Chemical Information System, the US National Library of Medicine, the US Environmental Protection Agency, the Gmelin Institute and the Beilstein Institute. There are, in addition, a considerable number of proprietary substance files that contain the collections of substances that have been accumulated by particular firms or laboratories; such files are normally highly confidential. These 'in-house' files typically are searched in much the same way as are public files. Many in-house files are searched with software developed within their particular company. Increasingly, however, in-house files are being managed and searched using software obtained from outside sources. Of the organizations mentioned in this section, both CAS, through its Private Registry service, and Questel, through 'in-house DARC', offer systems for managing proprietary substance collections. In addition to these, a number of firms have established in-house systems based on the MACCS software offered by Molecular Design Ltd, a US-based company. Another package for in-house use is called ORAC (see p. 84).

Chemical Abstracts Service
In 1965, CAS began operation of the Chemical Registry System. This system builds and maintains a file of chemical substances, the great majority of which have been selected as index entries to *CA*. Chemical structures are encoded in the form of atom–bond connection tables (see Chapter 6). Each structure is assigned an identifying number known as the CAS Registry Number®. By late 1991 the Registry System contained over ten million substances and was growing at the rate of about 10 000 substances per week. Substances in the file included organic and inorganic compounds, including polymers, minerals, alloys and mixtures. Several online files are based on data from the Registry System. Two of these, the files available from STN and Questel, provide for both structure-based and nomenclature-based searching. The others provide nomenclature searching; one of these, Dialog, augments the nomenclature with data fields which describe ring systems present in the corresponding structures.

STN International
The files on STN include the REGISTRY FILE®, which contains both structures and nomenclature from the Registry System. This file is searched using the Messenger software developed by CAS, which provides both identity, or 'full structure', searching and substructure searching, as well as nomenclature searching. Structure queries are entered as structure diagrams created with text commands or drawn on a graphics terminal or on a personal computer with graphics software. The user is allowed a considerable ability to describe alternatives in the desired substructure. Retrieved substances can be shown as structure diagrams either online or printed offline. CAS supplies a software package called STN Express for use on personal computers. With STN Express, the user can construct a structure query offline, and then go online and transmit the query to STN for searching. Within STN, the user can take Registry Numbers retrieved from the REGISTRY FILE® and use these to search the CA FILE® or other

STN files containing Registry Numbers.

Questel
Using software called DARC, originally developed by Professor J.E. Dubois at the University of Paris, Questel offers a file, also called DARC, which is based on the REGISTRY FILE®. Capabilities offered by the DARC system are very similar to those offered in STN, even though the underlying search algorithms are significantly different. Within the DARC system the user can restrict the search to one of several subfiles. These include: EURECAS, which contains all structures from the Registry except polymers; POLYCAS, which includes the polymer structures; and UPCAS, which includes monthly updates to EURECAS and POLYCAS.

Dialog Information Services
Dialog offers a family of files based on nomenclature from the Registry System. Each of these files includes Registry Numbers, chemical names (*CA* Index names and synonyms), and molecular formulae, augmented with ring-system information, The files include: CHEMZERO, which contains substances which are cited in the Registry System for which there are no references to *CA* from 1967 forward; CHEMSIS, which includes substances for which there is no more than one reference in any *CA* five-year Collective Index; CHEMNAME, which includes substances with two or more references; and CHEMSEARCH, with substances from updates to the file.

Orbit Search Service
This offers a file called CHEMDEX, searched with the Orbit software, which provides searching of nomenclature for substances entering the Registry System from 1972 forward.

Data-Star
This offers a nomenclature file called CHEMICAL NOMENCLATURE, which covers nomenclature for substances added to the Registry System for 1967 forward.

Chemical Information Systems Inc.
During the 1970s, a group of agencies of the US government, including some from the National Institutes of Health (NIH) and the Environmental Protection Agency (EPA), developed an innovative online information system called the NIH-EPA Chemical Information System (CIS). This system contained a number of different files sponsored by government agencies, most of them containing various kinds of numeric data about chemical substances, including files of mass spectra, carbon-13 nuclear magnetic resonance spectra, toxicity data, etc. Records for a substance in any file could be addressed by CAS Registry Number. (These numeric files are discussed later in this chapter in the

section on numeric data files, p. 90.) Structures, represented by connection tables, and names for the substances in the system's files were placed in a file called the STRUCTURE AND NOMENCLATURE SEARCH SYSTEM (SANSS). In the early 1980s a decision was made to discontinue government operation of the CIS in favour of private commercial operation. As a result of this decision the CIS is now operated by Chemical Information Systems, Inc., a subsidiary of Fein-Marquart Associates. User support for CIS services is available in the UK from Fraser Williams Scientific Systems of Poynton, Cheshire.

Currently SANSS contains records for some 350 000 substances, including those from the various CIS files, from other files produced by government agencies and other sources, and from the Toxic Substances Controls Act Inventory of the US Environmental Protection Agency (see below). SANSS can be searched by structure, nomenclature or molecular formula in much the same way that files based on the CAS databases can. Structures can be 'drawn' online using a series of commands corresponding to structural units, or they can be prepared using a personal computer-based software package called Super Structure. There are a number of differences on points of detail between searching SANSS and searching CAS-based files; the searcher trained in the use of one cannot simply begin searching the other without taking time to study and understand the points of difference. The CIS is an important information source and, at the time of writing, the most comprehensive source of numeric data about substances publicly available.

National Library of Medicine
The US National Library of Medicine (NLM) provides a series of online files related to various aspects of medicine and related areas. Among these is a large bibliographic file called TOXLINE, based on a variety of sources containing information related to toxicology. A file called CHEMLINE contains over 1 million substances from TOXLINE and other NLM databases and also from the Toxic Substances Control Act Inventory of the Environmental Protection Agency. Substances are represented in CHEMLINE by CAS Registry Numbers, systematic names, synonyms and molecular formulae. The records of many substances also include data about the ring systems present.

Environmental Protection Agency
The Environmental Protection Agency has produced a database containing information on approximately 60 000 compounds of commerce covered by the US Toxic Substances Control Act. This file, called the TSCA Initial Inventory, is available through Orbit, Dialog and the Life Science Network. It includes CAS Registry Numbers, preferred name, synonyms, and molecular formulae.

Gmelin Institute

The *Gmelin Handbuch der anorganischen Chemie*, which traces its history back to 1817, is a comprehensive treatise on inorganic chemistry published by the Gmelin Institute (see Chapter 8). The handbook's volumes contain information on more than 300 000 substances. These include organometallics, covalently bonded non-organometallics and 'classical inorganics', including solid solutions, minerals and alloys. Two components of the *Handbuch* can be searched online as the GMELIN FORMULA INDEX (GFI) file available via STN International. The GFI file contains information in English from 1924 to the present. The 'Formula Index' contains molecular formulae, element counts, molecular weights and the title of the particular *Handbuch* volume that contains data about a specific substance. The 'Complete Catalog' provides citations, including titles and abstracts, to the handbook volumes.

Beilstein Institute

The *Beilstein Handbuch der organischen Chemie*, which is discussed in detail in Chapter 9, contains evaluated numeric and factual data about more than three million organic compounds. Early in the 1980s, the Beilstein Institute began vigorous efforts to computerize their production operations and also to develop files for online searching based on the contents of the *Handbuch*. BEILSTEIN ONLINE is available through Dialog, Orbit and STN International. The initial implementation occurred late in 1988, with the complete file available by 1991. The publicly available Beilstein files include a structure file, providing capabilities for identity searching and substructure search and a factual file, with capabilities for data searching. The latter contains data including physical properties, molecular formulae, literature citations and comments about preparations and reactions. (The structure file is also expected to be available for in-house use.) The searcher locates substances of interest by searching the structure file, and then looks for data about those substances in the factual file, using the Beilstein Registry Number assigned by the Beilstein Institute. In many cases CAS Registry Numbers are also present. The introduction of online access to Beilstein data constitutes a major step forward in making older chemical information available online.

Markush structures

The term Markush structure comes from the development of US patent law (see Chapter 11 for a more detailed discussion of patent information sources). It refers to a generic structure of the type that is frequently claimed in a chemical patent. The following example is typical. It contains fixed groups, such as the pyridine ring, variable groups, such as 'R^1' and 'R^2', groups which have several alternatives such as 'alkyl (C_{2-8})', groups which have a virtually unlimited number of alternatives such as 'heterocyclyl', and logical relationships such as 'if R^1=H then R^2=Me'.

Base Portion

$R^2 \underset{(CH_2)_n - OH}{\overset{N - R^1}{\diagdown\diagup}}$

Variable Portion

- R^1 = H, OMe, Alkyl (C_{2-8})
- R^2 = Cl, Me, Heterocyclyl
- n = 0 - 4
- If R^1 = H then R^2 = Me

The ability to search for Markush structures is of great importance to the chemical industry, particularly for a firm which is preparing a new patent application or determining if a particular compound or patent claim will infringe a patent already granted. Adequate methods for storing and searching Markush structures have been a particularly difficult challenge in the development of chemical information systems. A Markush structure handling system must store these substances in a way in which all of the variability described in the patent is correctly preserved and from which the structure and all of the variables can be displayed on output. The search capability must be able to match structural units which are described at different levels of specificity. For example, a query containing an ethyl group should match a Markush structure which has a group described as 'alkyl' in the desired position.

Until 1988, no files of Markush structures meeting the above criteria were publicly available. The online services that did cover Markush structures did so through the use of 'fragment codes' corresponding to structural units which may or must be present in a Markush structure. Available services included the Derwent CPI file (available only to Derwent subscribers), the CLAIMS file of IFI/Plenum, available via Dialog, Orbit and STN, and the APIPAT file from American Petroleum Institute, available via Orbit and STN. A pioneering system for Markush structure handling developed in Germany is the GREMAS system, which encodes Markush structures by an elaborate fragment code; searches may be input topologically. The encoding of GREMAS is done by the IDC organization in Frankfurt, Germany, which is sponsored by several European chemical companies; searches are available to the member companies.

Since 1988 two major vendors have begun offering files of Markush structures. STN offers MARPAT, produced by CAS, which contains more than 40 000 Markush structures from more than 18 000 patents contained in the *CA File*. Questel offers MPHARM, which contains Markush structures from patents related to pharmaceutical research, and WPIM, which includes Markush structures from patents covered in specified sections of the *World Patents Index* produced by Derwent.

Chemical reaction databases

The science of chemistry is concerned with transformations between elements or compounds, as well as the structures and properties of individual compounds. However, the historical development of both printed and online information tools for chemical information has emphasized organization according to individual compounds rather than reactions, although several printed compilations of reactions have been prepared, including *Organic Reactions* and *Organic Syntheses*, published by John Wiley, and *Synthetic Methods of Organic Chemistry* (Theilheimer) (see Chapter 10). Online files of chemical reactions are now beginning to appear, and are likely to play a much larger role in computer-based chemical research in coming years than has been the case up till the present.

A reaction search can be thought of as a double structure search, in which the searcher specifies the structure of both reactant and product, either in terms of specific structures or in terms of substructures. Sometimes the searcher will not specify both ends of the reaction, but only the desired product or reactant structure. The searcher who wishes to prepare a particular substance, for example, may only describe a structure or substructure for the product, without constraining the transformations which may be involved in preparing that product. In the following example, the searcher has asked for reactions which lead to formation of the structural unit:

$$\text{C} - \text{CH}(\text{OH}) - \text{C}(=\text{C}) - \text{C}$$
Ring or Chain

One reaction which satisfies that query is the following:

$$\text{Ph-CO-CO-Ph} \rightarrow \text{Ph-CH(OH)-CO-Ph}$$

It was recognized early in the development of reaction searching capabilities that reaction searching was likely to lead to many false retrievals, i.e. cases where the specified substructures occurred in the reactant and in the product, but the actual reaction was not a transformation between them; they are simply artifacts in a different transformation. This led to the concept of identifying 'reaction sites', in other words of tagging those atoms and bonds that actually take part in a reaction. If retrieved reactions are displayed with the reaction sites identified, then the searcher can readily eliminate undesired retrievals through visual inspection. Even more useful is the ability to specify reaction sites as part of the search query, so that the only desired reactions will be retrieved.

The chemist who is searching a file of reactions is very often looking for methods that can be used to prepare a structure of interest. Such a search is likely to require not only skill in the mechanics of using the search system, but also imagination in stating queries that will elicit reactions of interest. The structure that is the chemist's real target may very well not be in the database, especially if it is thought not to have been reported previously, so the searcher must use substructure search techniques to find analogous reactions that are relevant to the problem at hand. There is at present no substitute for good chemical knowledge in selecting queries that will lead to information of interest without wasting inordinate time or resources in going down paths that are not useful. Ultimately, it can be hoped that expert systems will be developed to help the searcher with this particularly sophisticated type of searching.

The development of online reaction files has until recently emphasized services available only to subscribers, particularly the CRDS file described below, or developed for in-house use, including REACCS, ORAC and SYNLIB. Recently these have been joined by the CASREACT file from CAS.

Chemical Reactions Documentation Service (CRDS)
The CRDS file, prepared by Derwent Publications Ltd, is available via the Orbit Search Service to subscribers only. The file corresponds to the *Journal of Synthetic Methods* and follows the format of Theilheimer's *Synthetic Methods of Organic Chemistry*. Searching the file involves the use of the fragment codes developed by Theilheimer.

REACCS files
Molecular Design Ltd (mentioned above in connection with their MACCS software) have developed a software package for storing and searching chemical reactions, called REACCS, which is marketed to private firms for handling their proprietary files of reactions. They have also established a database division which prepares files of published reactions which can be searched with REACCS. Files available include:

THEILHEIMER — more than 40 000 reactions from the Theilheimer series.

JSM — 3000 literature reports, with some 4000 individual reactions, from the *Journal of Synthetic Methods*.

CURRENT LITERATURE FILE — interesting reactions abstracted from leading journals.

CURRENT CHEMICAL REACTIONS — published since 1979 by the Institute for Scientific Information, and now available in REACCS format, contains some 28 000 reactions.

ORAC

ORAC is a software package developed at the University of Leeds for managing and searching databases of chemical reactions. It is marketed by ORAC Ltd, primarily for use with in-house collections. ORAC is also available as a 'public database' containing more than 15 000 reactions extracted from the published literature by academic scientists.

SYNLIB

The SYNLIB file and SYNLIB software have been developed by Professor Clark Still of Columbia University, working in conjunction with Smith, Kline & French Corporation. The file now contains some 25 000 reactions from literature concerned with organic synthesis. SYNLIB is marketed by Distributed Chemical Graphics of Philadelphia.

CASREACT

The CASREACT® file, which became available online via STN International in early 1988, includes reactions from documents covered by the organic chemistry sections of *Chemical Abstracts*. The file covers the literature since 1985. At the time of its introduction it contained some 250 000 reactions, including both single-step and multi-step reactions; it now contains more than one million reaction steps from over 60 000 documents. CASREACT can be searched by reactant or product structures or by CAS Registry Numbers. Reaction sites can be specified, as can reaction roles including catalysts, reactants, products, reagents and solvents. Searches can be carried out for either single- or multi-step reactions.

Biomacromolecular databases

The growth of interest in the field of biotechnology in recent years has led to the development of a highly specialized type of chemical substance database. These databases, often referred to as 'sequence databases', contain representations of protein sequences and nucleic acid sequences. The sequences are made up out of amino acid residues, in the case of protein sequences, and of nucleotide residues, in the case of nucleic acid sequences, and frequently contain chains of hundreds or even thousands of residues. The resulting sequence records thus are primarily very long strings made up of the one- or three-letter amino acid symbols, or of the four letters G, C, A and T for the common nucleotides.

The searching of sequence databases has some characteristics both of text searching and of structure searching. Logically, it resembles structure search, in that one wishes either to do identity searching, i.e. determining whether a given sequence is already on file, or to do substructure searching, so as to find sequences on file which contain a desired unit. One might first think of searching a sequence file as a simple matter of matching alphabetic strings, but string matching is not adequate for this type of searching. This is true for a variety of reasons, among which are the imprecision of the methods used to determine sequences and the virtual impossibility of error-free data entry for such long chains of symbols. As a result, some novel software approaches have been developed for searching and displaying sequences. A technique called 'homology searching' is often used to search a sequence database and produce a ranked list of sequences on file which most closely resemble the query sequence. Display techniques have been developed which help the user assess the degree to which a retrieved sequence is a valid answer, and what the portions of similarity and difference are between that sequence and the query.

Existing sequence databases generally contain a few thousand sequences. The growth of these databases is likely to accelerate in coming years because of growing commercial interest in products of biotechnology (see Chapter 14) and because of the increasing power of automated equipment for determining sequences. At the same time, it can be expected that much more powerful tools for search and display will be developed. Most importantly, the decision of the US government to fund an effort to sequence the human genome in the course of the next 15 years will lead both to rapid growth in sequence data becoming available and to development of new methods for searching them, for example by the use of expert systems. The main sequence databases at present are the following:

GENBANK — compiled by the Los Alamos National Laboratory, contains more than 20 000 nucleic acid sequences. Each record includes 16 data items, including such information as bibliographic record, annotations showing such things as the start and stop points of coding regions, and source organisms.

EMBL NUCLEOTIDE SEQUENCE DATA LIBRARY — compiled by the European Molecular Biology Laboratory, contains more than 40 000 nucleic acid sequences.

PROTEIN IDENTIFICATION RESOURCE of the US National Biomedical Research Foundation — includes a protein sequence database of more than 25 000 proteins.

CAS REGISTRY FILE — contains more than 160 000 polypeptides which can be searched using the shortcut symbols for the amino acid residues.

With the exception of the Registry File, much of the use of these databases is done in-house by organizations who have acquired copies for this purpose. Online access is available for non-profit organizations through a network in the USA called BIONET, or through Intelligenetics, Inc., or the National Biomedical Research Foundation. The academic network of other countries, such as JANET in the UK, are linked to the US academic network BITNET through gateways, and one can communicate via JANET and BITNET through to other US networks (see Chapter 17).

Crystal structure databases

Crystal structures constitute another specialized type of chemical structure whose searching has many of the characteristics of numeric data searching discussed below. A few important crystal structure databases have become available online. These include the following databases.

CAMBRIDGE STRUCTURAL DATABASE — compiled by the Cambridge Crystallographic Data Centre, contains crystallographic data for more than 55 000 organic and organometallic structures, including all such substances whose crystal structures have been determined since 1935. Available online via the Canadian Scientific Numeric Database Service (CAN/SND) under the file name CRYSTOR, and in the UK via JANET.

CRYSTMET (NRCC Metals Crystallographic Data File) — produced at the National Research Council of Canada, contains crystallographic and bibliographic data for metallic structures determined by diffraction methods, with more than 6000 entries. Available online via CAN/SND and, in Germany, via Inkadata.

INORGANIC CRYSTAL STRUCTURE DATABASE, produced by the Institute of Inorganic Chemistry of the University of Bonn, contains data on more than 24 000 crystal structures for inorganic compounds. Available online via CAN/SND (as CRYSTIN) and STN (as ISCD).

NIST CRYSTAL DATA IDENTIFICATION FILE, produced by the US National Institute for Standards & Technology (formerly the National Bureau of Standards), contains data on more than 100 000 chemical structures. Available online via CAN/SND (as CRYSTDAT) and Inkadata (as NBSCD).

Sources of chemical substances

Frequently, an important practical question for the chemist is where a particular substance may be acquired. Databases which have been developed to answer this question include the following:

CHEMQUEST — formerly known as the FINE CHEMICALS DIRECTORY, this is produced by Molecular Design Ltd and is available via the Orbit Search Service. It contains names of suppliers for more than 130 000 substances, derived from over 50 catalogues.

CSCHEM, produced by Directories Publishing Co. and available via

STN, contains over 110000 records derived from the publication *Chem Sources USA*. A related file, CSCORP, provides addresses and other information about suppliers.

JANSSEN, produced by Janssen Chimica and available via Questel, contains information on more than 11 000 compounds from the Janssen catalogue.

Properties Data

A considerable amount of data about the properties of chemical substances — spectral data, thermophysical data, toxicity data, safety data, etc. — is available online. However, a much more complex situation exists for properties databases than for bibliographic or structural data. For the latter types of data, most online searching involves a small number of source databases which are searched through a few major online vendors. In the case of properties data, however, there are dozens of databases that have been developed by a number of organizations over a number of years, and they are available online through several different services, sometimes with search methods varying greatly from one service to another or even from one file to another. As interest in online use of properties data increases, the next few years can be expected to see the consolidation of a number of databases into larger services, or 'databanks', which will provide for searching of a variety of different data with consistent protocols and search languages. The largest databank up to the present is the Chemical Information System (CIS) mentioned above. It will be seen in the following paragraphs that this system already contains a number of the most important property databases.

Property databases are used in several ways. One is *data retrieval*, i.e. the searcher enters some identification of a substance (name, Registry Number, etc.) and asks for data about that substance. For example, the searcher might go to a file of safety data with the name of a common commercial substance looking for safety data about that substance, or to a mass spectral file looking for its mass spectrum. A second use of property data is *substance identification*; the searcher has one or more properties for an unknown substance and wishes to identify that substance. This is a common use of such data as mass spectra and nuclear magnetic resonance spectra. A third use is *data estimation*, which requires the availability of specialized software for use in conjunction with the database. Data estimation means the use of a computer program to predict data for a substance which is not on file, based on the data that are on file. One example of this, called 'spectrum estimation', uses data from a file of carbon-13 nuclear magnetic resonance data to predict the spectra of structures not on file, thus helping the user verify a hypothesis as to the structure of an unknown substance.

The process of searching a file of numeric data has special problems which must be provided for in the search system being used. One of these is the question of numeric ranges. A user asking for a substance melting at 150°C, for example, does not mean precisely 150.0000°C but rather some reasonable range about 150°C. The search system thus must store numeric values, and process numeric queries, such that the user can either specify numeric ranges, such as 147–153, or have default ranges imposed, and have such queries interpreted in a way that will satisfy his or her intent. Another key problem with numeric data is that of units. Either the user must know what units are used in the file being searched (for example, whether temperatures are Celsius, Kelvin or Fahrenheit), or else the search system must include a capability for converting the numeric values input by the user into those in which the data are stored.

The following paragraphs describe a number of important files of numeric data for chemical substances that are available at the time of writing. The reader may be assured, however, that this is a constantly changing area, and new service offerings will soon cause these lists to be outdated; the user should therefore consult database listings such as those of Gale Research mentioned earlier, or should seek current literature from database producers or vendors. This list does not attempt to cite all available data files, but is limited to a selection of files of interest in chemistry that are available from major vendors.

The databases discussed below include the categories of encyclopaedias and handbooks, spectral data, thermophysical data, toxicology and material properties.

Encyclopaedias and handbooks

These databases are, in general, the computer-readable equivalents of existing published data compilations. Two major handbooks, *Gmelin* and *Beilstein*, are discussed in Chapters 8 and 9 respectively, and also are discussed in this chapter in terms of structure searching. Both, however, are important not only as tools for substance identification, but as important sources of a variety of evaluated data. They include such data as physical properties, spectral data, comments about preparative methods, etc. Now that these handbooks are fully available online, they will constitute the largest online sources of data in terms of numbers of substances included.

In addition to *Gmelin* and *Beilstein*, some encyclopaedias of chemical data have become available online. These include the following:

Dictionary of Organic Compounds and *Dictionary of Organometallic Compounds* are available together via Dialog and the Life Science Network. The file contains information on more than 190000 substances, including physical properties, uses and reactions. The printed versions are described in Chapter 10.

Kirk-Othmer Encyclopedia of Chemical Technology and *Encyclopedia of Polymer Science and Engineering* are available in online form via BRS, Dialog, Data-Star and the Life Science Network. These files contain the full text of the encyclopaedia articles, including citations, tables and abstracts. They include a variety of information on chemical technology, manufacture and distribution of chemicals, etc. The printed versions are described in Chapter 4.

Merck Index (see Chapter 4) is a publication containing over 10 000 monographs describing some 30 000 chemicals, available online via BRS, CIS, Dialog, the Life Science Network and Questel. The files contain full text, including information on such areas as preparations, properties, pharmacological actions and toxicity.

Spectral data
Spectral data are available in several online files, including files of mass spectra, carbon-13 nuclear magnetic resonance (NMR) spectra and infrared spectra. (See also Chapter 7 for printed versions.)

Mass spectra

MSSS (Mass Spectral Search System) — produced by the US National Institute of Standards and Technology, available via CIS and Questel, contains mass spectra for more than 40 000 substances.

WMSSS (Wiley Mass Spectral Search System) — produced by Cornell University and John Wiley, available via CIS and Inkadata; contains approximately 130 000 mass spectra from the Wiley *Registry of Mass Spectral Data*.

MASS SPECTROMETRY BULLETIN, a bibliographic database on mass spectrometry and related subjects, produced by the Royal Society of Chemistry (RSC) and available via ESA-IRS. The same unit at the RSC, the Mass Spectrometry Data Centre, also contributes data to the MSSS (above) and publishes the *Eight Peak Index of Mass Spectra* (see Chapter 7).

Nuclear magnetic resonance spectra

C13-NMR/IR, available online via STN, includes over 102 000 NMR spectra for 75 000 compounds; these spectra are derived from a variety of sources, including the files of BASF and other organizations and several journals.

CNMR (Carbon-13 Nuclear Magnetic Resonance Search System), a file of more than 11 000 published NMR spectra compiled by the Netherlands Information Combine and available via CIS.

NMRLIT, a bibliographic file of some 41 000 citations related to NMR; it is produced by the US National Institutes of Health, is available via CIS and covers the years from 1964 to 1984, but it is no longer updated.

Infrared spectra

C13-NMR/IR, available online via STN, includes over 16500 IR spectra for 15000 compounds compiled by BASF.

IRSS (Infrared Search Systems) — a file of about 3000 complete spectra prepared by the Boris Kidric Institute and available via CIS.

SPIR (Search Program for Infrared Spectra) — a file of over 140000 spectra from several commercial collections and literature through 1975, collated by the American Society for Testing and Materials and available via the Canadian Numeric Scientific Database Service (CAN/SND).

Thermophysical properties

Databases related to thermophysical or thermodynamic data typically incorporate both straightforward files of numerical data values and also capabilities for predicting property values for unknown substances or for conditions different from those measured values on file. Some of these files, particularly those which are more heavily computational in nature, are available only through organizations that do not otherwise offer files of chemical interest; addresses for these are included in Appendix 2. Important files include the following.

CHEMTRAN — Produced by ChemShare Corporation and available via ChemShare and the General Electric Information Services Co. Provides data on physical properties of several hundred substances and the ability to calculate vapour–liquid equilibria.

DETHERM-SDR and DETHERM-SDC (Dechema Thermophysical Property Data Bank — Data Retrieval System and Data Calculation System). These files come from Dechema (Deutsche Gesellschaft für Chemisches Apparätewesen Chemische Technik und Biotechnologie), and are available via Inkadata. SDR contains data, with citations and abstracts, to literature on thermophysical properties of about 3000 industrially important compounds and mixtures. SDC contains data for calculating properties and phase equilibria of about 500 chemicals and mixtures in the fluid state.

DIPPR — textual information and numeric pure component physical property data for commercially important substances, including about 900 records from 1982 to the present. Produced by the Design Institute for Physical Property Data (DIPPR) of the American Institute of Chemical Engineers and available via STN.

JANAF THERMOCHEMICAL TABLES — produced by the National

Institute of Standards and Technology and available via STN. Contains evaluated data on thermochemical properties for some 1100 substances, including inorganic substances and also substances with one or two carbon atoms.

THERMO — available via CIS and STN, made up from the NIST Tables of Chemical Thermodynamic Properties (US National Institute of Standards & Technology), which contain data for over 14 000 inorganic substances and one- and two-carbon organic substances, and also the Thermodynamics Research Center Data, from the Texas A & M University Thermodynamics Research Center, which contains up to 15 values for each of about 400 organic and inorganic substances.

THERMODATA — produced and made available via Thermodata — provides critically evaluated thermodynamic data for more than 3000 inorganic substances.

Toxicology and environmental effects
See also Chapters 13 and 15. These include:

AQUIRE (AQUATIC INFORMATION RETRIEVAL DATA BASE) — produced by the US Environmental Protection Agency and available via CIS. Contains over 100 000 records on effects of over 5000 substances on fresh water and salt water organisms.

CCRIS (CHEMICAL CARCINOGENESIS RESEARCH INFORMATION SYSTEM) — produced by the US National Institutes of Health and available via CIS, DIMDI and NLM. Contains over 1200 records on more than 1000 substances, including carcinogenicity, mutagenicity and tumour promotion.

CESARS (CHEMICAL EVALUATION SEARCH AND RETRIEVAL SYSTEMS) — produced by the State of Michigan Department of Natural Resources and available via CIS. Contains toxicological data on about 200 substances, with up to 120 data items per substance.

CHEMICAL EXPOSURE — produced by the Science Applications International Corp., and available via Dialog. Contains over 25 000 citations to over 1800 chemicals that have been identified in human and animal biological media.

CHEMLIST — produced by the American Petroleum Institute and available via STN. Contains information on some 83 000 substances from the US Toxic Substances Control Act Inventory.

CTCP (CLINICAL TOXICITY OF COMMERCIAL PRODUCTS) — produced by the Dartmouth Medical School and available via CIS. Contains chemical and toxicological information on over 20 000 commercial products derived from about 3000 chemicals.

DERMAL ABSORPTION DATA BASE — produced by the US Environmental Protection Agency and available via CIS. Contains about 3000 records on health effects of about 600 chemicals administered to human and animal subjects via the dermal route.

ENVIRONMENTAL FATE — produced by the US Environmental Protection Agency and available via CIS. Contains more than 8000 records on the fate of some 450 chemicals released into the environment.

ENVIRONMENTAL FATE DATA BASES — produced and made available by the Syracuse Research Corp. Contains three interrelated databases of information on the environmental fate of organic compounds.

GENETIC TOXICITY — produced by the US Environmental Protection Agency and available via CIS. Contains mutagenicity data on over 2600 substances.

GIABS (GASTROINTESTINAL ABSORPTION DATA BASE) — produced by the US Environmental Protection Agency and available via CIS. Contains citations and reports on test conditions on more than 2400 substances administered to animal or human subjects.

ISHOW (Information System for Hazardous Organics in Water) — produced by the US Environmental Protection Agency and available via CIS. Contains data on more than 5400 substances manufactured or used in the Great Lakes watershed.

PHYTOTOX — produced by the University of Oklahoma Department of Botany and Microbiology and available via CIS. Contains about 70 000 records on the toxic effects of organic substances on terrestrial vascular plants.

RTECS (REGISTRY OF TOXIC EFFECTS OF CHEMICAL SUBSTANCES) — produced by the US Department of Health and Human Services, National Institute for Occupational Safety and Health, available via CIS, DIMDI and the National Library of Medicine. Contains over 135 000 unevaluated toxicological measurements for about 80 000 chemicals.

Properties of Materials and Products
These include:

AGROCHEMICALS HANDBOOK — produced by the Royal Society of Chemistry and available via Data-Star and Dialog. Contains information on chemical products used in agriculture.

CORROSION — available via Orbit Search Service. Contains corrosion resistance data describing the effects of about 2400 corrosive substances.

MATERIALS PROPERTY DATA NETWORK (MPD) — available via STN, the MPD comprises a set of databases which can be searched with a menu that helps in selecting the databases of interest. MPD databases include AAASD (aluminium standards and data), MARTUF (steel toughness), ALFRAC (aluminium fracture toughness), MDF (alloys), MHS (aircraft materials), PLASPEC (data on commercially available plastics) and STEELTUF (steel toughness).

POLYMAT — produced by the Deutsches Kunststoff Institut and available via STN. Contains properties and characteristics of about 7000 plastics and moulding materials.

Future Trends

At the beginning of this chapter it was noted that the dozen or so years prior to the preparation of the present edition of this book had been marked by a dramatic shift in patterns of use of chemical information. In particular, the growth in use of online databases has reached the point where most major sources of chemical information are now available online, and use of these databases is widespread in both industrial and academic laboratories. The coming years can be expected to be equally marked by changes, though not necessarily of the same sort. While prediction is always a hazardous enterprise, it seems reasonable to expect that the change in chemical information use during the 1990s will be marked by factors such as those noted below.

Development of consolidated database systems

Systems such as those of STN, Questel and, for in-house use, Molecular Design provide capabilities to search databases both of chemical structures and of data for those substances, including reactions, bibliographic data and property data. Ready accessibility to property data, however, has lagged behind the development of bibliographic and structure databases. The coming years can be expected to see the development of database systems in which improved search software provides ready access to a variety of substance data, as well as the ability to search and manipulate that data much more easily than at present. Such software will provide consistent treatment of all of the substance data within a system, while still respecting the particular characteristics and requirements of specialized types of data.

Improved ease of use of search systems

The user of today's online systems is faced with the need to acquire a fair amount of knowledge about the search software and the databases involved, and the details of this knowledge vary significantly from one system to another. Important advances can be expected in the development of user-friendly software, and particularly in the application of techniques of artificial intelligence, such as expert systems, to help the searcher in locating the information that is really needed to support his or her research activities.

In-house searching
The power of personal computers available at reasonable prices, and of powerful advanced workstations, have made a great deal of computing power available for use within the average research laboratory. Search software of considerable sophistication can be expected to become available for these computers; some of this software will have been adapted from that now in use on the large online systems. There will thus be an increasing trend towards buying or leasing databases for in-house searching rather than searching at an outside vendor's computer accessed via long-distance lines. It is not yet clear what balance will develop between use of databases on vendors' computers versus in-house use, or when it will be appropriate to use one mode or the other. One trend already evident is that of making relatively small numeric databases available for use at local sites; for example, databases of nuclear magnetic resonance and mass spectra are frequently incorporated into the computers that are part of the sophisticated analytical spectrometric instruments themselves.

Optical storage media
A great deal of attention is currently being given to the use of optical storage media as part of information systems. Attention has been particularly focused on the format known as CD-ROM (for 'compact disk-read only memory'), since these can be produced relatively inexpensively by the plants that manufacture compact disks for audio systems. These disks can hold substantial amounts of data, although several disks are likely to be required for files of significant size; for example, it has been estimated that a single volume of *Chemical Abstracts* would require at least three CD-ROM disks. It is not yet clear what applications of this technology will find a marketplace among users of chemical information, either in terms of size of file or of potential application; as a search medium, as an archival medium for long-term storage, etc. It may be the case that the CD-ROM format is not optimal for applications related to searching and that other types of optical disk holding larger amounts of data will prove more useful. Applications in which the optical disks are used to store data for display may be especially important; for example, a searcher might use an online structure search system to identify substances or reactions of interest and then access an optical disk system to obtain high-quality diagrams of the retrieved items.

Structure search capabilities

As discussed above and in Chapter 6, identity search and substructure searching of large chemical structure files have become widespread during the last decade. The next several years will see the increasing use of some types of searching that are currently still largely the subject of research or, at most, used on relatively small in-house files. Areas in which advances will be seen will include three-dimensional searching, similarity searching and biosequence searching.

Three-dimensional searching
The recent development of sophisticated molecular modelling programs, such as the CONCORD program developed at the University of Texas, makes it possible to generate large files of three-dimensional coordinates from two-dimensional connection tables. The ability to search these files for structures containing groups of atoms with distances and angles specified is of widespread interest, especially in the pharmaceutical industry.

Similarity searching
This is a concept which can be applied either to two-dimensional or to three-dimensional structures. It generally means a process in which a known structure is posed as a query against a structural database, and some similarity metric is used to identify those structures on a file which are most similar to the query.

Biosequence searching
Searching of files of biological macromolecules such as proteins and nucleic acids is already available on relatively small files. However, the amount of available sequence data is growing at a rate that will eventually overwhelm the currently used search techniques. The next few years can be expected to bring much faster and more sophisticated methods for searching such files.

Appendix 1. Publicly available databases and vendors

	BRS	CANSND	CIS	Data-Star	Datacent	Dialog	DIMDI	ESA-IRS	Inkadata	NLM	Orbit	Questel	STN	Life Sci
Bibliographic Files														
Chemical Abstracts														
CA File													x	
CAOLD													x	
CAPreviews													x	x
CA Search	x			x		x								x
Patent files														
APIPAT						x							x	x
CLAIMS						x					x		x	
INPADOC						x					x		x	
WPIM												x		
Other bibliographic files														
ANALYTICAL ABSTRACTS				x		x							x	x
BIOTECHNOLOGY								x						
CHEMICAL ENGINEERING AND BIOTECHNOLOGY ABSTRACTS				x		x					x			
CURRENT BIOTECHNOLOGY ABSTRACTS				x		x					x		x	
EMTOX						x	x				x		x	x
JICST-E								x			x		x	
TOXLINE	x		x				x	x		x	x			
Full-text Files														
CJVCH													x	
CJACS													x	
CJOAC													x	
CJRSC													x	
CJWILEY			x			x							x	
DRUG INFORMATION FULLTEXT	x												x	x

98 Online searching for chemical information

	BRS	CANSND	CIS	Data-Star	Datacent	Dialog	DIMDI	ESA-IRS	Inkadata	NLM	Orbit	Questel	STN	Life Sci
Structure and nomenclature files														
Files based on CAS Registry														
Registry File													x	
DARC												x		
CHEMZERO						x								
CHEMSIS						x								
CHEMNAME						x								
CHEMSEARCH						x								
CHEMDEX														
Chemical Nomenclature				x										
Other substance files														
Beilstein						x							x	
CHEMLINE							x							x
Gmelin Form Ind											x		x	
SANSS			x								x			
TSCA Inventory						x					x		x	
Markush structures														
CAS Marpat													x	
MPharm											x	x		
Reactions														
CASREACT													x	
CRDS														
Crystal structures														
Cambridge (CRYSTOR)		x												
CRYSTMET		x							x					
Inorganic Crystalline Structures		x												
NIST Crystalline Structures		x											x	

Sources	BRS	CANSND	CIS	Data-Star	Datacent	Dialog	DIMDI	ESA-IRS	Inkadata	NLM	Orbit	Questel	STN	Life Sci
CHEMQUEST											×			
CSCHEM												×		
JANSSEN													×	× × ×
Properties data														
Encyclopaedias and handbooks														
BEILSTEIN						×					×		×	
CHAPMAN & HALL						×								
KIRK-OTHMER	×					×								
MERCK INDEX	×		×			×						×		
Spectral data														
C13 NMR/IR			×											
CNMR		×												
IRSS			×											
MSSS			×											
NMRLIT			×									×		
SPIR			×						×					
WMSSS									×					
Thermophysical														
DETHERM-SDR, SDC			×										×	
DIPPR													×	
JANAF													×	
THERMO													×	

100 Online searching for chemical information

	BRS	CANSND	CIS	Data-Star	Datacent	Dialog	DIMDI	ESA-IRS	Inkadata	NLM	Orbit	Questel	STN	Life Sci
Toxicology, environment														
AQUIRE			x										x	
CCRIS			x				x			x				
CESARS			x			x								
CHEMICAL EXPOSURE											x			
CTCP			x						x					
DERMAL			x											
ENVIRONMENTAL FATE			x											
GENETIC TOXICITY			x			x	x			x				
GIABS			x											
ISHOW			x											
PHYTOTOX			x											
RTECS			x											
Materials														
AGROCHEMICALS														
CORROSION														
POLYMAT														

References and suggested reading

Ash, J.E., Chubb, P.A., Ward, S.E., Welford, S.M. and Willett, P. (1985), *Communication, Storage and Retrieval of Chemical Information*. Chichester: Ellis Horwood.

Directory of Online Databases, published in full twice a year, in January and July, with supplements in April and October. Detroit: Gale Research.

Hawkins, D.T. (1985), A review of online physical sciences and mathematics databases. Part 2: Chemistry, *Database*, **8**(2), 31–41.

Maizell, R.E. (1987), *How to Find Chemical Information: A Guide for Practising Chemists, Educators and Students*, 2nd edn, New York: John Wiley.

Warr, W.A. (1987), Access to chemical information: a review, *Database*, **10**(3), 122–126.

Appendix 2. Addresses

The following are addresses for online search services and vendors cited in this chapter.

BRS Information Technologies/Search Service, 8000 Westpark Drive, McLean, VA 22102, USA

CAN/SND, Canada Institute for Scientific and Technical Information, National Research Council Canada, Ottawa, Ontario K1A OS2, Canada

Chemical Information Systems, Inc. (CIS), 7215 York Road, Baltimore, MD 21212, USA

ChemShare Corporation, P.O. Box 1885, Houston, TX 77001, USA

Data-Star, Radio Suisse Ltd, Schwarztorstrasse 61, CH-3000 Berne 14, Switzerland or Plaza Suite, 114 Jermyn Street, London SW1Y 6HJ, UK

Datacentralen, Landlystvej 40, 2650 Hvidovre, Copenhagen, Denmark

Dialog Information Services, Inc.

    USA: 3460 Hillview Avenue, Palo Alto, CA 94304

    UK: Learned Information Ltd/Dialog, Woodside, Hinksey Hill, Oxford OX1 5AU

Australia: InSearch Limited/Dialog, P.O. Box K16, Haymarket, Sydney, NSW 2000

Canada: Micromedia Limited/Dialog, 158 Pearl Street, Toronto, Ontario M5H 1L3

Japan: Kinokuniya Co. Ltd, P.O. Box 55, Chitose, Tokyo 156

DIMDI, Weisshausstrasse 27, P.O. Box 420580, 5000 Köln 41, Germany

ESA-IRS (European Space Agency, Information Retrieval Service), ESRIN, Via Galileo Galilei, 00044 Frascati, Italy

Fraser Williams Scientific Systems, London House, London Road South, Poynton, Cheshire SK12 1YP, UK

Inkadata, c/o FIZ Karlsruhe, 7514 Eggenstein-Leopoldshafen 2, Germany

Molecular Design Ltd, 2132 Farallon Drive, San Leandro, CA 94577, USA

National Library of Medicine, 8600 Rockville Pike, Bethesda, MD 20894, USA

Occupational Health Services, Inc., 400 Plaza Drive, P.O. Box 1505, Secaucus, NJ 07094, USA

ORAC Ltd, 8 Blenheim Terrace, Leeds LS2 9HD, UK

Orbit Search Service

USA: Orbit Search Service, Inc., 8000 Westpark Drive, McLean, VA 22102

UK: Orbit Infoline Ltd, Achilles House, Western Avenue, London W3 0UA

Germany: Orbit Infoline GmbH, Hammerweg 6, D-6242 Kronberg

Australia: Orbit Search Service, Locked Bag 49, Botany, NSW 2019

Japan: Usaco Corporation, 13-12 Shinbashi 1-chome, Minatoku, Tokyo, 105

Questel

France: 83–85 Boulevard Vincent-Auriol, 75013 Paris

USA: Questel Inc., 5201 Leesburg Pike, Suite 603, Falls Church, VA 22041

STN International

USA:	2540 Olentangy River Road, P.O. Box 02228, Columbus, OH 43202
UK:	c/o Royal Society of Chemistry, Thomas Graham House, Science Park, Milton Road, Cambridge CB4 4WF
Germany:	Postfach 2465, D-7500 Karlsruhe 1
Japan:	c/o Japan Information Center of Science and Technology, 5-2 Nagatacho 2-chome, Chiyoda-ku, Tokyo 100
Australia:	CSIRO Information Resources Unit, 314 Albert Street, East Melbourne, Victoria 3002

Syracuse Research Corporation, Merrill Lane, Syracuse, NY 13210, USA

Tech Data, Information Handling Services, Department 438, 15 Inverness Way East, P.O. Box 1154, Englewood, CO 80150, USA

Technical Database Services Inc., 10 Columbus Circle, Suite 2300, New York, NY 10102, USA

CHAPTER SIX

Chemical structure handling by computer

M.D. COOKE

Introduction

Since the previous edition of this book, the orders-of-magnitude increase in performance/price ratio of computers has revolutionized the availability to end-users of means of handling chemical structures in the ways most natural to chemists. Few serious systems now offer anything less than queries and search results displayable in the standard language of chemists, the structure diagram. More recently, and in line with the increasing recognition of the importance of the third dimension in determining the properties of a molecule, the commercial emphasis has shifted to include 'three-dimensional database management systems'.

This brief review can provide only broad understanding of the principles underlying the main classes of chemical structure handling systems available to chemists. Recent publications should be consulted for further details on the theory (Ash *et al.*, 1985) and on current developments (Warr, 1988). Likewise, with recent summaries available (e.g. Meyer *et al.*, 1988), it makes no attempt to be comprehensive for all the software available: in any case, the situation and the program names and specifications change too rapidly!

Linear notations

Despite the availability of graphical input and display (see below), the Wisswesser Line-formula Notation (WLN) (Smith and Baker, 1976) and the more recent SMILES (Simplified Molecular Input Line Entry System) (Weininger *et al.*, 1989) retain their clear devotees.

```
WLN    = Z2R CQ DQ
SMILES = c1cc(O)c(O)cc1CCN
```

dopamine

Linear notations share the advantages of a minimal number of keystrokes needed for input, and the corresponding low demands on disk space. SMILES itself has an extra feature promoting its continued use, in being the external communication language for the MedChem (addresses are given in Appendix 1) set of chemical software. The CCT module of this program suite allows two-way communications with the connection table (CT) systems which underlie the more standard software representations detailed below. Daring achieves a similar conversion to CT for WLN.

Nevertheless, and despite the substructure searching facility provided in MedChem's associated MERLIN module, there remain serious drawbacks in the use of such linear notations. Although SMILES is relatively easy to encode (the user needs recourse to far fewer rules than the corresponding WLN set), any ready comprehension of the resulting alphabetic strings is still beyond the wit of all but the notation specialist. Also, SMILES cannot yet handle key structural features, such as stereochemistry or non-standard bonds. This prevents its use in any application where a unique and exact representation is essential.

Two-dimensional graphical systems

General principles
The distinguishing feature is that both input and display are in the familiar line-diagram universally used by chemists. Behind the simple appearance lies, however, a complex set of algorithms which enable the structure to be stored (a) accurately, (b) compactly, and (c) yet with sufficient information to enable fast retrieval from databases of over 100 000 compounds.

In the major commercial in-house systems (MACCS/REACCS, DARC-SMS and ORAC/OSAC) these objectives are achieved by the use of the following.

Connection table (CT)
The CT, in essence, is a list of all the atoms in the molecule; for each atom the associated data could consist of:

- the atomic type (C, O, N, S, etc.)
- the bond type to, and atom number of, each atom to which it is bonded
- the associated charge (if any)
- the x, y, z-coordinates
- the stereochemistry (if any)

It may be realized that, in the example given, hydrogen atoms need not be specifically identified, since their positions can later be recalculated (by the logic of valencies).

Although the CT is sufficient to enable searching to be complete, it is not a compact form of information, and searching will be inefficient. Further information is therefore held as SEMA.

SEMA (Stereochemically Extended Morgan Algorithm) code (Wipke and Dyott, 1974)

This is a unique hash-coded value determined from the CT. The presence or absence of the corresponding SEMA code for an exact-match query molecule enables near-instantaneous search response even from a large (over 100 000 compounds) database. In a further refinement an ACMF (Augmented Connectivity/Molecular Formula) value may also be stored.

Screens

These are to enable efficient substructure searching. For every registered compound a number of bits will be set or cleared depending on the presence or absence of certain key structural features. Examples of these could be ring size, heteroatoms, heteroatoms in rings, specific functional groups, or chain length. The optimal design of such a screening system will depend on the nature of the structure file, the nature of queries likely to be posed and the size of the file. The object is to ensure an even population of set bits over perhaps some 2000 such screens.

The bits set/cleared for the query fragment are compared against those for the registered compounds — and a reduced subset (typically less than 5 per cent) of the database is screened off; in a second and slower phase, the CT for each potential candidate is matched atom-by-atom against the query structure, thereby generating true 'hits'.

Graphic display

This display may be generated directly on a special high-resolution graphics terminal, or the required graphics may be emulated on a lower-resolution screen (as on a PC) by the use of special emulation software.

Structure-handling software

The software handling structure input may take CTs generated externally, or input directly to the computer under conventional cursor/keyboard/mouse control. As aids in this latter process:

1. Templates of common structural fragments (e.g. rings, amino acids, functional groups) may be retrieved by the user from pre-formed files.
2. Chemical intelligence may be employed by the software to highlight invalid valencies (e.g. pentavalent carbon atoms).

3. Displayed (part-) structures may be rotated, zoomed, stretched, etc., to provide the appearance of a particular conformation/view desired by the user.
4. Input structures may be automatically 'refreshed' or 'cleaned' to provide normalized bond lengths and angles; atom crowding and bond overlap is minimized (Shelley, 1983).
5. Stereochemical centres can be defined in a range of techniques (R/S labelling, wedge/dotted bonds, etc.).
6. Certain 'gross features' may be recognized, such as the presence of an aromatic ring from the input of one Kekulé structure.
7. Heteroatom-bonded hydrogens may be suppressed or made explicit in the display.
8. Other shortcuts may be available, such as group truncations, superatoms and polymer representation.

Chemical reactions

As a minimum, the computer handling of chemical reactions involves the storage of the reactant molecule(s) and the product molecule(s), with similar screening data as indicated above. Additionally, specific reactant(s) and product(s) are linked by reference to a particular reaction, and their status as reactant or product in this reaction is also recorded.

One further feature, atom-to-atom mapping, is however also essential to ensure that 'false hits' are eliminated. Without such direct mapping, a search for (for example) the oxidation of a ketone to yield an ester (and using the search technique of substructure search for ester in the product and substructure search for ketone in the reactant) would pull out the following incorrect type of reaction:

Atom-to-atom mapping ensures that both the carbon and the oxygen atoms in the ketone group in the reactant are mapped to the corresponding C and O atoms in the ketone group in the product — and in the search technique indicated above, the further rider must be added by the software handling the search query that when apparent substructural matches are found, each atom in the reactant should be found atom-mapped to the corresponding fragment found in the product.

Most programs use intelligence during database creation to 'automap' the input reactions, but the algorithms used are not perfect. Symmetry and rearrangements can result in false mapping, and the database inputter may therefore expect to have to adjust the mapping in some 5 per cent of the reactions.

Handling similar molecules and reactions

The substructural search techniques described so far allow the retrieval of related molecules which exactly meet the query structural requirements. Increasingly, however, chemists have recognized the need for help in generating ideas. In structural terms, this may mean 'fuzzy searching', where the results of a search are ranked by the degree to which molecules meet the query structure. The computer application of this area is due to the seminal work of Adamson and Bush (1975).

In practice, one simple implementation could be to define that minimum fraction (the Tanimoto coefficient) of the structural bit screens that must match. Better matching of a 'find' to what a chemist truly regards as a similar structure also requires that the criteria used in the comparison are differentially weighted.

Potential problems

Since chemists themselves can often disagree on a 'correct' representation, it will come as no surprise that the same problem areas can cause difficulties in the computer environment. Typically the following problems have been solved in one way or another — but rarely in all ways to satisfy all potential users all of the time.

Aromaticity

The problem is simply illustrated by thiophen. Depending on whether the ring bonds of thiophen are considered as a diene (cf. *Chemical Abstracts* convention, also followed by MACCS), or aromatic ($4n+2\pi$ electrons, followed for example by the ORAC/OSAC system) will affect whether a match is made or missed in a search for thioenols.

Other non-conventional bonds

These are exemplified by π-allyl or transition-metal carbonyls, and special input conventions may be required. In principle the chemist should be able to enter any of the following:

$Fe(CO)_3$ $Fe(CO)_3$ $Fe(CO)_3$ $Fe^+(CO)_3$

and 'hits' should be obtained on the general internally held representation.

Tautomers

A classic test case is presented by 4-hydroxypyridine or 4-pyridone:

110 *Chemical structure handling by computer*

The ORAC/OSAC solution is to store a registered molecule in all tautomeric forms, as generated by its intelligent rule-base. A query for either tautomer (or corresponding substructure) will then be successful. The success of this approach depends on the skill of the designers of the rule-base to ensure that all possible tautomers are in fact generated; 1,3,5-trihydroxybenzene fails at the time of writing.

The approach adopted by MACCS is radically different. Under the tautomer search option, a match is first made for the skeleton ignoring bond-orders. A further constraint is that target and search skeletons contain the same number of hydrogen atoms, and if so, a tautomeric match is assumed. This technique can clearly lead to 'false drops', for example few chemists would recognize a significant *enol* component in a normal saturated ester:

$$RCH_2CO_2Et \equiv R-CH=C\begin{smallmatrix}OH\\OEt\end{smallmatrix}$$

Stereochemistry
Stereochemistry is usually considered by the software as an atom-centred property. Biphenyls with 2, 2', 6, 6'-substituents sufficiently bulky to cause restricted rotation possess chirality about a bond and cannot at present be distinguished by the conventional CT system.

Other problems arising from 'perspective' are well described by Wipke (1987).

Superatoms and repeating units
Even for relatively small molecules, screen clutter and common parlance prompt the need for chemists to specify groups of atoms by shorthand methods. All major systems now allow not only input but also optional display of peptides in the form H-ala-gly-cys-OH. To enable accurate substructure searching, however, the internal CT is based on the full expanded structure.

Similar principles apply to polymers, expressed as for example

But in all cases, even though the shorthand may be brief, the maximum molecule size may still be limited by the software to perhaps 255 specified atoms (non-hydrogens and explicit hydrogens).

Atom-to-atom mapping
A chemist searching for N-oxidation of a pyrimidine might expect to find the following reaction

$$\text{pyrimidine} \xrightarrow{H_2NOH} \text{pyrimidine N-oxide}$$

Because the reaction actually involves a ring-opening and insertion of a different nitrogen atom (that in hydroxylamine), the two one-nitrogen atoms should not be mapped to each other, and a search strategy based on the formation of a N-O bond will therefore fail.

Multistep reactions
In even a simple three-step reaction (A → B → C → D) there are six transformations implicit, each of which should be retrievable by a searcher. The brute technique is to store each transformation as a separate reaction — but this has the obvious disadvantages of extra database demands and extra input; the application of computers would be to make some of the implicit steps (such as A → C in the above scheme) automatically generated by the computer.

A further complication arises from two individual reactions (e.g. A → B and B → C) which are registered in different databases at different times. Will a user be able to retrieve the theoretically possible reaction of A → C?

It seems true to report that although ideas abound (e.g. Blower *et al.*, 1987; Johnson, 1987), no fully satisfactory commercially implemented solution to multistep reactions yet exists.

Many of these examples exemplify the extreme care required by users in understanding the nature of the underlying databases. Search strategies may need to be modified accordingly to avoid (or make use of) these particular features.

Three-dimensional graphical systems

Unlike the essentially topological representations described above, it is a *sine qua non* that a three-dimensional structure system must be based on precise x,y,z-coordinates for each atom.

The systems available fall into the following broad (and often overlapping) categories.

1. *Simple display of stored three-dimensional structures.* These programs

usually allow the display to be viewed from any angle; the appearance can be ball/stick, solid, charge-density, etc. Impressions of depth can be achieved by colour density, and by spot-highlighting of solid spheres.

2. *Systems designed for complex pattern searching.* Such software will have additional keys (or screens) allowing searching based on, for example, specified distance, bond-angle or torsion-angle criteria.

3. *Systems designed to 'create' three-dimensional databases for use in (1) or (2).* The underlying software, much of it in the public domain through QCPE, covers the areas of molecular mechanics, quantum mechanics, *ab initio* and semi-*ab initio* calculations, and conformational analysis.

The information perspective

Viewed from the information perspective it is, however, the database systems (1) and (2) which are of most importance. Arguably, the most reliable data come not from theoretical calculations (3), but from actual crystal data.

The Cambridge Structural Database contains some 90 000 structures (1990); the data are evaluated for their chemical consistency (75 per cent apparently need correction of one sort or another), and the resulting LASER database can then be searched using QUEST software, and displayed using either CCDC's GSTAT/PLUTO combination, or other molecular modelling software, such as Sybyl.

Different problems are associated with the handling by computers of macromolecules, such as the proteins and nucleic acid structures given in the Brookhaven Protein Databank (Bernstein *et al.*, 1977); here the size of the stored molecules dictates that some simplification of the tertiary structure is essential if the resulting displays are to convey any meaning. Two techniques are commonly used:

1. All but the amino acid backbone is stripped off, and the resulting chain depth-cued, and/or

2. Discrete regions are analysed for the secondary protein structure — descriptive pictures of α-helix or β-sheet can then be graphically displayed (usually using static shaded-image on raster graphics) in the form of coloured cylinders and vectors (Lesk and Hardman, 1982).

In-house databases cannot easily rely on such crystal structure data, and for large databases quick conversion of two-dimensional to one or more 'theoretical' three-dimensional structures is accomplished using programs such as CONCORD (Rusinko *et al.*, 1989) or COBRA. The resulting three-dimensional connectivity can then be analysed, and bit screens set up to reflect distance data between two points in space; such points are usually defined as key atoms in the molecule. Much work is in progress to improve the search performance of such complex three-dimensional files (cf. Jakes *et al.*, 1987). Typical industrial products based on these principles are 3DSEARCH (Sheridan *et al.*, 1989), ALADDIN (Van Drie *et al.*, 1989), and MACCS-3D. One drawback to this approach is that, in general, only one or two conformations of any particular molecule are generated and stored; although the chosen conformation(s) may be of lowest energy, it/they may or may not be the most appropriate for the desired pharmacophoric searching.

In a recent innovation, the ChemDBS-3D product allows the storage of multiple conformations (say within 5–10 kcal of the energy minimum). This can be achieved within one 32-bit word per screen used; each bit represents one distance range, and is set if any one of the conformations contains the appropriate distance between a pair of defined centres. By restricting the definition of such centres to only a few chemical properties (such as H-donor, H-acceptor, charge, or ring centroids — rather than atom types), the total conformation range can be accommodated in a very limited number of very compacted three-dimensional screens. The immediate benefit is that the resulting database demand is dramatically reduced.

Searches are likewise screened, and potential hits identified, using the Ullman algorithm (see Brint and Willett, 1987); exact matching follows if the regenerated conformers exactly match the original query (Murrell and Davies, 1991).

Conventions and front-ends

Conventions are required if different programs are to exchange data with the minimum need for conversion routines. Various standards have emerged, notably:

> The MOLFILE format, reflecting the dominant position of Molecular Design Ltd (MDL) not only in world in-house installations but also in its own range of products from mini/mainframe to PC;

> The Standard Molecular Data (SMD) format (Barnard, 1989), the impetus for which might be said to be in part due to a desire originally by the non-MDL world to communicate with each other.

Neither convention yet has any official status comparable to the IUPAC statements on chemical nomenclature. Until adopted by, or at least given the blessing of, more than one of the major national standards groups and/or publishing groups, the formats will always remain 'tentative'. Software houses are then understandably reluctant to commit themselves to the import/export interfaces to a possibly variable 'standard'.

Front-ends address the related problem which occurs with presenting a user-acceptable interface to the various distinctly different chemical information systems to which any one user might wish to have access. As well as handling issues like a common command language, enabling auto-dialling, and providing usage statistics for accounting purposes, such 'front-ends' allow chemists/users to:

Use graphical input, while the actual access to the host system may have been devised with a traditional (and unfriendly) command language;

Pre-formulate or pre-process queries locally, thus saving on the communications and host time;

Capture search output in machine-readable form for further local processing.

Some of the major front-ends meeting one or more of these criteria include:

STN Express (for CA Online);

MOLKICK (for STN, Dialog and Questel) compiles a created structure to the appropriate search strings, i.e. CAS, DARC or ROSDAL (Beilstein);

DARC CHEMLINK (for DARC in-house, Markush DARC or Generic DARC).

References

Adamson, G.W. and Bush, J.A. (1975), A comparison of the performance of some similarity and dissimilarity measures in the automatic classification of chemical structures. *Journal of Chemical Information and Computer Science*, **15**, 55–58.
Ash, J.E., Chubb, P.A., Ward, S.E., Welford, S.M. and Willett, P. (1985), *Communication, Storage and Retrieval of Chemical Information*. Chichester: Ellis Horwood.
Barnard, J.M. (1989), Draft specification for revised version of the Standard Molecular Data (SMD) format. *Journal of Chemical Information and Computer Science*, **30**, 81–96.
Bernstein, F.C., Koetzle, T.F., Williams, G.J.B., Meyer, E.F., Brice, M.D., Rodgers, J.R., Kennard, O., Shimanouchi, T. and Asumi, M. (1977), *Journal of Molecular Biology*, **112**, 535.
Blower, P.E., Jr, Chapman, S.W., Dana, R.C., Erisman, H.J. and Hartzler, D.E. (1987), In Warr, W.A. (ed.) *Graphics for Chemical Structures: Integration with Text and Data*, ACS Symposium Series 341, pp. 399–407. Washington DC: ACS.

Brint, A.T. and Willett, P. (1987), Pharmacophoric pattern matching in files of 3D chemical structures: comparison of geometric searching algorithms. *Journal of Molecular Graphics*, **5**, 49–56.

Jakes, S.E., Watts, N., Willett, P., Bawden, D. and Fischer, J.D. (1987), Pharmacophoric pattern matching in files of 3D chemical structures: evaluation of search performance. *Journal of Molecular Graphics*, **5**, 41–48.

Johnson, A.P. (1987), In Warr, W.A. (ed.) *Graphics for Chemical Structures: Integration with Text and Data*, ACS Symposium Series 341, pp. 297–302. Washington DC: ACS.

Lesk, A.M. and Hardman, K.D. (1982), *Science (Washington, DC)*, **75**, 539.

Meyer, D.E., Warr, W.A. and Love, R.A. (1988), *Chemical Structure Software for Personal Computers*. Washington DC: ACS.

Murrell, N.W. and Davies, E.K. (1991), *Conformational Freedom in 3-D Databases* (in press).

Rusinko III, A., Skell, J.M., Balducci, R., McGarity, C.M. and Pearlman, R.S. (1988), *CONCORD: A Program for the Rapid Generation of High-quality Approximate 3-Dimensional Molecular Structure*. St. Louis, MO: The University of Texas at Austin and Tripos Associates. See also Rusinko III, A. et al. (1989), *Journal of Chemical Information and Computer Science*, **29**, 251–254.

Shelley, C.A. (1983), *Journal of Chemical Information and Computer Science*, **23**, 61.

Sheridan, J.P., Nilakantan, R., Rusinko III, A., Bauman, N., Haraki, K.S. and Vankataraghavan, R. (1989), 3DSEARCH: A system for three-dimensional substructure searching. *Journal of Chemical Information and Computer Science* **29**, 255–260.

Smith, E.G. and Baker, P.A. (1976), *The Wiswesser Line-Formula Chemical Notation*, 3rd edn. Cherry Hill, NJ: Chemical Information Management.

Van Drie, J.H., Weininger, D. and Martin, Y.C. (1989), ALADDIN: An integrated tool for computer assisted molecular design and pharmacophore recognition from geometric, steric, and substructural searching of three-dimensional structures. *Journal of Computer Aided Molecular Design*, **3**, 225–251.

Warr, W.A. (ed.), (1987), *Graphics for Chemical Structures: Integration with Text and Data*, ACS Symposium Series 341. Washington, DC: ACS.

Warr, W.A. (ed.), (1988), *Chemical Structures: the International Language of Chemistry. Proceedings of the Conference at Noordwijkerhout, Netherlands*. Springer-Verlag. Corresponding *Proceedings* for the 1990 Conference are in press.

Weininger, D., Weininger, A. and Weininger, J.L. (1989), SMILES. 2. Algorithm for generation of unique SMILES notation. *Journal of*

Chemical Information and Computer Science, **29**, 97–101. Reference 1 in this paper (Weininger, D., 1988, Journal of Chemical Information and Computer Science, **28**, 31ff) gives the original publication.

Wipke, W.T. and Dyott, T.M. (1974), Stereochemically unique naming algorithm. Journal of the American Chemical Society, **96**, 4825–4834.

Wipke, W.T. (1987), In Warr, W.A. (ed.) *Graphics for Chemical Structures: Integration with Text and Data*, ACS Symposium Series 341, pp. 1-8. Washington, DC: ACS.

Appendix 1: Addresses
Cambridge Crystallographic Data Centre, University Chemical Laboratory, Lensfield Road, Cambridge CB2 1EW, UK

ChemDBS-3D: Chemical Design Limited, Unit 12, 7 West Way, Oxford, UK

COBRA: Oxford Molecular Ltd, Terrapin House, South Parks Road, Oxford OX1 3UB, UK

Daring: Fraser Williams (Scientific Systems) Ltd, London House, London Road South, Poynton, Cheshire SK12 1YP, UK

MACCS: The Molecular ACCess System; also MACCS-3D. See address for REACCS

MedChem Project: Pomona College, Claremont, CA 91711, USA

ORAC (Organic Reactions Accessed by Computer): ORAC Ltd, 18 Blenheim Terrace, Leeds LS2 9HD, UK

OSAC: Organic Structures Accessed by Computer. See address for ORAC

Quantum Chemistry Program Exchange, Indiana University, Bloomington, IN 47401, USA

Questel, 83–85 Boulevard Vincent-Auriol, 75013 Paris, France

REACCS (The REactions ACCess System): Molecular Design Limited, 2132 Farallon Drive, San Leandro, CA 94577, USA; or 10 Armstrong Mall, Southwood Summit Centre, Farnborough, Hants. GU14 0NR, UK

Sybyl: Tripos Associates, St. Louis, MO, USA

Section B: Pure Chemistry

CHAPTER SEVEN

Standard tables of physico-chemical data

R.T. BOTTLE

Introduction

Physical chemistry often provides the theoretical basis of our understanding of phenomena associated with either organic or inorganic compounds or biochemical systems. Increasingly it is the basis of investigational techniques used in other branches of chemistry. Thus most books about physico-chemical methods fit into other specific subject chapters of this book. This chapter contains only a small section at the end dealing with general and practical treatises, selected monographs and texts. Physical chemistry allows us to quantify relationships and thus the bulk of the chapter is concerned with compilations of data obtained by physico-chemical techniques and/or required in physico-chemical calculations.

A literature search for the best numerical value of a physical property of a substance or material can be a most time-consuming operation. There are many reasons for this, and perhaps the important ones are, firstly, that many physical properties are determined as a means to an end (for example, identification) or as a step in the determination of another property, such as density, in a step towards kinematic viscosity determinations. Many physical properties of materials are ill-defined; the rigidity of a polymer, or the electrical conductivity of an electrolyte, needs careful definition before numerical values can be assigned to it. In the early years of physical chemistry and of chemical physics, new work was published in journals, etc., with a very limited circulation. Thus many details of physical properties are deeply buried in the literature, and effort, patience and time are required to retrieve them. For this reason, the standard books of tables are invaluable.

The first edition of Kaye and Laby's *Tables of Physical and Chemical Constants* (1911) was prepared like that of many other compilations by individuals having a strong interest in the properties of matter. Nevertheless, the compilation of a book of tables nowadays is beyond the capacity of any one person, and it is now usual to have an editorial board who in turn delegate the collection and scrutiny of data to specialists. These data are afterwards edited and published in a series of volumes. Such a system means that publication occurs when the data are ready. Therefore, it is not possible to follow any order of publication, and this occurs over a period of many years. Often supplementary and additional volumes are published (cf. the 5th edition of Landolt-Börnstein's *Tabellen*).

The compiling of data on physical properties, and their critical scrutiny and editing, is today a formidable and complex task which can only be solved by collaboration of workers from all over the scientific world and continuous publication year by year. It is in this way that such reference books as the *International Critical Tables, Landolt-Börnstein's Zahlenwerte und Funktionen aus Physik, Chemie, Astronomie, Geophysik und Technik*, etc., have been created, and in recent years the databases of chemical properties.

Tables of this type possess three essential characteristics. Each volume, or section, has a concise introduction giving the basic physical facts upon which the data are based. This is followed by the numerical data and finally by a complete series of references to the original literature. They give the critical researcher sufficient information to estimate the accuracy and status of the figures quoted and sufficient references to enable setting about making a personal determination if wished. All these features are present in any good quality tables.

Making a search

A search for numerical data from within books of tables can be practised at several levels of competence. Scientists who use all their scientific training, experience and imagination while making a search, will be richly rewarded. The information collected will be much more than the numerical data. The scientist will be aware of physical properties not known, or imperfectly understood, and may see how to combine the numerical data into a new and useful picture which may offer an explanation that has eluded previous workers and searchers. His or her knowledge of physical principles and experimental techniques will undoubtedly be extended.

Time is always saved if a little thought is exercised before a search is begun. It is advisable to write down, in detail, exactly what is sought, the definition of the unit in which it is expressed, the order and accuracy required, and the division of physics to which the property belongs. Often a physical property is sought which is not normally expressed directly but which is a function of two or more physical properties that can be given values.

Standard tables of physico-chemical data 121

A mathematical training is useful, but a basic knowledge of physics is essential, for those who wish to make a search for the numerical value of physical properties. It is not so much the knowledge of the technique of mathematics as an understanding of mathematical philosophy which enables a search to be conducted quickly and systematically. When the question has been clearly defined, the next stage is to read the introduction to the appropriate section of the book of tables and if necessary the subject matter in any standard physics text. When an understanding of the subject has been obtained, then a beginning may be made upon standard books of tables.

A useful tip is to ask oneself the question, 'Do I believe that the data will have been published (with sufficient accuracy for my purpose) before 1924?' If the answer is 'yes', then go straight to the index of the widely available *International Critical Tables*.

Generalized critical tables

International Critical Tables (7 vols and index, E.W. Washburn, ed., McGraw-Hill, 1926-33) were prepared under the auspices of the International Research Council and the National Academy of Sciences by the National Research Council of the USA under US editorship. They were broadly based upon the International Tables already published annually (see below).

The subject matter is divided into 300 sections, each section being critically examined by the best contributor available at that time. The tables satisfy all the criteria given above and the literature up to 1924 was examined in great detail. Nevertheless after 1924 a time-consuming search must be made in *Chemical Abstracts*, etc. (unless one can use *Landolt-Börnstein*). These tables have been well indexed with full cross-references. This allows easy reference to all subjects and materials and the index is thus the normal method of entering the tables. In a given table chemical compounds are arranged by formula according to a set of key numbers for the elements called 'Standard Arrangement'. (This is explained in Vol. I, page 96 and Vol. III, page viii. In most cases one can find the required compounds quite quickly without bothering with the 'Standard Arrangement'.) The bold print numbers in the bibliography following each table are a journal code which will be found at the end of each volume. The tables are quite easy to use and are probably the most suitable source of reference up to 1924.

Landolt-Börnsteins Zahlenwerte und Funktionen aus Physik, Chemie, Astronomie, Geophysik und Technik (Springer) is the sixth edition of an internationally known set of tables published between 1950 and 1980. A brief synopsis in English is appended to this chapter to help those not fluent in German to select the right volume quickly. The introductory sections of all volumes and parts are comprehensive: indeed, if a basic knowledge of physical principles is assumed a reader can use the introductory sections as a sound textbook of physics. The literature references are exhaustive and allow the reader to examine the original data upon which the tables are based and to repeat the work if necessary. These tables satisfy all the criteria given in the introductory section of this chapter and are probably the most accurate published data on physical constants. The first edition in 1883, of just 250 pages, was the first collection of physical constants with literature references for each value given. The fifth edition (1923) and its six volumes of supplements (1927–36) were, like the earlier editions, known as the *Tabellen* and often purchased by individual scientists. The number of pages in each edition shows exponential growth with time. One could forecast that a seventh edition would run to over half a million pages. Not surprisingly, it was decided around 1960 that there would be no seventh edition, but that the data of the sixth edition would be supplemented by a new series of volumes of a narrower subject field, published as the need arises and when the data are accumulated. This is called *Zahlenwerte und Funktionen aus Naturwissenschaften und Technik: Neue Serie/ Numerical Data and Functional Relationships in Science and Technology: New Series*, and now runs to over 130 volumes. At least part of the material may be made available in non-print format in the future. It started in 1961 under the general editorship of K.-H. Hellwege; currently O. Madelung is the chief editor. Each volume has its preface, table of contents and introductory chapters in English and German. The volumes are divided into seven groups: I, Nuclear and Particle Physics; II, Atomic and Molecular Physics; III, Crystal and Solid State Physics; IV, Macroscopic and Technical Properties of Matter; V, Geophysics and Space Research; VI, Astronomy, Astrophysics and Space Research; VII, Biophysics.

A *Comprehensive Index for the 6th Edition 1950–1980 and New Series 1961–1985* (300 pages) in both English and German was published in 1987. A paper bound *Comprehensive Index 1990* (only in English) covers the sixth edition and *New Series* volumes to the end of 1990. A gratis floppy-disk version enables one to search by keyword. Updated versions will probably appear every two years. A free *Guide to Landolt-Börnstein Data* uses the keywords in chapter headings in the sixth edition and *New Series* to lead one to the right volumes(s).

Kaye and Laby's Tables of Physical and Chemical Constants (Longmans). The 15th edition (1986) of these *Tables* has been published under the guidance of an editorial board of physicists drawn mainly from the National Physical Laboratory (NPL), who in turn collected critical contributions. SI units have been used since the 13th edition. (SI units are a consistent set based on the m, kg, s and amp set up by the

Standard tables of physico-chemical data

11th General Conference of Weights and Measures, 1960.) The data given satisfy all the criteria in the introductory section and cover a wide range of properties in just over 400 pages. The subjects covered are general physics, chemistry, atomic physics and mathematical tables.

Handbooks

This is a well-defined class of tables that contains a wealth of data which have been carefully compiled and edited in an effort to select material to meet the needs of scientific workers who lack the facilities of large technical libraries, which are often not conveniently near manufacturing centres. As a result every effort is made to select the most reliable information and to record and print it with accuracy. In many instances editions are prepared every year and in any case an editorial board is continuously editing, adding data and removing obsolete data from such handbooks. As a result, in the case of handbooks which have passed in a large number of editions, the data offered are accurate and of topical value to all scientific workers.

It is today a practice to publish special handbooks devoted to each well-known industry. Over 3400 were listed in *Handbooks and Tables in Science and Technology* (R.H. Powell, ed., 2nd edn, Onyx Press, 1983). To list all such handbooks would not be appropriate here. The names and outline of contents of the important general handbooks of data followed by more specialized ones are, however, given below.

Perry's Chemical Engineers' Handbook (6th edn, McGraw-Hill, 1984) is an authoritative reference book covering comprehensively the field of chemical engineering as well as important related fields. A considerable amount of data has been taken from the *International Critical Tables* but this has often been rearranged and recalculated in units used by engineers. The first three editions of *Perry* were edited by the late John H. Perry and the sixth edition was edited by the late R.H. Perry and D.W. Green. The following sections of physical data are covered:

Units and conversion tables; physical properties of pure substances: specific gravity, melting point, boiling point: vapour pressure of pure substances and solutions: dissociation pressures: densities of pure substances and aqueous inorganic solutions: thermal expansion: Joule-Thomson effect, critical constants, compressibilities. Latent heats of pure compounds, specific heats and thermodynamic properties: freezing points and elevation of boiling points: thermodynamic properties and chemical reaction kinetics: flow of fluids in pipes and channels; viscosity data; technology of fluid dynamics; heat transmission by conduction and convection; radiant heat transmission; heat transfer in evaporation; diffusional operations, distillation, solvent extraction and gas absorption, equilibrium relationships; distillation and sublimation; gas absorption and equilibrium data; solvent extraction and dialysis; thermodynamic properties of moist

124 *Standard tables of physico-chemical data*

air; electrochemical equivalents.

Handbook of Chemistry and Physics (D.R. Lide, ed, 72nd edn, Chemical Rubber Co./Blackwells, 1991):

Mathematical tables, numerical tables; physical constants of elements, inorganic compounds, organic compounds, alloys, plastics; thermodynamic constants of elements, oxides, hydrocarbons; thermal expansion, vapour pressure, heat conductivity; acoustics; velocity of sound and sound absorption; electrical characteristics; units and conversion factors; miscellaneous basic physical data; sources of data.

The first edition of this well-known but no longer cheap laboratory companion for chemists and physicists was published in 1914. Recently, new editions have been issued annually and with the 44th edition a considerable enlargement occurred. A much cheaper students' edition is also published. Also from the same publisher is *Handbook of Data and Organic Compounds* (R. C. Weast, ed., 2nd edn, 9 vols, 1988). CRC produce some 300 handbooks covering the physical and life sciences; a *Composite Index for CRC Handbooks* (3 vols or CD-ROM, 3rd edn, 1991) is now available. Annual supplements in print or CD-ROM format are planned. Although colloquially known as *The Bible* or *Weast*, the *Handbook of Chemistry and Physics* is often referred to as the *Rubber Handbook*; unfortunately a recent publication of the Swedish Institution of Rubber Technology now uses the title *Rubber Handbook* (1990). (This supersedes *Gummiteknisk Handbok*, 7th edn, 1985.)

Handbook of Chemistry (J.A. Dean, ed., 13th edn, McGraw-Hill 1985):

Physical properties of elements, minerals, organic compounds, industrial materials; miscellaneous tables of specific properties, solubility, density, electrical properties refractivity, crystal structure, hydrometry, vapour pressure, thermal properties, surface tension, viscosity, compressibility and expansion; numerical tables.

This is known as *'Lange's Handbook'* and like the *'Chemical Rubber Handbook', International Critical Tables, 'Landolt-Börnstein'*, etc. contains 'inverted' tables where the arrangement is according to the magnitude of a particular physical property and not by substance. Such indexes of melting and boiling points are well known for identifying compounds, but indexes of densities, refractive indexes, etc. are also useful. A particularly useful collection of such tables is contained in *Handbook of Tables for Organic Compound Identification* (3rd edn, Chemical Rubber Co., 1966).

Standard tables of physico-chemical data

In the *Longman Dictionary of Physics* (A. Isaacs and H.J. Gray, eds, 3rd edn, Longmans, 1991) the data are alphabetically arranged, carefully edited and critically surveyed.

The newer encyclopaedias and dictionaries (see Chapter 4) often contain physical data which are not available elsewhere. Because many of the contributions are written by authorities on their subject who include hitherto unpublished information with the usual criteria of accuracy. Since their arrangement is usually alphabetical, this permits easy and quick reference for physical data. Somewhat different in arrangement is *A Physicist's Desk Reference* (H.L. Anderson, ed., 2nd edn, Adam Hilger, 1989). This contains a useful collection of formulae, data and definitions.

Many series, monographs and reviews on organic chemistry contain sections, sometimes 50 pages or more in length, giving tabulated data. A number of these and other sources of data on organic compounds, including spectra, composition tables and other analytical data are listed in Chapter 10. The most comprehensive source of information in English on organic compounds is *Dictionary of Organic Compounds (DOC)*; its data are, however, often limited to melting and boiling points but it includes references (see p. 164). *Beilstein* (Chapter 9), *Gmelin* and *Mellor* (Chapter 8) should also be remembered as printed sources of data on organic and inorganic compounds.

Physical properties databases

Physico-chemical data were an early constituent of computerized factual databases (sometimes referred to as databanks) from the early 1960s. Several of these were described at a symposium, the proceedings of which were published in the February 1967 issue of *Journal of Chemical Documentation*. (Also of historical interest are the punched card systems used three or four decades ago. An example is *Documentation of Molecular Spectroscopy*, 1957–72, where 2500 spectra were issued on double edge-notched cards.) Although some databanks are derived from and update earlier printed equivalents, a number are not and exist without a printed equivalent.

Most chemical databanks contain physico-chemical data. A number are described in other chapters; for example those covering health and safety are discussed in Chapter 13. It is reasonable to assume that most of the data which can be associated with specific compounds can be retrieved via the GMELIN database for inorganic and organometallic compounds (Chapter 8) or from BEILSTEIN ONLINE for organic compounds (Chapter 9). These German 'twins' give us access to data from the early nineteenth century to about 1990 (but with the older material evaluated in the light of recent knowledge) on nearly five million compounds. Data determined since 1967 can be accessed via CAS files (Chapter 5) which by 1991 covered about ten million compounds.

A large number of quite specific databanks are available and the more important ones are discussed in Chapter 5. These include ten for thermophysical properties, four crystallographic ones and three each on IR, NMR and mass spectra. MARTINDALE ONLINE (Chapter 4) is an important source of both physical and medical data on pharmaceuticals. Thus this chapter deals mainly with printed sources of data but provides cross-references to their equivalent database formats where applicable.

Although designed as a predictive tool in chemical hazard evaluation, *CHETAH* (2nd edn, ASTM) can also predict heats of reaction, heat capacities and entropies. Its databank and associated programs allow four hazard criteria to be combined to give an overall hazard rating, in addition to predicting impact sensitivity from molecular structure.

Clerc (1987, 1990) has proposed a general mathematical model for spectral identification processes. Increasingly such models will be incorporated into databank searching software. If more information is required on systems for storing physico-chemical data than is given in Chapters 5 and 6, a good account is available in *Chemical Structure Information Systems: Interfaces, Communication and Standards* (W.A. Warr, ed., ACS Symposium Series, 1989). It is, however, probable that the artificial intelligence approaches to structure identification based on spectral databanks have not yet been optimized (Levy, 1991).

Specialized compilations

A wide range of chemical materials is being produced by industries in ever-increasing quantity and diversity. In order to promote the use of such materials, data — particularly on physical properties — are collected and often edited and published by the manufacturers. Almost invariably they have information available upon enquiry (see *Finding and Using Product Information: From Trade Catalogues to Computer Systems* (R.A. Wall, ed., Gower, 1986).

As well as industrial laboratories, government laboratories also produce and collect physico-chemical data. Prominent in these activities is the US National Institute of Standards and Technology (NIST), formerly the National Bureau of Standards (NBS). When the name change occurred in 1988, there was also a corresponding change in its journal titles. The *Journal of Research of the NIST*, as it is now called, often contains physico-chemical data. NIST is also one of the co-sponsors of the American Institute of Physics's *Journal of Physical and Chemical Reference Data* (1972–) which contains critically evaluated compilations of physical and chemical properties. Cumulative property and materials indexes now appear quinquennially. A supplement to volume 14 of this journal has been separately published by ACS as *JANAF Thermochemical Reference Tables* (M.W. Chase, C.A. Davies, J.R.

Downey, D.J. Frurip, R.A. McDonald and A.N. Syverud, eds, 3rd edn, 2 vols, 1985). This provides data for 1800 substances.

Up to 1959 the NBS published *Circulars*, two of which are invaluable to physical chemists: *C500 Selected Values of Chemical Thermodynamic Properties* (F.R. Rossini *et al.*, eds, 1952, 1268 pp.) and *C510 Tables of Chemical Kinetics, Homogeneous Reactions* (1951, 731pp., Supplement 1, 1956, 472pp.). In *C500* 'best' values are given for 'all inorganic compounds and organic compounds up to 2 carbon atoms' of heats and free energies of formation, entropies, heat capacities, heats and temperatures of fusion, vaporization and sublimation, etc. together with extensive literature references. The material in *C500* was comprehensively revised in *TN270-3* to *TN270-8* (1968–75). An earlier *Circular C461 Selected Values of Properties of Hydrocarbons* (1947), was revised in 1964 (3 vols).

C510 is a critically evaluated compilation of rates and rate constants and the data are arranged in order of increasing complexity of the key reactant. Many literature references are given and the above compilations satisfy the criteria set out earlier. NBS *Monograph 34* (vol. 1, 1961 and vol. 2, 1964) is a further supplement to *C510*. Gevantman and Garvin (1973) have produced a comprehensive listing of compilations of kinetics data for the NBS. Kinetics data are also contained in *Tables of Bimolecular Gas Reactions* (A.F. Trotman-Dickenson and G.S. Milne, NSRDS-NBS 9, 1967). The NBS also published compilations of X-ray data.

From 1964 to 1983, NBS published 72 reports covering physical properties of over 40 000 mainly organic compounds collected from US government agencies in the *National Standard Reference Data Series* (NSRDS). *Standard Reference Data Publications* (G.B. Sherwood, ed., USGPO, 1985) includes materials, property and author indexes to these.

In the 1960s the International Council of Scientific Unions (ICSU) set up its Committee on Data for Science and Technology (CODATA). CODATA has produced numerous publications. One is an *Inventory of Data Resources in Science and Technology* (UNESCO, 1982) which lists 652 organizations in 90 countries as well as a number of printed data sources. This bibliography is complemented by an irregular serial, *CODATA Bulletin* (Pergamon Press, 1969–).

IUPAC has sponsored the collection of physico-chemical data since 1910. This work was the basis of *International Critical Tables*. Since the late 1930s the practice was established of critically evaluating the data and for each volume to cover a specific field with a self-contained bibliography. These were known as *Physico-Chemical Selected Constants (New Series)*. In the 1950s the *Chemical Data Series* commenced. An example is no. 130, *Handbook of Thermodynamics and Transport Properties of Alkali Metals* (R.W. Ohse, ed., Blackwells, 1985) which runs to nearly a thousand pages. (For the IUPAC *Solubility Data Series*,

see p. 129.)
Two Russian collections of thermodynamic data are *Termodinamichyeskiye Svoistva Individual'nyikh Vyeshchyestvo* (Izdat. Akad. Nauk SSSR, 1962) and *Termodinamichyeskiye Konstanti Vyeshchyestvo* (V.P. Glushkov, ed., 1965–80). The former appeared in two volumes, the first of which dealt with theory and calculations of thermodynamic properties and contained 3292 literature references; the second contains tables of thermodynamic properties, mainly of simple molecules. The latter appeared in 10 parts and covered heats of formation, heat capacities, entropies, etc.

Elsevier's *Physical Sciences Data* series (1978–) now covers most fields, though a number of the volumes are just indexes to the literature. No. 41, *Handbook of Electrolyte Solutions* (V.V.M. Lobo and J.L. Quaresoma, eds, 2 vols, 1989), covers data on activity coefficients, densities, viscosities, etc. It appears to supersede Lobo's *Electrolyte Solutions: Literature Data on Thermodynamic and Transport Properties* (Universidad do Coimbra, 1979). *Nonaqueous Solution Chemistry* (O. Popovych, Krieger, 1990) usefully updates a recently neglected area. Activity coefficients are comprehensively dealt with in K.S. Pitzer's *Activity Coefficients in Electrolyte Solutions* (2nd rev. edn, CRC Press, 1991).

Electrokinetic data for electrochemical reactions, potentials, response constants, etc. are contained in the 10 volumes of the *CRC Handbook Series in Inorganic Electrochemistry* (L. Meites *et al.*, eds, 1980–). It is arranged by element (alphabetically by symbol). Applied aspects of electrochemistry are dealt with in *Battery Reference Book* (T.R. Crompton, Butterworths, 1990) or by a group of German authors in *Battery Technology Handbook* (H.A. Kiehne, ed., Dekker, 1989).

Several of the older classics, such as W.M. Lattimer's *Oxidation Potentials* (2nd edn, Prentice-Hall, 1952) or H.S. Harned and B.B. Owen's *Electrolyte Solutions* (3rd edn, Reinhold, 1958) are still useful sources of electrochemical data. The *Atlas of Electrochemical Equilibria in Aqueous Solution* (M. Porbaix, Pergamon Press, 1966) which is a translation of the 1963 French edition, is a unique source of information on such equilibria. Mainly polarographic data are given in *Electrochemical Data* (L. Mietes, P. Zuman *et al.*, Wiley; vol. A 1974).

Buffers for pH and Metal Ion Control (D.D. Perrin and B. Dempsey, Chapman and Hall, 1979) gathers widely scattered data and includes small computer programs for calculating buffer compositions.

Some of the *Advances in Chemistry Series* (ACS) are compilations of data. Nos 6 and 35 (1952 and 1963) dealt with *Azeotropic Data*; these have been updated and replaced by no. 116 (1973). This includes five-component systems in the 17 000 systems covered and has a formula index. A 1961 compilation by A.W. Francis provides data on *Critical Solution Temperatures* (no. 31) for over 6000 systems. No. 18 is a com-

Standard tables of physico-chemical data 129

pilation of the *Thermodynamic Properties of the Elements* (1956) by D.R. Stull and G.C. Sinke.

Solubilities figured high in a list of data most frequently sought by US chemists (Weisman, 1967). They were also high on a list of desirable subjects for comprehensive critical compilations. There is little reason to believe that solubility data are markedly less important today. The classic work in this field is *Seidell's Solubilities*. The fourth edition was completed in 1966 by the American Chemical Society's publication of Vol. II (compounds of elements from K to Z) of *Solubilities of Inorganic and Metallorganic Compounds* (W.F. Linke, ed.). Volume I (1959) and the earlier editions were published by Van Nostrand. It satisfies the criteria set out in the introductory section of this chapter and covers the literature up to 1956. IUPAC *Solubility Data Series* has been published by Pergamon since 1979. The first 38 volumes were edited by A.S. Kertes; the editor is now J.W. Lorimer. Volume 43 was published in 1990 and Volumes 19 and 38 are indexes to the preceding volumes. A Russian work (edited by V.V. Kaprov in 1961) was revised and translated as *Solubilities of Inorganic and Organic Compounds* (H. and T. Stephen, eds, Pergamon Press, 1963). Volume I (two parts) deals with binary systems and both aqueous and non-aqueous solvents. Volume II (three parts) covers multicomponent systems. Volume III (three parts, 1979) on multicomponent systems covered the literature to 1965 and was edited by H. Silcock. Slightly broader in scope but not comprehensive is the *CRC Handbook of Solubility Parameters and other Cohesion Parameters* (A.F.M. Barton, 2nd edn, CRC, 1991); it contains 2900 references.

W.G. Moffatt's unique collection of element pair mixtures has been made generally available in loose-leaf binders as *Handbook of Binary Phase Diagrams* (5 vols, General Electric Co. — now Genium Publishing, 1979–). Although element pairs and their binary compounds are shown graphically, other two-component systems of compounds are not. Elsevier's *Physical Sciences Data* series (see above) provides a literature source book on phase diagrams for compounds in vol. 10, *Phase Diagrams* (J. Wisniak, two parts, 1981) and its supplement, vol. 27 (1986).

The *CRC Handbook of Radiation Chemistry* (Y. Tabata et al., eds, 1991) is an ample source of data on sources, dosimetry, etc. in this area and is well referenced. It also contains an outline of experimental methods and theoretical considerations and covers free radicals, polymerization and other radiation-induced reactions.

The International Union of Crystallography has been active for over fifty years in compiling data; it brought to publication *International Tables for Crystallography* (T. Hahn, ed., 2nd edn, 1987–). Volume A on symmetry also contains a detailed theoretical background. Computerized data compilations have been available for over twenty years from Cambridge Crystallographic Data Centre (see Chapter 6). NBS was also a prolific collector of such data. Its NSRDS (see p. 127) produced *Crystal Data: Determination Tables* (various editors, 3rd edn, 6 vols, 1972–83). This is a publication of NBS and the Joint Committee on Powder Diffraction Studies and covers 43 000 organic and 34 000 inorganic compounds. The NIST CRYSTAL DATA IDENTIFICATION file contains 160 000 structures (see Chapter 5).

Spectroscopic data

Spectroscopy affords a good example of both fundamental investigations of phenomena and the application of the techniques as a tool for analytical or structural determinations by organic and inorganic chemists, biochemists, etc. Although most of the data compilations are mainly used for the latter purpose, they are dealt with here since they are largely numerical or graphical in form. For databanks of NMR, IR and mass spectra, see the review by Warr (1991).

A useful basic introduction (and some data) is provided by R.M. Silverstein and G.C. Bassler's *Spectrometric Identification of Organic Compounds* (4th edn, Wiley, 1981), which covers infrared, ultraviolet, nuclear magnetic resonance and mass spectra. Further details of other books dealing with practical and interpretative aspects are given in the physical methods section of Chapter 10.

Compilations of infrared spectra rank high in any list of physical data frequently sought by chemists. Fourier-transform infrared is now one of the fastest growing analytical techniques. This is also reflected in its literature with the appearance of works such as *Practical Fourier Transform Infrared Spectroscopy* (J.R. Ferraro and K. Kristinan, eds, Academic Press, 1990) and *Fourier Transform Infrared Spectroscopy: Applications to Chemical Systems* (J.R. Ferraro and L.J. Basile, eds, Academic Press, 4 vols, 1978–85).

Collections of infrared spectra come in all sizes. Most monographs provide some data (see also pp. 184–5). An up-to-date descriptive text accompanies the data in the *Handbook of Infrared and Raman Characteristic Frequencies of Organic Molecules* (D. Lin-Vien, N.B. Colthup, W.G. Fately and J.G. Grassell, Academic Press, 1991). Several collections are in loose-leaf binders and some were issued in microform, e.g. *Infrared Spectra of Selected Chemical Compounds* (R. Mecke and F. Langenbucher, Hayden, 1966). All the 1900 simple organic or inorganic compounds in the Mecke collection are of known structure. There are alphabetical, formula, molecular and serial number indexes and an index of chemical classes.

The *Aldrich Library of FT-IR Spectra*, compiled by C.J. Pouchert (2 vols, 1985) selected 11 000 spectra from the files of a leading American laboratory chemicals supplier. The *Merck FT-IR Atlas* (K.G.R. Pachler,

Standard tables of physico-chemical data 131

F. Matick and H.V. Greenwich, VCH, 1988) contains spectra for 3000 compounds.

Even larger and more expensive are the *Sadtler Reference Spectra Collection*, published by Sadtler Research Laboratories, Philadelphia. The 12 Standard Collections are updated regularly and included 59 000 prism plus 48 000 grating IR, 48 000 ultraviolet and 32 000 nuclear magnetic resonance (NMR) spectra for 30 000 compounds by 1980. Since then some 80 000 compounds have been added. There are also Special Interest Collections and Commercial Compounds Spectra covering drugs, polymers, surfactants, etc. These include:

Infrared Spectra Handbook of Adhesives and Sealants, 1987
Infrared Spectra Handbook of Common Organic Solvents, 1983
Infrared Spectra Handbook of Inorganic Compounds, 1984
Infrared Spectra Handbook of Intermediates, 1984
Infrared Spectra Handbook of Minerals and Clays, 1982
Infrared Spectra Handbook of Priority Pollutants and Toxic Chemicals, 1981
Infrared Spectra Handbook of Atlas of Polymer Additives, 3 vols, 1981
Infrared Spectra Handbook of Monomers and Polymers, 1981
Infrared Spectra Handbook of Rubber Chemicals
Infrared Spectra Handbook of Surfactants

All are in loose-leaf binders or in microform. Each spectrum shows the purity and source of the sample and technique used and is identified by its serial number in the several indexes. These are: (a) Total Spectra Index, which has four parts (Alphabetical, Serial Number, Molecular Formula and Chemical Class); (b) Infrared Spec-Finder (based on wavelength of strongest band); (c) Ultraviolet Spectra Locator (based on λ_{max} values); and (d) NMR Chemical Shift Index.

An updated index to Sadtler spectra was produced in 1980 and a supplement covering infrared, ultraviolet and NMR spectra issued from 1981 to 1988 has also appeared.

The Sadtler Collections, together with 25 000 infrared and ultraviolet spectra from the old *Documentation of Molecular Spectroscopy* system (1957–1972) plus the API/TRC, Koblenz Society, Infrared Committee Japan and Aldrich (see above) collections and spectra from the literature, about 200 000 infrared spectra, have been indexed by the American Society for Testing Materials (ASTM) with alphabetical and molecular formula lists. Database tapes are also available. Thus, if it is available, a search of latest edition of the ASTM's *Spectral Index* should reveal if spectra for the compound one is seeking have been published. There is, of course, considerable overlap in the compounds covered by the various collections.

Specialized analytical handbooks often contain infrared spectra. The US Environmental Protection Agency compiled a loose-leaf *Manual of*

132 Standard tables of physico-chemical data

Chemical Methods for Pesticides and Devices (1976–) which contains some 400 infrared spectra. *Clarke's Isolation and Identification of Drugs in Pharmaceuticals and Post-mortem Material* (A.C. Moffat, ed., 2nd edn, Pharmaceutical Press, 1986) gives ultraviolet, infrared and mass spectra in addition to chromatographic data for over 1300 drugs. Also in this area is *Infrared Reference Spectra* (British Pharmacopoeia Commission, 1980–) which is updated by supplements.

Several collections of mass spectra are available. Probably the largest is the CD-ROM edition of the *Registry of Mass Spectral Data* (F.W. McLafferty and D.B. Stauffer, 5th edn, 1989) which is derived from the *Wiley/NBS Registry of Mass Spectral Data* (7 vols, 1988–89) by the same compilers. It contains data for 118 144 compounds from the collections of the NIH, the Environmental Protection Agency and the NBS/NIST. The Mass Spectrometry Data Centre published *Eight Peak Index of Mass Spectra* (4th edn, 1991) for 66 720 compounds. MSDC has produced *Mass Spectrometry Data Sheets* since 1966 and also a guide to current literature, *Mass Spectrometry Bulletin* (1966–). The MSDC was originally at AWRE, Aldermaston, but is now at the Royal Society of Chemistry, Cambridge. A large Russian collection is produced by their Academy of Science's Novosibirsk Institute of Organic Chemistry. As it consists mainly of spectra with the names of compounds in both English and Russian, it can be quite easily used even if one has virtually no knowledge of Russian, *Atlas Mass Spektrov — Organicheskikh Soedinyenii* has run to 18 volumes so far. The same Institute has also produced a 35-volume set of infrared spectra — *IK Spektryi* (1967–) and has more recently started publishing its ultraviolet spectra collection, *UF Spektryi* (1986–). *Mass and Abundance Tables for Use in Mass Spectrometry* (J.H. Beynon and A.E. Williams, Elsevier, 1963) is a computer compiled aid for identifying ions (containing C, H, N or O only) up to a mass number of 500.

In the 1950s and 1960s, a number of literature indexes to spectra were published. Their main use now must be as a checklist for compilation editors, as the majority of the spectra so indexed will have already been trawled into the major collections. Equipment manufacturers (notably Varian Associates and JEOL for NMR) have also published spectra collections. These are now also to be found, for example, in the Sadtler collections.

General texts and treatises

There are many undergraduate textbooks, though few excellent ones. Since the 1960s there has been an increasing tendency to fragment the subject and to issue books confined to certain aspects such as thermodynamics, kinetics, etc. Most of the general texts are of US origin and it should be remembered that they normally use a different set of thermodynamic symbols and electrochemical sign conventions from UK and European chemists.

The US texts dominated the post-war era, firstly with S. Glasstone's *Textbook of Physical Chemistry* (2nd edn, Macmillan, 1948) and then with W.J. Moore's *Physical Chemistry* (5th edn, Longmans, 1972). This dominance, at least in the UK, was ended by P.W. Atkins's outstanding text, *Physical Chemistry* (4th edn, OUP, 1990). This edition made good a number of gaps, such as colloids and macromolecules, which were observable in earlier ones. The *McGraw-Hill Physical Chemistry Source Book* (1988) is a collection of over 200 articles which is useful for quick reference or for supplementing textbooks.

There is unfortunately still no up-to-date general treatise. Academic Press published *Physical Chemistry: An Advanced Treatise* (H. Eyring, D. Henderson and W. Jost, eds, 1967–75) in 11 volumes but it lacked any information on surface chemistry. Physical chemistry is, however, abundantly covered by monographs on a narrow well-defined topic which sometimes form part of a series. An example is the Elsevier series *Studies in Physical and Theoretical Chemistry* (1980–) where vol. 73 was published in 1991. As with textbooks, fragmentation of the subject is again evident. Several exhaustive treatments of topics are given in multivolume series. They are normally very good sources of physicochemical data as well as of the theoretical background. Electrochemistry is even covered by two such series, *Comprehensive Treatise of Electrochemistry* (J.M. O'Bockris et al., eds, 10 vols, Plenum, 1981–85) and the *Encyclopedia of Electrochemistry of the Elements* (A.J. Bard, ed., Dekker, 1975–86). The latter includes inorganic compounds in the first 10 volumes, whilst organic compounds are covered in vols 11–15 (1978–85). Elsevier's *Comprehensive Chemical Kinetics* (31 vols, 1969) was originally edited by C.H. Bamford and C.F.H. Tipper but latterly by R.G. Compton.

Out-of-date monographs and treatises can to a limited extent be brought up to date through the *Annual Review of Physical Chemistry* (1950–) or Wiley's *Advances in Chemical Physics*. Of the RSC *Specialist Periodical Reports* in this area, few have reasonably recent volumes. These are *Catalysis* (1989), *Nuclear Magnetic Resonance* (1991) and *Spectroscopic Properties of Inorganic and Organometallic Compounds* (1991).

A good guide to most practical techniques on physical chemistry is *Physical Methods of Chemistry* (B.W. Rossiter and J.F. Hamilton, eds, 2nd edn, 9 vols, Wiley, 1985). This is derived from A. Weissberger's classic work *Techniques of Chemistry*, which has been through several editions; see Chapter 10. As many of the basic *non-automated* methods have changed little in the past two or three decades, the earlier editions of Weissberger or *Physico-chemical measurements* (J. Reilley and W.N. Rae, 3 vols, Van Nostrand, 1948–54) contain some still useful information. Recent teaching aids for practical chemistry are the videos produced by the RSC, ACS, Open University, etc., for example the RSC's 40-minute video *Interpreting Infrared and NMR Spectra*.

References

Clerc, J.T. (1987), in *Computer-enhanced Analytical Spectroscopy*, H.L.C. Meuzelaar and T.L. Isenhour, eds, Plenum Press, pp. 145–162.

Clerc, J.T. (1990), in *Computer Methods in UV, Visible and IR Spectroscopy*, W.O. George and H.A. Willis, eds, RSC, pp. 13–24.

Gevantman, L.H. and Garvin, D. (1973), *International Journal of Chemical Kinetics*, **5**, 213.

Levy, G.C. (1991), *Spectroscopy*, **6** (4), 50.

Warr, W.A. (1991), *Chemometrics and Intelligent Laboratory Systems*, **10**, 279–292.

Weisman, H.M. (1967), *Journal of Chemical Documentation*, **7**, 9.

Appendix: Contents of the 6th Edition of Landolt-Börnstein's Tables

Volume 1 — Atomic and molecular physics (5 parts)

Part 1 (1950). Atoms and ions

The system of units; length, mass, time, mechanical, electrical, thermal, photometry; relationship between electrical and magnetic units. The basic constants of physics, velocity of light, Planck's constant, gravitation constants; proton, α-particle. Wavelength of atomic spectra; Faraday effect and other effects due to the external shell of electrons.

Part 2 (1951). Molecular structure

Atomic distances and structure; energy of chemical handling; dissociation of di- and poly-atomic molecules; oscillation and rotation of molecules; the restraint of molecular rotation.

Part 3 (1951). The external electron ring

Band spectra of diatomic and polyatomic molecules; light absorption of solutions in visible and ultraviolet; energy of ionization; optical rotation; electrical moments of molecules; electrical and optical polarization of molecules; magnetic moments; diamagnetic polarization; quantum yield in photochemical reactions.

Part 4 (1955). Crystals

Symmetry, crystal class, space groupings; lattice type structure and dimensions of crystals; ion and atomic radii; lattice energy; internal variations; electron emission of metals and metalloids; X-ray and electron and high frequency spectra of crystals; lattice distortion and absorption in alkali; halogen compounds.

Part 5 (1952). Atomic nucleus and elementary particles

The energy of the atomic nucleus; hyperfine structure of spectral lines; naturally radioactive atoms; rotation of the nucleus and light quanta; cosmic radiation.

Volume 2 — Properties of matter in the various states of aggregation (10 parts)

Part 1 (1971). Thermal mechanical state

SI units; T, V and P measurement; hardness; natural isotopes; densities and compressibilities of solutions; non-ionized solutions.

Part 2. Equilibria other than melting point equilibria

Part 2 (a) (1960). Vapour condensed phase equilibria and osmotic phenomena

One-component system; vapour pressures of pure substances; densities of coexisting phases of pure substances; effect of pressure on melting and transition points; liquid crystals. Multicomponent systems; vapour pressures of mixtures; heterogeneous equilibria; colligative properties.

Part 2 (b) (1962). Solution equilibria 1

Multicomponent systems (contd): solution equilibria of (a) gases in liquids, (b) gases in solid and liquid metals, (c) solids and liquids in liquids (solubilities of organic substances in water, etc.; solubilities of inorganic substances in water, etc.).

Part 2 (c) (1964). Solution equilibria 2

Solubilities of organic substances in organic liquids; equilibria in systems with several immiscible phases.

Part 3 (1956). Melting point equilibria and interface phenomena

Melting point diagrams of metal alloys; binary and ternary systems of inorganic compounds; reciprocal salt pairs; silicates; melting point diagrams of organic systems and inorganic organic systems.

Part 4. Calorimetry (1961)

Molar heat capacities, entropies, enthalpies and free energies and their temperature dependence tables for calculating thermodynamic functions from known molecular vibrations Joule-Thomson effect; magnetothermal effect with paramagnetic salts at very low temperatures: thermodynamic functions of mixtures and solutions (including heats of absorption and neutralization).

Part 5 (a) (1969). Transport phenomena 1. Viscosity and diffusion

Viscosity of gases and liquids; diffusion of gases and liquids.

Part 5 (b) (1968) Transport phenomena 2. Kinetics, homogeneous gas equilibria

Dynamic constants (contd); thermal conductivity; thermal diffusion in gases and liquids; reaction velocities in gases and solids; homogeneous gas equilibria.

Part 6 (1959). Electrical characteristics 1

Electrical conductivity of metals; ionic conductivity in solids (crystals); Hall effect and transistor effect; conductivity transition point; photoconductivity; piezoelectric, elastic, dielectric constants of some systems: dielectric properties of crystals, crystalline solids, glass, synthetic materials (Volume 4, Part 3, 658); crystalline liquids, pure liquids, aqueous solutions, non-aqueous solutions; gases. Thermoelectric effects; Peltier effect; Thomson effect; photoemission and secondary electron emission effects.

Part 7 (1960). Electrical characteristics 2. Conductivity of electron systems

Conductivity of molten salts; pure liquids; conductivity, transport numbers and ion conductivity of aqueous electrolytic solutions; transport numbers; electrolytes in heavy water; conductivity in non-aqueous solutions — organic and inorganic; electrophoretic mobilities and electrokinetic potentials; e.m.f. of reversible and irreversible cells in aqueous and non-aqueous solutions; e.m.f. of cells in molten salts; electrolytic dissociation of constants; acid–base indicators; buffer mixtures and compositions.

Part 8 (1962). Optical constants

Metals and alloys; non-metallic substances; glasses and plastics; optical and magneto-optical characteristics of liquid crystals; liquids; gases.

*Part 9 (1962). Magnetic Characteristics 1

*Part 10 (1967). Magnetic characteristics 2

---

*These volumes have tables of contents and introductory material in both English and German.

Volume 3 (1952) — Astronomy and geophysics (1 part)

Types and location of instruments; location and time constants; frequency and occurrence of elements; solar system; magnitude of radiation; movement of stars; star systems and special star types; galactic mist. The shape of the earth; force of gravity; minerals and rocks; earth's magnetic field; oceanography; hydrography; meteorology; external atmosphere.

Volume 4 — Basic techniques (4 parts)

Part 1 (1955). Natural materials and mechanical properties

Basic technical and physical units; mass, length, time, area, volume, frequency, velocity; units of force, pressure, energy, work; electrical units; atomic weights; hydrometry and pyknometry. Natural and artificial building materials; cement, mortar, stone, wood, paper, cellulose. Fibrous materials: ceramics, glass, natural and synthetic rubber. Friction and viscosity: static and dynamic friction and viscosity flotation; acoustics.

Part 2. Metallic materials

Part 2 (a) (1963). Fundamentals, Testing methods, Ferrous materials

Characteristics of metals; test methods (investigation of mechanical properties); non-destructive testing; steel production; standards and physical characteristics for ferrous metals; special chemical and physical properties; ferrous and oxygen-free alloys.

Part 2(b) (1964). Sintering materials, Heavy metals

Powder metallurgy and sintering materials: W, Rh, Ta, Mo, Nb, V, Cr, Co, Ni, Mn, noble metals, Cu, Sb, Zn, Pb, Bi, Sn.

Part 2(c) (1965). Light metals, Special materials, Semiconductors, Corrosion

Ti, Be, Al, Mg, Li, Rb, Cs; liquid metals, reactor materials, U, Pu, Zr, Hf, Th, rare earths, semiconductors, solders, enamel, hard solders, welding, cutting and extrusion of metals; metal adhesion; corrosion behaviour of materials (arranged by corrosive agent).

Part 3 (1957). Electrical, optical and X-ray techniques

Technical conductivity of solid and liquid materials; thermo-elements; discharge through gases; insulating materials — synthetic and natural; magnetic properties; illuminations; light sources; luminous substances; light filters; photosensitive materials. X-ray techniques, circuits and dispositions. General and medical X-ray techniques: X-ray spectra; α–radiation; fine line structure of X-ray spectra.

138 *Standard tables of physico-chemical data*

Part 4. Determination of thermal and thermodynamic properties

Part 4(a) (1967). Methods of measurement, Thermodynamic properties of homogeneous materials
Thermometry; hygrometry; thermodynamic properties of gases, vapours, liquids and solids.

Part 4(b) (1972). Thermodynamic properties of mixtures, Combustion, Heat transfer
Thermodynamic equilibria of mixtures; adsorption. Fuels and combustion. Thermal conductivity.

Part 4(c). Absorption equilibria of gases in liquids

Part 4(c1) (1976). Absorption in liquids of low vapour pressure
Water, D_2O, organic liquids, oils, HF, NH_3, etc.; aqueous solutions.

Part 4(c2) (1980). Absorption in liquids of high vapour pressure

CHAPTER EIGHT

Inorganic and nuclear chemistry
J.A. TIMNEY

Introduction

This chapter concentrates on the sources available to those searching for information in inorganic chemistry (including organometallic chemistry) and the somewhat smaller field of nuclear chemistry. It is interesting to compare the growth and decay of the two areas over the last 30 years. Inorganic chemistry has experienced an impressive and sustained renaissance. The amount of research (measured by the number of papers) continues to rise in both academic and industrial laboratories. The growth in inorganic chemistry has been most dramatic in relation to the ever-expanding chemistry of the d-block elements. One feels that if this rich vein ever begins to thin out then the chemistry of the f-block elements will become more prominent. Either way, the growth in inorganic chemistry appears assured. In contrast to this, nuclear chemistry appears to be in something of a trough. This is, partly, because it has ceased to be distinct and is not really in need of its own literature. This causes articles and reviews in the subject to be disguised, more often than not, as inorganic or physical chemistry. It may be that nuclear chemistry as a separate entity (and not to be confused with the thriving science of radiobiochemistry) slowly vanishes from the scene.

Inorganic chemistry

Initial searching

One can not over-emphasize the value of *Chemical Abstracts* (covered in Chapter 3), at any stage of literature searching. Like any tool, this becomes much more useful as the user becomes more familiar with its workings. *Chemical Abstracts* apart, the outstanding text for the researcher looking for general information in inorganic chemistry is *Advanced Inorganic Chemistry* (F.A. Cotton and G. Wilkinson, 5th edn,

Wiley Interscience, 1988). This text, designed primarily as an undergraduate text book, contains references to review articles and many original papers across the whole spectrum of inorganic chemistry. Earlier editions contained a great deal of theoretical discussion which has now been replaced by largely experimental work. The book is broken down into four parts dealing with basic principles, the chemistry of the main group elements, the transition elements and, finally, a survey of selected areas (mainly organometallic chemistry and bio-inorganic chemistry). It is a reflection of the trends in modern inorganic chemistry that the chemistry of the transition elements dominates this book.

Outstripping Cotton and Wilkinson in scope, if not in readability, is *Gmelins Handbuch der anorganischen Chemie* (Gmelin), a vast compilation of all the literature relevant to the inorganic chemist. The *Handbuch* began in 1819 with the publication of the *Handbuch der theoretischen Chemie* in three volumes, with Leopold Gmelin as the author/editor. The second and third editions were similarly titled and arranged. The fourth edition appeared as the *Handbuch der Chemie* and was intended to cover all types of chemistry. The fifth edition omitted organic chemistry, which was taken over by *Beilstein*. In 1871 the sixth edition, now titled the *Handbuch der anorganischen Chemie* appeared with Karl Kraut as the editor. Kraut was also at the helm for the seventh edition and this version is often referred to as the 'Gmelin-Kraut Handbuch'. In 1922 the eighth edition appeared as *Gmelins Handbuch der anorganischen Chemie*. Now, Gmelin usually occupies a prominent position in any collection of reference books devoted to inorganic chemistry. The Hauptband (main work) has been augmented across the years by the publication of Ergänzungsbände (supplements). The numerous supplements and extensions that have appeared over the years tend to give Gmelin an uneven, disorganized look. Do not be deceived. Gmelin is distinctly user-friendly and contains hundreds of valuable tables and much graphical material. The classification system is given in Table 8.1. A compound is located within a particular volume via the constituent element having the highest system number. Thus, $CaCl_2$ would be located under system number 28 (calcium) rather than number 6 (chlorine).

In keeping with modern trends, Gmelin (more properly the Gmelin Institut für anorganische Chemie der Max-Planck-Gesellschaft zur Forderung der Wissenschaften, Frankfurt, Germany) produces a database that is available through the Scientific and Technical Information Network (STN International, P.O. Box 2465, D-7500 Karlsruhe 1, Germany). Under the title GFI (GMELIN FORMULA INDEX), this vast store of information cross-references the appearance or use of a compound with its location in the *Handbuch*. There are in excess of 3×10^5 compounds in the database and it is updated twice a year. A video on *Gmelin* is available from Springer-Verlag, who publish for the Gmelin Institute.

Table 8.1. The Gmelin classification

No.	Element	No.	Element	No.	Element
1	Noble gases	25	Caesium	49	Niobium
2	Hydrogen	26	Beryllium	50	Tantalum
3	Oxygen	27	Magnesium	51	Protactinium
4	Nitrogen	28	Calcium	52	Chromium
5	Fluorine	29	Strontium	53	Molybdenum
6	Chlorine	30	Barium	54	Tungsten
7	Bromine	31	Radium	55	Uranium
8	Iodine	32	Zinc	56	Manganese
9	Sulphur	33	Cadmium	57	Nickel
10	Selenium	34	Mercury	58	Cobalt
11	Tellurium	35	Aluminium	59	Iron
12	Polonium	36	Gallium	60	Copper
13	Boron	37	Indium	61	Silver
14	Carbon	38	Thallium	62	Gold
15	Silicon	39	Lanthanides	63	Ruthenium
16	Phosphorus	40	Actinium	64	Rhodium
17	Arsenic	41	Titanium	65	Palladium
18	Antimony	42	Zirconium	66	Osmium
19	Bismuth	43	Hafnium	67	Iridium
20	Lithium	44	Thorium	68	Platinum
21	Sodium	45	Germanium	69	Technetium
22	Potassium	46	Tin	70	Rhenium
23	Ammonium	47	Lead	71	Transuranic elements
24	Rubidium	48	Vanadium		

Other concise sources of information which should be useful to the inorganic chemist are Mackay and Mackay's *Introduction to Modern Inorganic Chemistry* (4th edn, Blackie, 1989), *Vogel's Textbook of Quantitative Inorganic Analysis* (5th edn, J. Bassett *et al.*, Longman, 1989), and *Organotransition Metal Chemistry* (A. Yamamoto, John Wiley, 1986). Although now somewhat out of date, the five volumes of *Comprehensive Inorganic Chemistry* (A.F. Trotman-Dickenson, executive editor, Pergamon, 1973) are still worth searching if the literature spans a number of decades. More recently, Wilkinson, Stone and Abel have produced *Comprehensive Organometallic Chemistry* (Pergamon, 1982) which, as the title suggests, is a useful source book for those searching for work in organometallic chemistry.

If the information sought is about simple compounds, rather than research work, the classic **CRC Handbook of Chemistry and Physics** (see Chapter 7) is strongly recommended. This contains a wealth of data relevant to any chemist and is a must for any library in this field. Another source of information about compounds rather than their usage is the catalogues produced by the producers of fine chemicals (see Chapter 12). A good example of this type of source is the *Catalogue Handbook of Fine Chemicals* produced by the Aldrich Chemical Co. Ltd. In addition to a basic data, this considerable (1990–91 edn, 2150 pp.) text provides references to infrared spectra, NMR spectra and safety data.

Finally, in this section, researchers are encouraged to work their way through the *Nomenclature of Inorganic Chemistry: Recommendations 1990* (G.J. Leigh, ed., IUPAC Commission on the Nomenclature of Inorganic Chemistry). Although this does not make for light reading, it is well worth the effort.

Reviews

After an initial search in, for example, Cotton and Wilkinson or Gmelin, the next stage of information gathering will probably focus on the many review articles that are published. These are often written by an expert in a particular field who attempts to draw together many pieces of research into a coherent whole. Apart from the references (often to be counted in the hundreds) that such reviews make available, the workers and institutions that are active in a particular field are identified, which becomes very useful later on.

The most useful type of review is typified by the *Specialist Periodical Reports* (SPRs) published annually by The Royal Society of Chemistry. These are published annually (if the field is active enough) and seek to report all published work in specific areas. For example, SPR 13 is devoted to Organometallic Chemistry and its constituent chapters cover virtually every aspect of the subject. Less exhaustive in their treatment of advances in inorganic chemistry, but useful nonethe-

Inorganic and nuclear chemistry 143

less, are the *Annual Reports of the Progress of Chemistry* of The Royal Society of Chemistry, which have been published since 1904.

Inorganic chemistry is often featured in publications of a general nature such as: *Accounts of Chemical Research* (American Chemical Society); *Angewandte Chemie* (International Edition in English; VCH Publishers); *Chemical Reviews* (American Chemical Society); *Chemical Society Reviews* (The Royal Society of Chemistry); *Monatshefte für Chemie* (Springer-Verlag, Wien).

There is no real short cut to finding information in these publications. Their contents pages need to be scanned for likely articles. However, reviews which are written specifically for inorganic chemists are certainly not scarce. These include: *Advances in Inorganic Chemistry* (Academic Press); *Advances in Organometallic Chemistry* (Academic Press); *Coordination Chemistry Reviews* (Elsevier); *Comments on Inorganic Chemistry* (Gordon and Breach); *Progress in Inorganic Chemistry* (Wiley Interscience); *Transition Metal Chemistry* (Chapman and Hall).

Having found a good review and examined the references contained within, it is well worth turning to the Author Index of the *Science Citation Index* (*SCI*) to find out who, if anybody, has used or developed published material (see Chapter 3). Major pieces of work tend to have a significant entry in the *SCI* and it is likely that the main personalities and institutions can be identified at this stage.

Journals

The major drawback related to review articles is the timelag between work being submitted for publication and its appearance in a review. This process can be as long as 18 months which is acceptable for background research, but useless for those requiring recent advances in a subject. The indispensable *Chemical Abstracts* has been mentioned elsewhere and we need not repeat the discussion. Suffice it to say that inorganic chemistry as a whole is well represented in *Chemical Abstracts* and a great deal of time can be saved by consulting the relevant sections before going on to specific papers in the specialist journals.

Searching the journals can be a tedious process (especially those issues that are recently published and have not been abstracted by *Chemical Abstracts*) but will produce information that is 6–12 months old (typical delay times between submission to a journal and publication). Papers relevant to the inorganic chemist can be found in a great number of publications. However, the major sources for publications in inorganic chemistry are: *Inorganica Chimica Acta* (Elsevier); *Inorganic Chemistry* (American Chemical Society); *Journal of Coordination Chemistry* (Gordon and Breach); *Journal of Organometallic Chemistry* (Elsevier); *Journal of the American Chemical Society*; *Journal of the*

Chemical Society, Dalton Transactions (The Royal Society of Chemistry); *Organometallics* (American Chemical Society); *Polyhedron* (Pergamon) (formerly the *Journal of Inorganic and Nuclear Chemistry*); *Zeitschrift für anorganische und allgemeine Chemie*.

Most of those listed above contain a section for communications or letters which have a much shorter timelag between submission and publication than a full paper. Very often, such short communications (such as those found in the *Journal of the Chemical Society, Chemical Communications*) herald the publication of a major paper and it is worth making routine checks for the appearance of the senior author's name in the appropriate index.

Quite often, inorganic chemistry shows up in journals which are published along national lines. As a researcher, it is useful to have access to the *Canadian Journal of Chemistry*, the *Bulletin de la Société Chimique de France* and *Russian Chemical Reviews* (a British Library Document Supply Centre cover-to-cover translation journal) since these countries have very active inorganic and organometallic chemists. The chance of a major piece of work appearing in other smaller national journals of chemistry is quite slight.

For those who do not have funds available to subscribe to dozens of journals, a cheaper and relatively fast alternative exists. The Institute for Scientific Information publication *Current Contents—Physical, Chemical and Earth Sciences* contains copies of the contents pages of, literally, hundreds of journals. Although the researcher ends up with nothing more than a title and the authors of a published paper, there is usually enough information available to take the search further. Most authors make the titles of their papers keyword-rich, so that computer searches can turn them up quite readily. This paucity of information is more than compensated for by the speed that one can scan through the appropriate literature. Along the same lines, the publishers of the *Journal of Organometallic Chemistry* produce the *JOM Pipeline* which gives a list of articles accepted for publication in future issues. If a paper looks particularly interesting, contact the senior author. It is an unusual scientist who does not reply to those who read his/her publications and preprints are, invariably, available.

Nuclear chemistry

It is important to emphasize from the outset that chemistry which involves the use of radioisotopes and ionizing radiations is likely to be subject to specific legislation which governs the scope and scale of any research. In the field of nuclear chemistry (unlike nearly all mainstream chemistry), knowledge does not give the researcher the right to experiment. To take the UK as an example, any institution using radioactive substances must register with the Department of the Environment (or with the Department of Education and Science in the case of schools) and nominate one or more 'competent persons' to assist in enforcing current regulations. These 'competent persons' will have undergone special training which will probably have included some guidance in information gathering in the relevant literature.

The researcher in nuclear chemistry should be warned that articles which are of military potential are unlikely to be released and those papers which are considered suitable for public consumption may find their way into a large number of journals. Many of the comments made earlier about inorganic chemistry are directly applicable to the nuclear chemist. Specifically, a considerable amount of useful information about radiation and radioisotopes is to be found in the *CRC Handbook of Chemistry and Physics* (72nd edn, D.R. Lide, ed., CRC Press, 1991).

Additionally, there is a considerable body of work being carried out under the general umbrella of organic chemistry and/or biochemistry and publications devoted to these areas will also need thorough scrutiny. As far as textbooks go, nuclear chemistry (as distinct from the applications of nuclear technology in biological sciences or physics) is well covered in *Nuclear Chemistry: Theory and Applications* (G.R. Choppin and J. Rydberg, Pergamon, 1980) and the second edition of *Essentials of Nuclear Chemistry* (H.J. Arnikar, John Wiley, 1987). Both have excellent coverage of the subject although the references contained within are somewhat limited. For a good text devoted to the actual chemistry of the actinides (rather than the effects of their radiation) there are three extensive works to choose from: *Actinides in Perspective* (N.M. Edelstein, ed., Pergamon, 1982); *Handbook on the Physics and Chemistry of the Actinides* (A.J. Freeman and C. Keller, eds, North-Holland, 1985); *The Chemistry of the Actinide Elements*, vols 1 and 2 (J.J. Katz, G.T. Seaborg and L.R. Morss, Chapman and Hall, 1986).

Much that is nuclear chemistry is now amalgamated within the boundaries of inorganic chemistry and there has been a definite shrinkage of the number of column inches devoted to nuclear chemistry as a 'stand-alone' subject. This trend shows up in the paucity of journals devoted to specific topics in the subject. The *Journal of Inorganic and Nuclear Chemistry* became *Polyhedron* in 1982 and, it must be said, has only occasional articles which might be termed 'nuclear chemistry'. Along the same lines, *Advances in Inorganic Chemistry and Radiochemistry* became abbreviated to *Advances in Inorganic Chemistry* (edited by A.G. Sykes, Academic Press) in 1987. Quite specifically for the nuclear chemist is the *Journal of Radioanalytical and Nuclear Chemistry* (Elsevier-Sequoia). Articles of interest often appear in the *Journal of the Less-Common Metals* and *Radiochimica Acta*. On the positive side, the Scientific and Technical Information Network (STN) carries a database called INIS (International Nuclear Information System) produced by the International Atomic Energy Agency (IAEA)

146 *Inorganic and nuclear chemistry*

in Vienna. There are at present 1.5 million entries in INIS dating from 1970. Twice-monthly updating is carried out to cope with the large number of citations. However, only a small fraction of INIS can be regarded as nuclear chemistry.

The older literature contained a vast number of wartime reports which had been declassified and thus made available to the public in the late 1940s and the 1950s. *Nuclear Science Abstracts* (1947–76) was produced by the US Atomic Energy Commission and is a better guide to this material than *Chemical Abstracts*. In 1976 when *Nuclear Science Abstracts* ceased publication INIS carried on the hard-copy service under the title *INIS Atomindex* (1976–).

CHAPTER NINE

Organic chemistry: the Beilstein Handbuch

T.C. OWEN, R.M.W. RICKETT AND R.T. BOTTLE

Nature, uses and limitations

The *Handbuch der organischen Chemie*, initiated by Friedrich Konrad Beilstein*, is the largest compilation of information on organic chemistry. It gives an exhaustive survey of over 1.5 million organic compounds which have been definitely characterized and also of naturally occurring organic materials which are of unknown structure. It is, therefore, a most voluminous work; in the current (fourth) edition, the main work (*Hauptwerk, HW*) comprises some 200 000 entries in 27 volumes covering the literature prior to 1910, and this is augmented by supplementary series of volumes, each corresponding in content to the conumerate volume in the *Hauptwerk*. The first supplement (*Erstes Ergänzungswerk, EI*) covers the literature from 1910 to 1919 and the second (*Zweites Ergänzungswerk EII*) from 1920 to 1929.

In 1958 the first volumes of the third supplement (*Drittes Ergänzungswerk, EIII*) covering the literature from 1930 to 1949, appeared. In 1972 volumes of the fourth supplement (*Viertes Ergänzungswerk, EIV*) covering the literature from 1950 to 1959, started to appear. As from Volume 17, these were combined into a single supplement (*EIII/IV*) with literature coverage from 1930 to 1959.

In 1984 a *Supplementary Series V / Fünftes Ergänzungswerk (EV)* commenced publication with the heterocyclic compounds of Volume 17. *EV* covers the literature from 1960 through 1979 and is in English. The production of *EV* is likely to keep the Beilstein Institut busy until the end of the century (Prof. R. Luckenbach, personal communication, May 1990); thus a sixth supplement covering the literature of the 1980s is

---

* Beilstein, a Russian who had studied in Germany, completed the first edition of his *Handbuch* in 1882 and subsequently produced two further editions. The fourth edition (and also supplements to the third edition, 1899–1906) were issued by the German Chemical Society until after the Second World War, when its preparation was vested in the Beilstein Institut, Frankfurt am Main. For further historical details, see E.H. Huntress (1938), *Journal of Chemical Education*, **15**, 303.

unlikely to appear until then, if ever. The literature is, however, being collected and evaluated and these 'excerpts' have been made available as BEILSTEIN ONLINE since 1988. The database is searchable on Dialog and at STN. By 1991 organic substances contained in *HW* and *EI* to *EIV* (covering the period 1830–1960) will have been entered into the BEILSTEIN DATABASE. The database is discussed in detail in *The Beilstein Online Database: Implementation, Content and Retrieval* (S.R. Heller, ed., ACS Symposium no. 436, 1990), but further details are given in Chapter 14. It is now supplemented by *Beilstein Current Facts in Chemistry*, a quarterly CD-ROM covering over 80 leading journals from 1990.

Beilstein itself is now an extremely expensive serial, each 'volume' consisting of several parts, sometimes costing more than £1200 each. Second-hand sets are often available from librarians unwilling to continue this commitment, and microform versions of the main work and first two supplements are available more cheaply.

The unique and important features of *Beilstein* are its comprehensiveness (information on all known compounds being listed) and its compilation of the information. In contrast, the comprehensive information in *Chemical Abstracts* is scattered, while the compiled information in the various dictionaries is far from complete.

Thus, a comprehensive survey of the literature (prior to 1960) on ethyl benzoate simply involves a few pages of Volume 9 of *HW, EI, EII, EIII and EIV*; extraction of the same information from other sources is a major operation involving considerable mental (and physical!) labour. It is our experience that information is often missed in such a search. However, *Beilstein* can never be up to date. No supplementary volume can ever appear until the whole of the period which it covers has elapsed and in practice a long delay has always occurred. *EIV* was not completed until 26 years after the end of 1959. The work, therefore, gives no guide to the recent literature. It is an information retrieval system for ascertaining the major physical and chemical properties, preparations, etc., of any organic compound reported in the literature during the period covered. It also gives a detailed guide to the extent of knowledge in a given field together with ample references to the original literature. The need to consult the early abstract literature, which may be difficult to access, incomplete and sometimes even misleading, is thus removed. The information so obtained is frequently sufficient for one's requirements, more recent data often being unnecessary.

Information listed

The information listed varies from a few lines to many pages, depending on the importance of the particular compound. A full but abbreviated account is given, including, as available: name(s); formula; structure(s);

Organic chemistry: the Beilstein Handbuch 149

history; occurrence; formation; preparation; physical, chemical and physiological properties; technology; analysis; addition compounds; salts; and conversion products of unknown structure.

Finding a compound in *Beilstein*

Indexes

The location of information in *Beilstein* was considerably simplified when the cumulated general indexes to the *Hauptwerk*, *EI* and *EII* were published in 1956. Location by name is alphabetical in the subject index (*Sachregister; EII*, Volume 28, parts 1 and 2). Location in the formula index (*Formelregister; EII*, Volume 29, parts 1, 2 and 3) is according to the Hill system (see p. 37). The entry against each formula is subdivided according to the names of the isomeric substances, each of which is followed by the appropriate volume and page numbers. Thus the entry **20**, 181, *I* 54, *II* 96, **22** *II* 619 against pyridine in the subject index and against C_5H_5N–pyridine in the formula index, indicates data on this compound on page 181 of Volume 20 of the *Hauptwerk*, page 54 of the same volume in the first supplement, page 96 of this volume in the second supplement, and page 619 of Volume 22 in the second supplement. Similar general indexes to *HW* and *EI* only were published in 1938–40 as volume 28 and 29 of the *Hauptwerk*. Care should be exercised in case the earlier indexes should be consulted in mistake for the later set. While the 1956 and subsequent indexes use the Hill classification system, earlier formula indexes use the Richter system, in which other elements follow C and H in a set sequence (O, N, Cl, Br, I, F, S, P). In addition to the general indexes, each volume has its own subject and formula index. These are of value in cases where the general indexes are anteceded (e.g. *EIII* to *EV*).

A new set of general indexes are scheduled for publication between 1991 and 1993 as Volumes 28 (*General-Sachregister*) and 29 (*General-Formelregister*). Costing over £9000 these will use German (IUPAC) nomenclature and cover *HW* plus *EI* to *EIV* in 23 sub-volumes.

The various guides to the use of *Beilstein* (see Bibliography) throw great emphasis on an understanding of the system of classification adopted in the construction of the work as a means of locating information. This is because general indexes did not exist prior to 1938, and only to the *Hauptwerk* and first supplement prior to 1956. Between these dates it was possible to locate entries in the second supplement for compounds previously reported in the earlier series by either of the following methods: the appropriate earlier volume number could be determined and the index to the appropriate volume in the second supplement consulted; or the system number (*Systemnummer*) for the compound could be observed in the earlier work and the same number located in the volume of the later series. The system numbers comprise a series of 4877 arbitrary units into which the subject matter is sub-divided and one can quickly find a specific compound by scanning the entries under the appropriate system number. Data on compounds first reported after 1920 could not be located in this way, and a knowledge of the classification system adopted by the editors was essential.

Compounds noted in earlier series may be located in the third to fifth supplements by utilizing the corresponding main work pagination (e.g. H123) which is printed at the top of each right-hand page. Other compounds can often be similarly located by looking for the pagination of closely related compounds. If this method fails, one must then locate the compounds by means of the system as described below.

Table 9.1. Functioning classes in the Beilstein classification system

1.	Stem nuclei (and non-functioning derivatives)*
2.	Hydroxy compounds*
3.	Carbonyl compounds*
4.	Carboxylic acids*
5.	Sulphinic acids
6.	Sulphonic acids*
7.	Seleninic and selenonic acids
8.	Amines*
9.	Hydroxylamines
10.	Hydrazines
11.	Azo compounds
12.	Hydroxyhydrazines
13.	Diazonium, diazo and isodiazo compounds
14.	Azoxy compounds
15.	Nitramines and isonitramines
16.	Triazanes
17.	Triazenes
18.	Hydroxytriazenes
19.	Azoamino-oxides
20.	Tetrazanes
21.	Tetrazenes
22.	Compounds with chains of more than four nitrogen atoms
23–28.	Organometallic compounds having carbon bonded directly to the metallic element, *except* salts and all compounds of the alkali metals

*It is useful to memorize the order of the groups in these classes in particular.

Organic chemistry: the Beilstein Handbuch 151

Table 9.2. Correspondence of volumes and system numbers with divisions and classes

Division	Sub-division	Class*	Sub-class	Volume	System number
I		1-3		1	1-151
I		4a		2	152-194
I		4b	Polyfunctional carboxylic acids	3	195-322
I		5 and over		4	323-449
II		1		5	450-498
II		2		6	499-608
II		3a	Mono- and polycarbonyl compounds	7	609-736
II		3b	Hydroxycarbonyl compounds	8	737-890
II		4a	Mono- and polycarboxylic acids	9	891-1050
II		4b	Hydroxy and carbonyl acids	10	1051-1504
II		5-7		11	1505-1591
II		8a	Monoamines	12	1592-1739
II		8a and b	Poly- and hydroxyamines	13	1740-1871
II		8b	Other amines	14	1872-1928
II		9 and 10		15	1929-2084
II		11 and over		16	2085-2358
III	1 cyclic O (S, Se, Te)	1-3a		17	2359-2503
III	1 cyclic O (S, Se, Te)	3b and over		18	2504-2665
III	2 or more cyclic O	All		19	2666-3031
III	1 cyclic N	1		20	3032-3102
III	1 cyclic N	2 and 3		21	3103-3241
III	1 cyclic N	4 and over		22	3242-3457
III	2 cyclic N	1 and 2		23	3458-3554
III	2 cyclic N	3a		24	3555-3633
III	2 cyclic N	3b and over		25	3634-3793
III	More than 2 cyclic N	All		26	3794-4187
III	Cyclic O and N	All		27	4188-4720
Subject index. Part I, A-G; Part II, H-Z				28	
Formula index. Three parts				29	
In 1938 the following volumes appeared for which there are no supplements.					
IV	Rubber, Gutta Percha, Carotenoids			30	4723-4723a
IV	Carbohydrates			31	4746-4767a

*'a' indicates unmixed functionality; 'b' indicates mixed functionality.

Use of the classification system

Any substance of known structure is entered at a position which is absolutely determined by that structure. Each structural formula is therefore its own index. Location is effected by applying the following three operations.

1. Examination of the structural formula (with manipulations as detailed below) determines to which division, sub-division, class, sub-class, etc., the substance belongs (see Table 9.1).

2. Inspection of Table 9.2 and application of the classification thus determined indicates the appropriate volume.

3. Random opening of the appropriate volume and examination of the entries on the page so selected will make it immediately apparent whether the desired compounds must appear before or after the entries inspected (provided that one knows the classification scheme). Adjustment and repetition of the inspection process soon leads to location of the desired compounds.

Perusal of a worked example will clarify the application of these operations. (The example should be re-examined in detail when the later section on the classification system has been studied).

4-Chloro-6-hydroxypyridine-2-carboxylic acid.

Inspection shows this to belong to Division III (heterocyclic), sub-division one cyclic nitrogen, latest 'functioning group' carboxylic acid, Class 4; Table 9.2 therefore indicates Volume 22. Random opening of this volume, at page 443, for example, reveals entries concerning aminoquinolines. Clearly this is the correct division and sub-division, but amines (Class 8) come later than the desired carboxylic acids, so pages are skimmed through in the direction of the front of the book until heterocyclic carboxylic acids appear (page 385). Again stopping at random, page 133 reveals dicarboxylic acids. These must come before hydroxycarboxylic acids, so pages are checked, somewhat more carefully, in the reverse direction until hydroxycarboxylic acids with three atoms of oxygen appear at page 190. Close inspection then reveals successively hydroxypyridine carboxylic acids, page 212, under which is 6-hydroxypyridine-2-carboxylic acid, page 213, and the desired 4-chloro-6-hydroxypyridine-2-carboxylic acid as a 'stem nuclear non-functioning derivative' thereof on page 214.

Organic chemistry: the Beilstein Handbuch

The classification system

Divisions

The four divisions comprise:
 Division I   Acyclic
 Division II Alicyclic
 Division III Heterocyclic
 Division IV Natural products not assigned in the above three divisions

The assignment of a compound to Division I, II or III depends on the derivation of its 'stem nucleus', which is obtained by replacing all atoms or groups attached to carbon by the appropriate number of hydrogen atoms, except where such replacement would require rupture of a heterocyclic ring.

Example

Compound	Stem nuclei	Division
$CH_2ClCH_2CH_2COOH$	$CH_3CH_2CH_2CH_3$	I
$CH_3CH_2COOCH_2CH_3$	$CH_3CH_2CH_3$ and CH_3CH_3	I
$C_4H_9SO_2SC_2H_5$	C_4H_{10} and C_2H_6	I

[Cyclohexane with CH$_3$, CH$_3$, NH$_2$ substituents] → [Cyclohexane with CH$_3$, CH$_3$] II

[4-Cl-phenyl-COOCH$_2$-phenyl(CH$_3$, NH$_2$)] → [benzene with CH$_3$, CH$_3$] and [benzene with CH$_3$] II

[Thiophene with CH$_3$] → [Thiophene with CH$_3$] III

[Pyranose sugar with CH$_2$OH, OH, OH, OH] → [Tetrahydropyran with CH$_3$] III

154 Organic chemistry: the Beilstein Handbuch

Frequently, stem nuclei corresponding to more than one division occur in one compound. In such cases the compound is assigned to the division corresponding to that stem nucleus which comes last in the classification. This *principle of latest position* is applied widely elsewhere in the classification system.

Example

Compound	Stem nuclei	Division
CH₃CH₂COO—⟨benzene⟩	CH₃CH₂CH₃ and ⟨benzene⟩	II
⟨2-NO, 5-Br, 1-COOCH₂—(3-pyridyl-5-NO₂)⟩ benzene	⟨toluene (CH₃)⟩ CH₃ and ⟨3-methylpyridine⟩	III

Sub-divisions

The stem nuclei of Divisions I and II contain only carbon and hydrogen, so that further classification depends on the type and number of functional groups present (see 'Functioning classes', below). In Division III, however, other atoms are present in the stem nucleus, necessitating the definition of a sequence of sub-divisions governing the order in which heterocyclic stem nuclei must be placed before further classification according to functional groups may be considered. The defined sequence considers first oxygen and then nitrogen heterocycles as follows:

1. Stem nuclei containing 1, 2, 3, 4 ... n oxygen atoms
2. Stem nuclei containing 1, 2, 3, 4 ... n nitrogen atoms
3. Stem nuclei containing 1, 2, 3, 4 ... n oxygen and 1 nitrogen
4. Stem nuclei containing 1, 2, 3, 4 ... n oxygen and 2 nitrogen, etc.

Heterocyclic stem nuclei containing the heavier elements of group 6 of the periodic table, sulphur, selenium and tellurium, are considered to arise by replacement of oxygen and are listed immediately following the parent stem nuclei containing that element. Heterocycles containing other cyclically bound atoms are considered after all O, N, S, Se, Te heterocyclics have been dealt with. In cases where stem nuclei corresponding to more than one sub-division occur in one compound, assignment is to the last possible sub-division in accordance with the principle of latest position in the system. (Note: where there is more than one hetero atom, these may be in one ring or distributed over several rings.)

Functioning classes
Having determined the gross order by division and sub-division according to stem nuclei, the order of consideration is next defined according to the functional groups attached thereto. The order of precedence is detailed in Table 9.1, the order of the first eight or ten classes therein being of overriding importance. Stem nuclear compounds having either no substituents or the seven 'non-functioning' substituents (those which contain no replaceable hydrogen atoms) -F, -Cl, -Br, -I, -NO$_3$, -NO$_2$, -N$_3$ only comprise Class 1 and are dealt with in order of decreasing saturation and increasing number of carbon atoms. Thus Volume 1 deals with Division 1, Class 1 in the order:

Hydrocarbons C_nH_{2n+2}

>Methane and its non-functioning derivatives
>
>Ethane and its non-functioning derivatives
>
>Propane and its non-functioning derivatives
>
>Butane and its non-functioning derivatives
>
>Similarly 2-methylpropane, pentane, 2-methylbutane, 2,2-dimethylpropane, hexane, . . . higher paraffins, each with its non-functioning derivatives

Hydrocarbons C_nH_{2n}

>Ethene and its non-functioning derivatives
>
>Propene and its non-functioning derivatives
>
>But-1-ene and its non-functioning derivatives
>
>Similarly but-2-ene, 2-methylpropene, pent-1-ene, etc.

In an analogous manner, Volume 5 deals with Division II, Class 1 in the order:

Hydrocarbons C_nH_{2n}

>Cyclopropane and its non-functioning derivatives
>
>Cyclobutane, methylcyclopropane, cyclopentane, methylcyclobutane, 1,1-dimethylcyclopropane, etc., each with its non-functioning derivatives.

Hydrocarbons C_nH_{2n-2}

Cyclopropene, cyclobutene, cyclopentene, methylcyclobutene, methylenecyclobutane, etc., each with its non-functioning derivatives.

After all known compounds having no functioning* groups have been entered in each division, those containing such groups are dealt with in order as in Table 9.1. The functioning groups of classes 2 to 22 all contain replaceable hydrogen and are joined to the stem nucleus by single bonds, carbonyl compounds being regarded in this connection

as *gem*-diols,

$$\begin{array}{c} \diagdown OH \\ C \\ \diagup \diagdown \\ OH \end{array}$$

Compounds containing a functioning group attached to a carbon atom which bears one or more other functioning or non-functioning groups of any kind are regarded as 'replacement derivatives' (see below). They appear under that carbonyl compound or carboxylic acid which results from replacement of all these groups by an appropriate number of hydroxyl groups followed by elimination of water from *gem*-diols as may be appropriate. The same applies to compounds in which nitrogen is joined to carbon by multiple valence, e.g.

$$CH_3CH\diagup^{NO_2}_{\diagdown OH} \quad , \quad CH_3CH\diagup^{SO_3H}_{\diagdown NH_2} \quad , \quad CH_3CH{=}NNH_2 \text{ and } CH_3CH\diagup^{NH_2}_{\diagdown NH_2}$$

will be all classed as derivatives of, and would be considered under

$$CH_3CHO \left(\text{i.e. } CH_3CH\diagup^{OH}_{\diagdown OH} \right), \text{Division I, Class 3}$$

$$CH_3CH\diagup^{OH}_{\diagdown OH} \quad \text{is a derivative of } CH_3CH_3, \text{Division I, Class 1.}$$

---

*Functioning must not be confused with the common term *functional*.

$CH_3CH_2NHNH_2$ and $CH_3CH_2SO_3H$ are functioning compounds of Division I, Classes 10 and 6, respectively.

It frequently occurs that different functioning or non-functioning groups occur, or may be conceived to occur, on different carbon atoms of a given stem nucleus. The principle of latest position is applied in such cases. Any non-functioning group, therefore, is considered after all derivatives of the functioning groups present; e.g. 2-amino-4-chlorobenzoic acid comes after all the esters, amides, etc. of 2-aminobenzoic acid.

Examples
Bromoethanoic acid is considered under ethanoic acid, Division I, Class 4, Volume 3, as a non-functioning derivative, just as bromoethane is considered under ethane.

1-Amino-2-hydroxyethanol, $HOCH_2CH_2NH_2$, is considered under ethylamine, Division I, Class 8, Volume 4.

Where several functioning groups occur, the latest functioning group is considered first, *followed by the others in the order in which they appear* in Table 9.1. Thus the order of appearance of the classes in any division (or sub-division of Division III), non-functioning derivatives being considered along with each parent functioning compounds, would be:

Stem nuclear compounds

Monohydroxy compounds

Dihydroxy compounds

Trihydroxy compounds, etc.

Monocarbonyl compounds

Dicarbonyl compounds, etc.

Carbonyl-hydroxy compounds - with 2 atoms of oxygen
- with 3 atoms of oxygen
- with 4 atoms of oxygen
  etc.

Monocarboxylic compounds

Dicarboxylic compounds

Tricarboxylic compounds, etc.

Carboxylic-hydroxy compounds - with 3 atoms of oxygen
- with 4 atoms of oxygen
- with 5 atoms of oxygen
etc.

Carboxylic-carbonyl compounds - with 3 atoms of oxygen
- with 4 atoms of oxygen
etc.

Carboxylic-hydroxy-carbonyl compounds - with 4 atoms of oxygen
- with 5 atoms of oxygen
etc.

Monosulphinic acids

Disulphinic acids, etc.

Hydroxy-sulphinic acids - with 1 OH and 1, 2, 3, 4, etc. SO_2H groups
- with 2 OH and 1,2,3,4, etc. SO_2H groups, etc.

Carbonyl-sulphinic acids - with 1 C=O and 1,2, etc. SO_2H groups
- with 2 C=O and 1,2, etc. SO_2H groups, etc.

Carbonyl-hydroxy-sulphinic acids - with 2 atoms of oxygen in the C=O and 1, 2, etc. SO_2H groups
- with 3 atoms of oxygen and 1,2,3, etc. SO_2H groups, etc.

Carboxylic-sulphinic acids — as for hydroxysulphinic acids

Carboxylic-hydroxy-sulphinic acids - with 3 atoms of oxygen and 1,2,3, etc. SO_2H groups
- with 4 atoms of oxygen, similarly, etc.

Carboxylic-carbonyl-sulphinic acids — as for carboxylic-hydroxy

Carboxylic-hydroxy-carbonyl-sulphinic acids — similarly.

While the number of variations obtainable in this manner is as limitless as the number of organic compounds which it is possible to envisage, it is obvious that the majority of compounds corresponding to the higher combinations will not be known. However, from a knowledge of the order in which the combinations will appear, it is possible to judge whether the entries on a page, selected by opening the appropriate vol-

Organic chemistry: the Beilstein Handbuch 159

ume at random, come before or after the desired compound. Repetition of the process leads to easy location of the desired entry. With a little practice, location by this means becomes easily as rapid and convenient as the use of the indexes, and is, of course obligatory where indexes are not available.

Further subclassification
Application of the classification system as above is invariably sufficient to locate any known compound quite easily. For completeness, it may be mentioned that each of these groups (e.g. acyclic monohydroxy compounds) is arranged in order of decreasing saturation (C_nH_{2n+2}, C_2H_{2n}, C_nH_{2n-2}, etc.), and then each group having a given degree of saturation (acyclic monohydroxy compounds, C_nH_{2n}) is grouped in order of increasing number of carbon atoms.

Functioning derivatives
Functioning derivatives are those compounds which arise formally by 'coupling' a functioning group with a 'coupling component'. In such coupling there is eliminated as water an H from the functioning group and an OH of an organic coupling component, or an H or OH from the functioning group and an OH or H of an inorganic coupling component. Such derivatives are located by 'uncoupling' followed by location of the parent functioning compounds.

A discussion of the order of precedence of such derivatives is beyond the scope of the present work and is unnecessary for the location of a given derivative. However, when both the parent functioning compound and the coupling component are organic, ambiguity frequently arises as to which is which. The principle of latest position is applied, the compound being regarded as a derivative of that one of its components which comes latest in the system.

Examples
1. *Ethyl benzoate.* Uncoupling gives ethanol and benzoic acid, which can be recoupled in two ways regarding either as the parent functioning compound and the other as the coupling component. Benzoic acid, however, comes later in the system than ethanol, so that this ester is regarded as a derivative of the acid.
2. *Phenylmethyl ethanoate.* By a similar process, and in contrast to the previous example, this ester is regarded as a derivative of benzyl alcohol, with ethanoic acid as the coupling component, the alicyclic (Division II) alcohol coming later than the acyclic acid.

3. N-Ethyl-4-chlorobenzamide, $ClC_6H_4CONHCH_2CH_3$. Whereas in examples (1) and (2) uncoupling on either side of the singly bonded ester oxygen gave the same components, in this case uncoupling the bond CO-N gives 4-chlorobenzoic acid and ethylamine, while uncoupling the N-CH bond gives 4-chlorobenzamide and ethanol. 4-Chlorobenzamide is not an index compound and must be further uncoupled, giving ammonia and again 4-chlorobenzoic acid. The latest of these is the acid, and it is under this compound that the substituted amide will be listed — not, however, as a derivative with ethylamine, which would require elimination of H instead of OH from an organic coupling component, but as a second-degree derivative from coupling with ammonia, elimination of H from inorganic coupling components being permitted, and then with ethanol.

Non-functioning and replacement derivatives of functioning compounds and functioning derivatives
These are obtained by substitution of stem nuclear hydrogen by non-functioning substituents and of functioning oxygen by sulphur, selenium or tellurium. They and their derivatives appear after the parent functioning compounds and their functioning derivatives have all been disposed of. Thus ethyl chlorethanoate is considered as a functioning derivative of chlorethanoic acid, which is itself a non-functioning derivative of ethanoic acid and comes after that acid and all its functioning derivatives have been dealt with. The ester is not regarded as a derivative of ethyl ethanoate.

With some practice, location of compounds by use of this system becomes easy. The above is, however, an abbreviated guide and in a few instances may not suffice. In such cases it is most instructive to locate the desired compound, or something closely related to it, by means of indexes and then to examine the surrounding text, whereupon the factors which decide its location are determinable. If further help is needed in locating compounds in *Beilstein*, use Runquist's programmed guide, Weissbach's guide or the Beilstein Institute's own guide *How to Use Beilstein* (gratis).

Two computer programs are also available. BEILSTEIN KEY runs under MS-DOS on any IBM-compatible PC and takes the user through a step-by-step analysis of the required molecule to identify the relevant volume. SANDRA (Structure AND Reference Analyser) permits the user to draw the structure using a mouse. It then analyses the structure and locates its position, usually within a few pages, in the *Handbuch*.

Another program from the Beilstein Institute is AUTONOM. This creates the IUPAC systematic name of an organic compound from its structural diagram on a PC.

Bibliography

Beilsteins Handbuch der organischen Chemie (1918), 4th edn, vol. 1 pp. 1–47. Heidelberg: Springer.

Deutsche Chemische Gesellschaft (1929), *System der organischen Verbindungen: Ein Leitfaden für die Benutzung von Beilsteins Handbuch.* Heidelberg: Springer.

Giese, F. (1986), *Beilstein's Index, Trivial Names in the Systematic Nomenclature of Organic Chemistry.* Heidelberg: Springer.

Hancock, J.E.H. (1968), *Journal of Chemical Education*, **45**, 336.

Heller, S.R., ed., (1991), *The Beilstein Online Database: Implementation and Retrieval.* ACS Symposium Series no. 436.

Heller, S.R. and Milne, W. (1989), *Beilstein Online, Reference Manual* (STN or Dialog versions). Heidelberg: Springer.

Huntress, E.H. (1938), *A Brief Introduction to the Use of Beilsteins Handbuch*, 2nd edn. New York: Wiley.

Runquist, O. (1986), *Beilsteins Handbuch: a Programmed Guide.* Burgess.

Toler, L. (1989), *Beilstein Online, Workshop for STN users.* Heidelberg: Springer.

Weissbach, O. (1976), *The Beilstein Guide* (translated by H.M.R. Hoffman). Heidelberg: Springer.

CHAPTER TEN

Organic chemistry: other reference works

A.G. OSBORNE

The volume of published work on organic chemistry continues to increase considerably. Since the last edition of this book was written a number of new dictionaries and reference books have been introduced and these publications now merit a separate division. The range of serials and data compilations has continued to increase, particularly in the area of regular reviews and updates of the literature.

With the implementation of the COSHH (Control of Substances Hazardous to Health) regulations in 1989, the literature of organic chemistry now contains many works dealing with all aspects of health and safety. These have not been included in this chapter; however, a comprehensive range of material is available from Croner Publications. The reader is also referred to Chapter 13.

Major reference works

Rodd

Chemistry of Carbon Compounds (E.H. Rodd, ed., 1st edn, Elsevier, 1951–62; *Rodd's Chemistry of Carbon Compounds* (S. Coffey, ed., 2nd edn, Elsevier, 1964–89); *Supplements to the 2nd Edition (Editor S. Coffey) of Rodd's Chemistry of Carbon Compounds* (M.F. Ansell, ed., Elsevier, 1973–). Rodd is a general introduction to organic chemistry and is complete in five volumes, each volume being in several parts. A General Index was published as Volume 5. A valuable theoretical interpretation of the properties and reactivity of each group of compounds is given in addition to the structural formulae, methods of preparation and chemical and physical properties of the individual compounds listed. Publication of a completely rewritten second edition began in 1964; the first part of Volume 4, dealing with heterocyclic compounds, was published in 1973, and this was to appear in 12 parts (4A–4L), but after the publication of eight of the parts, Dr Coffey sadly passed away in 1980.

Accordingly Volumes 4C and 4D, which appeared next, were jointly edited by Coffey and Ansell, and the final parts I and J (containing Chapters 42–47) were then combined and appeared in 1989 as volume IJ edited by Ansell only. Owing to the very long time involved in producing the volumes in this series, a range of supplements (edited by M.F. Ansell) was commenced in 1973 in order to keep this major work of reference up to date. Supplements to Volumes 1–3 and parts A/B/E–H/K/L of Volume 4 have now appeared, the last being Volume 4E published in 1990. Supplements to the remaining parts of Volume 4 (for which the second editions appeared much later) are planned to be produced in due course. Publication of the Index (Volume 5) to the second edition and the supplements has, however, been abandoned.

A series of second supplements to the second edition (M. Sainsbury, ed.) is now in progress, with Volume 1A/B published late in 1991. Recently a camera-ready format has been adopted to expedite publication.

Chemistry of Heterocyclic Compounds

Chemistry of Heterocyclic Compounds (A. Weissberger, ed., 1950–70, A. Weissberger and E.C. Taylor, eds, Interscience, 1970–). This set of monographs is intended to cover comprehensively the chemistry of the complete field of heterocyclic chemistry. It has been appearing at regular intervals since 1950; Volume 50 on bicylic azepines appeared in 1991. The series also includes several multi-volume works and the four-part work on pyridine and its derivatives (Volume 14), edited by E. Klingsberg (1960–64) and the supplement, edited by R.A. Abramovitch (1974–75), could be considered as a series in its own right. Several volumes have revised editions or supplements; that on pyridine–metal complexes appeared in 1985 and contained 8701 references.

Comprehensive Chemistry series

These are medium-sized works designed to fill the gap between the smaller and less informative texts and existing multi-volume treatises such as *Rodd* above. The following works, devoted to organic chemistry subjects, have appeared to date.

Comprehensive Organic Chemistry (D.H.R. Barton and W.D. Ollis, eds, 6 Vols, Pergamon Press, 1979). The contents of these volumes reflect all the important facets of modern organic chemistry. Mechanistic organic chemistry has been adopted to provide a constant and correlative theme. The emphasis throughout is placed upon the synthesis, properties and reactions of all the important classes of organic compounds.

Organic chemistry: other reference works   165

Comprehensive Organometallic Chemistry (G. Wilkinson, F.G.A. Stone and E.W. Abel, eds, 9 vols, Pergamon Press, 1982). These volumes bring together a survey of the entire body of research in organometallic chemistry for both the main groups and the transition elements.

Comprehensive Heterocyclic Chemistry (A.R. Katritzky and C.W. Rees, eds, 8 Vols, Pergamon Press, 1984). This presents a comprehensive account of fundamental heterocyclic chemistry, with the emphasis on basic principles and, as far as possible, on unifying correlations in the properties, chemistry and synthesis of heterocyclic systems and their analogous carbocyclic structures.

Comprehensive Polymer Science (G. Allen, ed., 7 Vols, Pergamon Press, 1988). This work is designed to set down the structure of the vast subject of polymer science and technology in such a way that scientists and teachers of polymer science, and workers in associated fields, can find an authoritative and comprehensive account of any topic of interest.

Comprehensive Medicinal Chemistry (C. Hansch, ed., 6 Vols, Pergamon Press, 1990). These volumes provide the rational design, mechanistic study and therapeutic application of chemical compounds. They cover general topics, enzymes and other molecular targets, membranes and receptors, quantitative drug design and biopharmaceuticals in full detail. Also included is a unique compendium of drug information covering over 5000 compounds.

Also recently published is *Comprehensive Organic Synthesis* (Pergamon Press, 9 Vols, 1991); see also p. 76.

Elsevier's Encyclopaedia of Organic Chemistry (F. Radt, ed., Elsevier, 1940–56; F. Radt, ed., Springer, 1959–69). Only Volumes 12, 13 and 14, dealing with condensed carboisocyclic compounds, had appeared when publication was suspended with the fourth supplement to Volume 14 in 1956. However, further supplements were published by Springer to bridge the gap until the steroid sections of *Beilstein* (Chapter 9) appeared. The complete work deals with bicyclic, tricyclic, tetracyclic and higher cyclic compounds, including triterpenes and steroids, and contains separate surveys of methods for the synthesis of the more important ring systems in addition to information on individual compounds.

Dictionaries and handbooks

Dictionary of Organic Compounds (J. Buckingham, ed., 5th edn, Chapman and Hall, 1982). The first four editions of this work were published by Eyre and Spottiswoode and edited by Sir Ian Heilbron and his collaborators. They were commonly referred to as *'Heilbron's Dictionary'* or *'Heilbron & Bunbury'*. The new, revised and enlarged 5th edition comprises five main-work volumes and two index volumes including a Compound Name Index with extensive cross-references and a Molecular Formula/CAS Registry Number Index. Supplements are issued annually, the ninth supplement being published in 1991. The fifth supplement is in two volumes, the second part representing a cumulative index for supplements 1–5 inclusive. The formula, source, some physical constants and chemical properties and literature references to preparation are given for each parent compound, together with a list of derivatives. Nomenclature in the new edition conforms to IUPAC recommendations, but continuity with the earlier editions has been preserved by use of cross-references, and many trade names (e.g. of drugs and insecticides) have been included.

Three new features of the fifth edition, which lists more than 150000 compounds within 46400 entries, include the listing of the CAS Registry Number for each compound as well as references to spectral data. Each compound is also provided with a unique DOC reference number (e.g. 6,6'-biquinoline: B-01378) to facilitate location. The DOC number consists of a letter of the alphabet followed by a five-digit number. For entries in the main work the first digit is invariably zero, whilst for entries in the supplements, the first digit refers to the supplement number. The information in the DOC is also available online via Dialog as file 303 (see Chapter 5).

Dictionary of Organophosphorus Compounds (R.S. Edmundson, ed., Chapman and Hall, 1988) is a companion to the *Dictionary of Organic Compounds* covering the whole field of organophosphorus chemistry; well over 20000 compounds are covered in 5200 entries. A valuable feature is the Type of Compound Index which classifies all the compounds in the dictionary by structural type (e.g. phosphonic acids, six-membered rings containing phosphorus, etc.).

The information in *Dictionary of Organometallic Compounds* (J. Buckingham, ed., Chapman and Hall, 1984) which comprises two volumes and an index, is arranged in the same general manner as the *Dictionary of Organic Compounds*. Compounds are initially filed by the metal element in alphabetical order of chemical symbol, and thence by formula. Annual Supplements (J.E. Macintyre, ed.) have also been issued; the fifth supplement (1989) comprises two volumes and contains an index to all previous supplements. Also available is a *Complete Structure Index to Supplements 1–5* (J.E. Macintyre, ed., 1990) which should greatly extend the value of the previously published supplements. No more supplements are to be issued; instead a second edition of this dictionary is planned for publication in 1994.

Organometallic Chemistry Sourcebooks (various editors, Chapman and Hall, 1985–) contain information derived from the *Dictionary of Organometallic Compounds* and provide carefully tailored information to workers in specialized areas, e.g. 'Organometallic Compounds of Ruthenium and Osmium', G.R. Knox, 1985. The older literature is also summarized in *Handbook of Organometallic Compounds* (N. Hagihara, M. Kumada and R. Okawara, eds, Benjamin, 1968).

Dictionary of Antibiotics and Related Substances (B.W. Bycroft, ed., Chapman and Hall, 1988) is a large single-volume work providing an authoritative survey of the antibiotic field. Previously, chemical data on these compounds were scattered throughout the literature. The arrangement and layout is again similar to that of the *Dictionary of Organic Compounds* and the volume provides an extensive range of preferred names and synonyms as well as giving the source and activity of the antibiotic. The dictionary also includes a Type of Compound Index in which all the approximately 6000 antibiotics included are classified under one or more of 46 type headings allowing rapid location of all related compounds.

Dictionary of Alkaloids (I. Southon and J. Buckingham, eds, 2 vols, Chapman and Hall, 1989) is a large two-volume dictionary and is the definitive work on this important class of compounds. It covers about 10 000 alkaloids within 5000 entries. The dictionary contains over 15 000 references altogether with some entries containing as many as 30 references.

Dictionary of Drugs (J. Elks and C.R. Ganellin, eds, Chapman and Hall, 1990) is a large two-volume reference work providing information on more than 6000 drugs. The entries include all those drugs listed in the British and American Pharmacopoeias, and the World Health Organization's List of International Nonproprietary Names. Each entry includes pharmacological actions and clinical use.

Dictionary of Steroids (D.N. Kirk, R.A. Hill, H.L.J. Makin and G.M. Murphy, eds, Chapman and Hall, 1991). This large two-volume work features over 15 000 steroids in 6000 entries. The coverage is virtually comprehensive for natural steroids, and extensive entries are provided for synthetic and semisynthetic derivatives. The extensive indexes are bound as a separate volume and include a Type of Compound Index and a Species Index. The indexes permit rapid access to the data by botanical or zoological name.

Also planned for publication in 1992 is the *Dictionary of Terpenoids* (R.A. Hill and J.D. Connolly, eds, Chapman and Hall) covering over 20 000 terpenoids in a three-volume work.

Dictionary of Natural Products CD-ROM (Chapman and Hall, 1992).

This natural products database contains chemical and physical data for 65 000 organic substances and includes all natural products contained in other Chapman and Hall dictionaries (DOC, antibiotics, alkaloids, drugs and terpenoids). The CD-ROM may be searched by text or substructure and operates in the popular Windows version 3.0 environment. Publication was expected in the spring of 1992.

Faraday's Encyclopaedia of Hydrocarbon Compounds (J.E. Faraday *et al.*, eds, Chemindex/Butterworths, 1945–69). Thirteen volumes have appeared in loose-leaf ring-binder format. Each volume contains the original compilation of data and numerous supplements. Detailed information on the synthesis, physical properties and analysis of hydrocarbons containing 1–14 carbon atoms is given.

Other reference works, including series publications

Organic Chemistry (A.T. Blomquist, ed., 1964–71; A.T. Blomquist and H. Wasserman, eds, Academic Press, 1971–) is a series of monographs on topics in organic chemistry; certain volumes have appeared with later editions. The following volumes have appeared so far:

1, Carbene chemistry; 2, Bridged aromatic compounds; 3, Conformation theory; 4, Carbanion chemistry; 5, Oxidation in organic chemistry; 6, Structure and mechanism in organophosphorus chemistry; 7, Ylid chemistry; 8, 1,4-Cycloaddition reactions; 9, Cycloaddition reactions of heterocumulenes; 10, Cyclobutadiene and related compounds; 11, Dehydrobenzene and cycloalkynes; 12, Organic functional group preparations;. 13, Ring-forming polymerisations, parts A and B; 14, Carboxylic *ortho* acid derivatives; 15, Organic charge-transfer complexes; 16, Nonbenzoid aromatics, Vols I and II; 17, Acidity functions; 18, Chemistry of indoles; 19, Chemistry of the heterocyclic *N*-oxides; 20, Isonitrile chemistry; 21, Conformational analysis; 22, Catenanes, rotaxanes and knots; 23, Reaction mechanisms in sulphuric acid and other strong acid solutions; 24, Carbon-13 NMR spectroscopy; 25, The isoquinoline alkaloids: chemistry and pharmacology; 26, Organic reactive intermediates; 27, Ring transformations of heterocycles, Vols I and II; 28, Organic synthesis with noble metal catalysts; 29, Polymer synthesis, Vol. I; 30, Total synthesis of steroids; 31, Sulphur ylides: emerging synthetic intermediates; 32, Anodic oxidation; 33, Transition metal organometallics in organic synthesis; 34, The chemistry of pyrroles; 35, Pericyclic reactions; 36, Prostaglandin research; 37, Reactions of organosulphur compounds; 38, Strained organic molecules; 39, Ozonation in organic chemistry; 40, Singlet oxygen; 41, Pyrolytic methods in organic chemistry; 42, Rearrangements in ground and excited states; 43, Thermal electrocyclic reactions; 44, Hydrocarbon thermal isomerisations; 45, Cyclophanes; 46, The proton: Applications to organic chemistry; 47, Hetero Diels–Alder methodology in organic synthesis.

Studies in Organic Chemistry (various editors, Elsevier, 1979–) is a series presenting detailed reviews of certain groups of compounds or reactions; for example Volume 12 (1982) was devoted to crown com-

Organic chemistry: other reference works 169

pounds, their characteristics and applications. Recent volumes have, however, tended to be devoted to conference proceedings, such as Volume 26 (1986) on 'New trends in natural products chemistry'. A separate series with the same title (*Studies in Organic Chemistry*, P.G. Gassman, ed., Dekker, 1973–5) amounted to only four volumes.

Organometallic Chemistry (P.M. Maitlis and F.G.A. Stone, eds, Academic Press, 1969–) is a series of monographs, unnumbered, and devoted to special topics in organometallic chemistry. Representative titles include 'The chemistry of organotin compounds' (Poller, 1970) and 'Organoborane chemistry' (Onak, 1975).

Chemistry of Functional Groups (S. Patai, ed., Wiley, 1964–) is a series intended to cover all aspects of the chemistry of the important functional groups in organic chemistry. Coverage is generally restricted to the more recent developments and to subjects inadequately covered by reviews. Several more than the 20 volumes originally planned have been published so far. Representative titles include: *Chemistry of Alkenes* (2 vols, 1964); *Chemistry of the Ether Linkage* (1967); *Chemistry of the Hydrazo, Azo and Azoxy groups* (2 parts, 1975); *Chemistry of Peroxides* (1983); *Chemistry of Organic Selenium and Tellurium Compounds* (2 vols, 1986). A recent extension includes a four-volume series on the *Chemistry of Organophosphorus Compounds* (F. Hartley, ed.). The earlier volumes (1964–85) of this highly successful publication have recently been released on microfilm.

Reactive Intermediates in Organic Chemistry (G.A. Olah, ed., Wiley, 1968–76) was a series of volumes, unnumbered, dealing with reactive intermediates in organic chemistry of which the following appeared: *Carbonium Ions* (5 vols, G.A. Olah and P. von R. Schleyer, 1968–76); *Radical Ions* (E.T. Kaiser and L. Kevan, 1968); *Nitrenes* (W. Lwowski, 1970); *Carbenes* (2 vols, R.A. Moss and M. Jones Jr., 1973–75); *Free Radicals* (2 vols, J.K. Kochi, 1973); *Halonium Ions* (G.A. Olah, 1975). Volumes dealing with carbanions and with arynes were planned for the series. A new series, *Reactive Intermediates: A Serial Publication* (M. Jones and R.A. Moss, eds., Wiley, 1978–) was then introduced to preserve the currency of the original works by covering the recent literature for each major type of reactive intermediate, the chapter headings of each issue being very similar to the titles of the volumes issued and planned for the earlier series. Another series with the same title *Reactive Intermediates* (R.A. Abramovitch, ed., Plenum, 1980–) also provides current in-depth reviews on the actions and interactions of these species.

In *Reaction Mechanisms in Organic Chemistry* (initially edited by C.D. Hughes, later C. Eaborn and N.B. Chapman, Elsevier, 1963–), the following monographs have appeared: 1, *Nucleophilic Substitution at a Saturated Carbon Atom* (1963); 2, *Elimination Reactions* (1963); 3, *Electrophilic Substitution in Benzenoid Compounds* (1965); 4, *Electrophilic Additions to Unsaturated Systems* (1966); 5, *The Organic*

Chemistry of Phosphorus (1967); 6, *Aromatic Rearrangements* (1967); 7, *Steroid Reaction Mechanisms* (1968); 8, *Aromatic Nucleophilic Substitution* (1968); 9, *Carbanions: Mechanistic and Isotopic Aspects* (1975); 10, *Mechanisms of Oxidation by Metal Ions* (1976).

Tetrahedron Reports on Organic Chemistry (D.H.R. Barton, ed., Pergamon Press, Vols 1–5, 1978) are volumes containing reviews reflecting the breadth of interest and endeavour of the modern organic chemist. Each volume comprises ten reports previously published in *Tetrahedron*. Publication of the separate collective volumes ceased after Volume 5; however, the Reports still continue to appear in *Tetrahedron*, e.g. no. 220 on 'Orthoquinodimethanes' in 1987 and no. 300 on 'Ninhydrin and ninhydrin analogs, synthesis and applications' in late 1991.

Conformational Analysis (E.L. Eliel, N.L. Allinger, S.J. Angyal and G.A. Morrison, Wiley, 1965) was a treatment of conformational analysis which included both basic principles and applications to natural-product chemistry. Also useful is the *Topics in Stereochemistry* series (*see* p. 175).

The Essential Oils (E. Guenther, ed., Van Nostrand, 1948–52) is a work complete in six volumes, the first two containing data on the chemical groups which comprise the major constituents of the essential oils. Each botanical family is given one chapter and the botany, geographical distribution and cultivation of the plant source are described in addition to the distillation, physical and chemical properties, production and industrial uses of the essential oils. The work is still useful and contains a large number of references to other sources of information.

Organic Sulphur Compounds (N. Kharasch, ed., Vol. 1, Pergamon Press, 1961); *Chemistry of Organic Sulphur Compounds* (N. Kharasch and C.Y. Meyers, eds, Vol. 2, Pergamon Press, 1966) was a series of well-documented articles on organic sulphur compounds which included chapters on the biochemical functions of selected compounds and studies on reaction mechanisms in addition to the synthesis and reactions of the more important groups of these compounds. Although further volumes in the series were planned, none have yet appeared.

Heterocyclic Compounds (R.C. Elderfield, ed., Wiley, Vol. 1, 1950, Vol. 9, 1967) is a series obviously less comprehensive than the much larger Weissberger (see above) but is nevertheless still useful. Each chapter discussed the chemistry of a group of heterocyclic compounds under the main headings of synthesis and reactions. The chapters are then grouped into volumes by general heterocyclic types — e.g. Volume 1: three-, four-, five- and six-membered polycyclic compounds containing one O, N or S atom; Volume 2: five- and six-membered polycyclic compounds containing one O or S atom.

The Chemistry of Natural Products (K.W. Bentley, ed., Interscience, 1957–65) was a series of monographs on topics in organic chemistry intended mainly for undergraduates; it is noted for the liberal use of structural formulae and reaction schemes, which are separated from the text; original literature references are given. Fields covered included alkaloids, carbohydrates, terpenes, vitamins and natural pigments.

Natural Products Chemistry (K. Nakanishi, T. Goto, S. Ito, S. Natori and S. Nozoe, eds; Vols 1–2, Academic Press, 1974–5; Vol 3, Oxford University Press and Kodansha University Science Books, 1983) are works attempting to bridge the gap between the organic chemistry textbook and the comprehensive treatise. Structure, syntheses and reactions of classes of natural products and of individual members are given. The presentation uses many structural formulae and much of the material is presented in terse lecture-note format.

Handbook of Naturally Occurring Compounds (2 vols, T.K. Devon and A.I. Scott, Academic Press, 1972–5) and *A Handbook of Alkaloids and Alkaloid-Containing Plants* (R.F. Raffauf, Wiley, 1970). These two handbooks contain data of interest to the natural-product chemist; names, formulae and literature references are given.

The Alkaloids (R.H.F. Manske *et al.*, eds, 1950–81; A. Brossi, ed., Academic Press, 1983–) is a monograph, initially intended to be complete in five volumes, giving a comprehensive survey of the field of alkaloid chemistry and pharmacology. The occurrence, separation and properties of the more important alkaloids are covered in addition to their structure, stereochemistry, synthesis and pharmacology. Volume 1 contains a chapter 'Alkaloids in the plant', and Volume 5 is heavily biased towards the pharmacology of the clinically important alkaloids. Volume 6 (1960) is a supplement to Volumes 1 and 2, bringing them up to date; Volume 7 is a supplement to Volumes 2–5; while Volume 8 (1965) covers the indole alkaloids. The practice of grouping chapters on similar alkaloids, either chemically or botanically, was abandoned after Volume 8. Volumes 9 onwards contain reviews of recent developments.

Ergebnisse der Alkaloid-Chemie bis 1960 (H.-G. Boit, Akademie Verlag, Berlin, 1961), a useful volume which reviewed the advances in alkaloid chemistry between 1950 and 1960 under the general headings of synthesis, degradations and reactions, and biosynthesis, is written in the style of *Annual Reports of the Chemical Society* but with many more structural formulae. There are many references to the original literature, and the less-common alkaloids are described in tables under the headings of melting point, refractive index, derivatives, sources and functional groups.

Steroid Reactions, an Outline for Organic Chemists (C. Djerassi, ed., Holden-Day, 1963), although all the examples and literature references are for steroids, is a book which is virtually an abstract of functional group chemistry. The structural formulae, reaction conditions (e.g. 70% AcOH, 100°C, 40 min) and the yield are given, and all reactions of the same type are grouped together. Some of the chapter headings are:

'Protection of carbonyl and alcohol groups'; 'Introduction of double bonds'; and 'Metal ammonia reductions of steroidal enones'.

The Proteins (H. Neurath, ed., 2nd edn, Academic Press, Vol. 1, 1963, Vol. 5, 1970) is a monograph on protein chemistry which was originally intended to be complete in five volumes, although the third edition is devoted to reviews of specific subjects. In the second edition the field of protein composition, structure and function was comprehensively covered under such headings as: 'Fractionation of proteins, amino acid analysis of polypeptides', 'X-ray analysis and protein structure', 'Genetic control of protein structure', 'Interaction of proteins with radiation (analytical)', and 'Structure and function of antigen and antibody proteins'. Volume 5 is devoted to metalloproteins. The third edition is edited by H. Neurath and R.L. Hill; Volume 1 appeared in 1975 and Volume 5 in 1982.

The Lipid Handbook (F.D. Gunstone, J.L. Harwood and F.B. Padley, eds, Chapman and Hall, 1986) is a useful reference compendium which covers the occurrence, isolation, chemical identification and technological applications of this class of natural products, from the publishers of *Dictionary of Organic Compounds* (see p. 166).

Medicinal Chemistry (A. Burger, ed., 3rd edn, Wiley-Interscience, 1970) is now called *Burger's Medicinal Chemistry* (M.E. Wolff, ed., 4th edn, Wiley-Interscience, 1979–81). The fourth edition of this work is in three parts. Part 1 features the basis of medicinal chemistry whilst Parts 2 and 3 contain monographs dealing with the various classes of drugs, with broad therapeutic classes being grouped into contiguous chapters.

Volume 1 of the three-volume *Handbook of Solvents* (I. Mellan, Reinhold, 1957–59) is itself called *Handbook of Solvents* and deals with pure hydrocarbons; Volumes 2 and 3 are called *Source Book of Industrial Solvents* and deal with halogenated hydrocarbons and with monohydric alcohols. The work has been written with the large-scale user in mind but will also be of use to the research worker.

Tables in *Industrial Solvents Handbook* (E.W. Flick, ed., 3rd edn, Noyes Data Corp., 1985) contain pertinent data concerning the physical properties and degrees of solubility of solutes, etc. Many of the data are presented graphically and a number of phase diagrams are included. The text is organized according to solvent functional group. A companion volume is the *Solvents Safety Handbook* (D.J. DeRenzo, ed., Noyes Data Corp., 1986), which is a compilation of data on 335 hazardous and frequently used solvents.

SOCMA Handbook, Commercial Organic Chemical Names (Chemical Abstracts Service, 1965) was produced jointly by the Chemical Abstracts Service and the Synthetic Organic Chemical Manufacturers Association. It provides a means by which the chemical composition or structure of 6300 industrial organic compounds can be identified from a variety of their trade names. There are sections dealing with pure compounds and also with mixtures and polymers.

Gardner's Chemical Synonyms and Trade Names (J. Pearce, ed., 9th edn, Gower Technical Press, 1987) is a widely used rapid source of information on trade names and synonyms for chemicals of commerce. First published in 1924, the ninth edition contains some 12000 new entries. The book is in two parts: Part 1 is an alphabetical listing of names, whilst Part 2 lists those manufacturers who have supplied information. The volumes were initiated by William Gardner and previous editions were published by Crosby Lockwood and the Technical Press.

Elsevier Monographs on Toxic Agents (E. Browning, ed., Elsevier, 1959–64). Several of the volumes in this series, edited by a former H.M. Medical Inspector of Factories, deal with organic compounds — e.g. *Toxic Aliphatic Fluorine Compounds* (F.L.M. Partison, 1959) and *Carcinogenic and Chronic Toxic Hazards of Aromatic Amines* (T.S. Scott, 1962). Also useful is another work, *Toxicity and Metabolism of Industrial Solvents* (E. Browning, Elsevier, 1965), which contains data for over 150 compounds with original literature references.

Selective Toxicity (A. Albert, 7th edn, Chapman and Hall, 1985) deals with the physical and chemical properties that are responsible for the selectivity of toxic agents. The volume contains over 3000 references to the literature. See, however, Chapter 13 for general aspects of health and safety.

Named reactions

The practice of naming reactions after the discoverer and not on systematic grounds has become widespread in organic chemistry. This system gives a quick code which is very convenient for specialists in the narrow field but is inconvenient for others searching the literature for a particular reaction. Several books list these named reactions in alphabetical order and give information about the reaction with references to the literature. Chemical dictionaries are sometimes useful as is the list given in the *Merck Index*. Details of three of the most useful collective works follow.

Name Index of Organic Reactions (J.E. Gowan and T.S. Wheeler, 2nd edn, Longmans, 1960) has 739 reactions listed with a brief description, equation and leading references. The book contains a Type of Reaction Index in addition to a General Index of Reactions, Reagents and Products.

Organic Name Reactions (H. Krauch and W. Kunz, Wiley, 1964) contains mechanistic details, yield and rate of reaction in addition to the usual description, formulae and list of references for 523 named reactions.

Name Reactions and Reagents for Organic Synthesis (B.P. Mundy and M.G. Ellerd, Wiley, 1988) is a concise compilation of the most popular, commonly used and widely known name reactions and reagents employed in synthetic organic chemistry.

Serial publications

Encyclopaedic publications having been considered in the section on reference works, the serial works considered here are mainly of the review type which appear at fairly regular intervals. These reviews are written by specialists in the particular field and are therefore authoritative. Extensive bibliographies are a valuable feature of these publications, which are therefore useful in literature searches. The following list is a representative selection.

American Chemical Society, Monograph Series (Reinhold, 1921–71; American Chemical Society, 1972–) is a well-established series of monographs on a variety of topics in pure and applied chemistry and is most conveniently considered here; it incorporates a number of volumes of interest to the organic chemist. Typical titles include: No. 147 *Organometallic Chemistry* (H. Zeiss, ed., 1960); No. 150 *Industrial Organic Nitrogen Compounds* (A.J. Astle, 1961); No. 159 *Formaldehyde* (J.F. Walker, 3rd edn, 1964); No. 170 *Regulation of Purine Biosynthesis* (J.F. Henderson, 1972); No. 173 *Chemical Carcinogens* (C.E. Searle, 1976); No. 178 *Aromatic Substitution by the SRN1 Mechanism* (R.A. Rossi and R.H. di Rossi, 1983); No. 182 *Chemical Carcinogens*, 2nd edn (C.E. Searle, 1984); No. 185 *Pharmacokinetics* (P.G. Wellings, 1986).

Specialist Periodical Reports (Chemical Society, 1967–80; Royal Society of Chemistry, 1980–) are volumes containing systematic and comprehensive review coverage of the progress of research in specific topics. The most widely expanding subjects are covered annually, while others appear biennially or less frequently. Certain titles, such as *Aromatic and Heteroaromatic Chemistry* and *Terpenes and Steroids*, are of importance to the organic chemist. Further details are given in Chapter 4.

MTP International Review of Science, Organic Chemistry (D.H. Hey, ed., Butterworths, 1972–76) has ten volumes in the first series of biennial reviews on organic chemistry, covering literature published in 1970 and 1971, and appeared in 1973. Each volume does not contain an index, these being collected in an additional index volume. Volume titles were: 1, *Structure Determination in Organic Chemistry*; 2, *Aliphatic Compounds*; 3, *Aromatic Compounds*; 4, *Heterocyclic Compounds*; 5, *Alicyclic Compounds*; 6, *Amino Acids, Peptides and Related Compounds*; 7, *Carbohydrates*; 8, *Steroids*; 9, *Alkaloids*; 10, *Free Radical Reactions*. The second series, covering literature published in 1972 and 1973, appeared in 1975–76. A biochemistry series has also been produced which could also prove of interest to the organic chemist. This series was discontinued in 1976.

In *Progress in Organic Chemistry* (various editors, Butterworths, 1952–73), each volume in the series contained a number of specialist articles on recent developments in theoretical and experimental organic chemistry. A wide range of topics was covered and natural products were well represented. Surveys on 'Salamander alkaloids', 'Acidic hydrocarbons', 'Synthesis of prostaglandins' and 'Chemiluminescence of organic compounds' appeared.

Organic Reactions (various editors, Wiley, 1942–) are volumes which appeared at roughly yearly intervals until Volume 26 (1977). After a pause they then resumed in 1982 with Volume 27, and now appear at rather erratic intervals, often with more than one volume being published in a single year. The volumes contain review chapters each devoted to a single reaction of wide applicability. Experimental conditions for examples of each reaction are given in addition to a survey of the scope and limitations. Volume 40 appeared in 1991 and included a chapter on 'The Pummerer reaction of sulphinyl compounds'.

Organic Reaction Mechanisms (various editors, Wiley, 1966–) is a series providing annual summaries of the progress of work on organic reaction mechanisms generally and fairly uniformly, and not just on selected topics. Each volume is divided into about 15 main chapter headings.

Annual Reports in Organic Synthesis (various editors, Academic Press, 1970–) is an organized annual review of information of use to the synthetic organic chemist, compiled by abstracting about 40 primary chemistry journals, a list of which is given. There is no index; instead an extensive table of contents is provided. The information is presented in the form of structural equations giving details of yield, applicability and original literature reference.

Advances in Physical Organic Chemistry (V. Gold, ed., 1963–76; V. Gold and D. Bethell, eds, 1977–86; D. Bethell, ed., Academic Press, 1987–) is a series of fairly specialized review articles directed mainly at the research worker or potential research worker in the field. The areas covered include both theoretical and practical aspects, as shown by two recent chapter titles: 'Charge density–NMR chemical shift correlations in organic ions'; 'Nucleophilic aromatic photosubstitution'. *Progress in Physical Organic Chemistry* (R.W. Taft et al., eds, Interscience, 1963–) provides more general reviews than those in the *Advances in Physical Organic Chemistry* series and should be of value to advanced students as well as non-specialists. Each chapter contains much tabulated numerical data.

Topics in Stereochemistry (E.L. Eliel et al., eds, Wiley, 1967–) is a series intended for the advanced student, the teacher and the active researcher to obtain great detail about stereochemical topics which are often only summarized in standard texts. The topics covered are aimed at both the organic and the inorganic chemist, and also deal with certain stereochemical aspects of biochemistry.

Progress in Medicinal Chemistry (G.P. Ellis and G.B. West, eds,

Butterworths, 1961–71; Elsevier, 1973–) is a series containing reviews written by specialists concerned with the development and study of new drugs, and is intended for the chemist, biochemist, pharmacologist and, to a lesser extent, the clinician.

Advances in Lipid Research (R. Paoletti and D. Kritchevsky, eds, Academic Press, 1963–) has reports on subjects including structural investigations, separations, metabolism and biosynthesis of lipids. The reports contained in this series are designed mainly for research workers and specialists in the field. *Progress in the Chemistry of Fats and Other Lipids* (R.T. Holman, W.O. Lundberg and T. Malkin, eds, 1952–63; R.T. Holman, ed., Pergamon Press, 1963–78) and *Progress in Lipid Research* (R.T. Holman, ed., Pergamon Press, 1978–) form a further series of reviews and monographs on the various aspects of lipid chemistry, including the reactions, syntheses and analyses of different classes of lipid.

Advances in Protein Chemistry (various editors, Academic Press, 1944–) has appeared almost annually since 1951 and contains reviews on topics of current interest in protein chemistry. Reviews have appeared on topics such as single proteins, groups of proteins, protein structure and experimental techniques in protein chemistry and are illustrated by the following examples: 'Phosphoproteins' (1974); 'Molecular orbital calculations on the conformation of amino acid residues of proteins' (1974); 'Membrane receptors and hormone action' (1976); 'Species adaptation in a protein molecule' (1984).

Fortschritte der Chemie organischer Naturstoffe/Progress in the Chemistry of Organic Natural Products (founded by L. Zechmeister, various editors, Springer, 1938–) is a valuable series of collected reviews covering the synthesis, reactions, stereochemistry and properties of a wide range of natural products, in addition to the use of physical methods in structure determination. The articles are contributed in English, French and German. Volume 55 appeared in 1989. A cumulative index for volumes 1–20 (1938–62) appeared in 1964.

Advances in Carbohydrate Chemistry (various editors, Academic Press, 1945–68) and *Advances in Carbohydrate Chemistry and Biochemistry* (various editors, Academic Press, 1969–) present a series of critical reviews on developments in carbohydrate chemistry covering biochemical, industrial and analytical aspects. The original title was altered after Volume 23 to the new title (commencing as Volume 24) to emphasize the coverage of biochemical topics. A recent innovation has been the inclusion of obituaries of prominent carbohydrate chemists.

Advances in Heterocyclic Chemistry (A.R Katritzky, ed., 1963–6; A.R. Katritzky and A.J. Boulton, eds, 1966–81; A.R. Katritzky, ed., Academic Press, 1982–) covers a large and rapidly expanding field. Each volume from Volume 12 (1970) contains a contents list for the earlier volumes. Supplement 1, entitled *The Tautomerism of Heterocycles*, appeared in 1976, updating the earlier reviews of this subject which originally appeared in Volume 1 and 2 (1963). Supplement 2, devoted to 'Pyrylium salts', was published in 1982.

Only three volumes appeared in the series *Advances in Alicyclic Chemistry* (H. Hart and G.J. Karabatsos, eds, Academic Press, 1966–71) which was a collection of specialist monographs in the field of alicyclic chemistry.

Advances in Organometallic Chemistry (F.G.A. Stone and R. West, eds, Academic Press, 1964–) contains authoritative reviews devoted to the complete field of organometallic chemistry; coverage also extends to the chemistry of complexes containing organic ligands. *Organometallic Reactions* (E.I. Becker and M. Tsutsui, eds, Wiley, 1970–75) and *Organometallic Reactions and Syntheses* (E.I. Becker and M. Tsutsui, eds, Plenum, 1977) were mainly concerned with the synthesis and reactions of selected categories of organometallic compounds. Many chapters contained tabulated examples using the same format as used in *Organic Reactions*. *Advances in Free Radical Chemistry* (G.H. Williams, ed., Logos Press, 1965–72; Elek, 1975; Heyden, 1980–) is a series similar in design to the other *Advances* series.

Laboratory techniques and synthetic methods (including analysis)

Advances in Organic Chemistry: Methods and Results (various editors, Interscience, 1960–) is a collection of review articles covering new techniques and preparative methods which are considered likely to be of wide application in the field of organic chemistry. Expertly written reviews cover the mechanism, scope and application of the reactions, in addition to the experimental details. Volume 9 appeared in two parts (1976 and 1979) and was entirely devoted to a survey of iminium salts in organic chemistry.

The fourth edition of *Methoden der organischen Chemie, Houben-Weyl* (E. Müller, and later H. Kropf, eds, Georg Thieme Verlag, 1952–), which is in German, deals comprehensively and critically with methods of all types, providing full experimental details for appropriate examples, in addition to the theoretical background. In this it follows closely the guiding principles laid down for the third edition but has been considerably expanded to deal with the new material available and to give, for the first time, an extensive coverage of the patent literature. The new edition is now complete in 15 volumes (many of which are in several parts) with an index published as Volume 16. Through the publication of 20 supplementary and additional volumes (which commenced in 1982), the fourth edition will be brought up to date and several additional classes of compounds not previously described will be covered. For technical reasons, the editors are publishing each volume or part as it becomes available and not in strict sequence. This, fortunately, is not too great an inconvenience as each part is virtually an independent monograph. The supplementary volumes are co-ordinated with the volumes of the fourth edition by cross-references since the volume numbers are not the same. Thus supplementary volume E3 (1983) devoted to aldehydes updated the original Volume 7 of the fourth edition which dated from 1954.

By 1987 supplementary Volumes E1–E5 and E11 had appeared and these were accordingly included in the index (Volume 16) to the fourth edition. The work when completed will contain sections on analytical methods and the physical and chemical investigation of structure in addition to syntheses. An explanatory booklet detailing the content and expected supplement publication schedule is available from the publishers.

Supplementary volume 12 on 'Organotellurium compounds' appeared in 1990 and was a landmark in the history of *Houben-Weyl* since it was the first volume to be published in English. In future volumes will appear in English wherever possible.

The reviews in Volume 1 of *Newer Methods of Preparative Organic Chemistry* (W. Foerst, ed., Vol. 1, Interscience, 1948; Vols 2–6, Academic Press, 1963–71) gave details of the scope and limitations, preparation of reagents and experimental data on a number of important reactions and reagents. It was a selection from a collection of articles published in Germany in 1944 and this translation was published in 1948 by consent of the Alien Property Custodian. Volumes 2–6 were translations of collections of reviews which originally appeared in *Angewandte Chemie*. These reviews were considerably expanded and numerous experimental procedures also included.

Organic Syntheses (various editors, Wiley, 1921–) is an annual series which gives satisfactory methods for the synthesis of individual organic compounds, and each method is checked in two independent laboratories before publication. Volumes 1–30 are out of print but six revised collective volumes have now been issued: I (Vols 1–9, 2nd edn, 1941); II (Vols 10–19, 1943); III (Vols 20–29, 1955); IV (Vols 30–39, 1963); V (Vols 40–49, 1973); VI (Vols 50–59, 1988). A useful feature of all volumes is the very comprehensive indexing. A Cumulative Index for Collective Volumes I–V was published in 1976. Volume 50 (1970) and all subsequent volumes contain brief references to unchecked procedures received for publication, giving equations and yields. Full details of these procedures, whether subsequently accepted or rejected for publication, may be obtained from the publishers. From Volume 64 (1986) a camera-ready format has been adopted.

In 1991 Pergamon Press published *Comprehensive Organic Synthesis* edited by B.M. Trost and I. Fleming. All nine volumes were published simultaneously in the second half of the year (Volume 9 is the index), to provide the most modern, complete and authoritative reference work on

the reactions and reagents of organic synthesis.
For *Synthetic Methods of Organic Chemistry* (W. Theilheimer, ed., Karger, Basel, 1948–81); *Theilheimer's Synthetic Methods of Organic Chemistry* (A.F. Finch, ed., Karger, Basel, 1982–), an English edition is available which abstracts new methods for the synthesis of organic compounds. The abstracts appraise the applicability of the method, show the number of steps and the yield and give references to the original paper or to a usable abstract for experimental details. The syntheses are arranged systematically in each volume by reaction type and author names are not used. The indexes include a name index, a reaction type index, a reagent list and a list of supplementary references. Volume 40 (1986) contains a cumulative index to Volumes 36–40; the next cumulative index will appear in Volume 45.

Technique of Organic Chemistry (A. Weissberger, ed., Interscience, 1945–70) is a thorough and comprehensive treatise in 14 volumes with a critical survey of techniques used in the organic laboratory, placing emphasis on theoretical background throughout but in no way neglecting reference to practical details. Topics covered included physical methods of organic chemistry, separation and purification, distillation, organic solvents, and the various techniques of chromatography and spectroscopy. *Techniques of Chemistry* (A. Weissberger, ed., Interscience, 1971–) is a treatise which is the successor to the *Technique of Organic Chemistry* (*TOC*) series, formed by combination with the companion series *Technique of Inorganic Chemistry*. The new series contains later editions of works of the previous series including: Vol. 1, *Physical Methods of Chemistry* (4th edn, previously Vol. 1 of *TOC*); Vol. 2, *Organic Solvents* (4th edn, previously Vol. 7 of *TOC*); Vol. 4, *Elucidation of Structures by Physical and Chemical Techniques* (2nd edn, previously Vol. 11 of *TOC*); Vol. 6, *Investigation of Rates and Mechanisms of Reactions* (4th edn, previously Vol. 8 of *TOC*); Vol 13, *Laboratory Engineering and Manipulations* (3rd edn, previously Vol. 3, Part 2 of *TOC*). In addition, new techniques have also been covered and the series has now reached Volume 21 (1990) on *Solubility Behaviour of Organic Compounds*.

Organic Synthesis (2 vols, V. Migrdichian, Reinhold, 1957) was a summary of the preparation and reactions of organic functional groups which gave a good survey of the older literature. *Synthetic Organic Chemistry* (R.B. Wagner and H.D. Zook, Wiley, 1953) is a single volume which summarized the most common syntheses of organic mono- and di-functional compounds. Very few experimental details are given but there are over 6000 references to original papers which contain explicit details. Each chapter is devoted to an individual functional group and many of the relevant compounds are listed in tables giving yield, m.p./b.p. and references.

Survey of Organic Syntheses (C.A. Buehler and D.E. Pearson, 2 vols,

Wiley, 1970–77) lists the principal methods for the synthesis of the main types of organic compounds. Each chapter is devoted to the creation of one functional group from another. The equation for each reaction is given and its value and limitations are discussed, together with basic theory. Specific examples are also given in sufficient detail to permit laboratory preparation.

Many methods have been devised for the synthesis of labelled compounds in order to increase the yield or to obtain maximum utilization of the isotopic reagent. *Organic Synthesis with Isotopes* (2 parts, A. Murray and D.L. Williams, Interscience, 1958) describes many of these methods in sufficient detail to be used in the laboratory. Part 1 deals with compounds of isotopic carbon, and Part 2 with compounds labelled with halogens, hydrogen, nitrogen, oxygen, phosphorus and sulphur.

Each volume in the series *Isotopes in Organic Chemistry* (E. Buncel and C.C. Lee, eds, Elsevier, 1975–) provides a number of reviews on specific topics such as 'Carbon-13 in organic chemistry' (Vol. 3) and 'Isotopic effects' (Vol. 6).

Volume 1 of *Reagents for Organic Synthesis* (L.F. and M. Fieser, 1967–86; M. Fieser and J.G. Smith, Wiley, 1988–) contains the structural formula, molecular weight, physical constants, preparation, purification, commercial suppliers and applications for 1120 reagents of use to the organic chemist. These are arranged in alphabetical order and literature references are given, the coverage being up to early 1966. Supplementary volumes 2–15 (1990) cover recent developments on previous reagents, with cross-references to the data in earlier volumes and also new reagents. These later volumes do not cumulate. A collective index for Volumes 1–12 has recently been published.

Synthetic Reagents (J.S. Pizey, ed., Ellis Horwood, 1974–) are volumes giving thorough coverage to the use of a selected group of very versatile reagents, and complement Fieser and Fieser. Among the reagents treated are: lithium aluminium hydride (Vol. 1); diazomethane (Vol. 2); sulphuryl chloride (vol. 4); and polyphosphoric acid (Vol. 6).

The original volume of *Compendium of Organic Synthetic Methods* (I.T. and S. Harrison, Wiley, 1971–) is a systematic listing of functional-group transformations designed for use by synthetic organic chemists. Reactions are classified on the basis of the functional group of starting material and product, and there is a very useful quick reference summary table. The information given is very terse: an equation, reaction conditions, catalyst, yield and abbreviated original literature reference(s). Volume 2 (1974) is a supplement to the original work, with a chapter on difunctional compounds. Both volumes also cover methods for the protection of functional groups. Subsequent volumes (various editors) provide additional information spanning three- or four-year periods; thus Volume 6 (1988) covered 1983–86.

Comprehensive Organic Transformations (R.C. Larock, ed., Verlag

Chemie, 1989) is a further guide to functional-group preparations, with many references to the literature and a valuable Transformation Index simply arranged under 'to' and 'from'.

Organometallic Compounds, Methods of Synthesis, Physical Constants and Chemical Reactions (M. Dub, ed., 2nd edn, Springer, 1966–75) is in three volumes. Volume 1 contains derivatives of the transition metals of Groups III to VIII of the periodic table; Volume 2 contains Ge, Sn and Pb; while Volume 3 is devoted to As, Sb and Bi. The second edition covers the literature from 1937 to 1964 and provides a comprehensive, non-critical source of information on organometallic compounds; a formula index is also available. Supplements to volumes 1–3 scanning the literature from 1965 to 1968 appeared in 1975, 1973 and 1972, respectively.

Industrial Chemicals (F.A. Lowenheim and M.K. Moran, 4th edn, Wiley, 1975) is a revision of the earlier work of Faith, Keyes and Clark. The manufacture (with flow diagram and details), properties, uses, economics, transportation requirements and plant sites are given for a large number of important industrial chemicals, including many organic compounds.

Purification of Laboratory Chemicals (D.D. Perrin and W.L.F. Armarego, 3rd edn, Pergamon Press, 1988) is in three main sections: Section 1 (Chapters 1 and 2) is a discussion, with theoretical background, of general purification techniques; Section 2 (Chapters 3 and 4) outlines the purification method for each of a large number of organic and metalorganic compounds; Section 3 (Chapter 5) discusses general principles of purification for use with compounds not listed in Section 2. Chapter 6 deals with the purification of biochemicals.

Handbook of Laboratory Waste Disposal (M.J. Pitt and E. Pitt, Ellis Horwood, 1985) is a practical manual dealing with all forms of laboratory waste disposal. Chapters are included on biological materials, radioactive substances and also the use of waste-disposal contractors. There are several useful appendices listing known carcinogens, deteriorating chemicals, etc.

Methods in Carbohydrate Chemistry (R.L. Whistler *et al.*, eds, Academic Press, 1962–80) is a series containing descriptions of reliable methods for the handling, preparation, reaction and analysis of carbohydrates, and the field covered ranges from simple monosaccharides to complex polysaccharides.

For *A Laboratory Manual of Analytical Methods in Protein Chemistry (including Polypeptides)* (P. Alexander *et al.*, eds, Pergamon Press, 1960–69), Volumes 1–3 (1960–61) contained reviews on the manipulation of proteins — e.g. their separation and isolation (Vol. 1); composition, structure and reactivity (Vol. 2); and determination of size and shape of protein molecules (Vol. 3). The later volumes contain reviews of recent developments and new techniques.

Tables of Resolving Agents and Optical Resolutions (S.H. Wilen, University of Notre Dame Press, 1972) is a collation of resolving agents and resolutions. It is tabulated and contains many references to the original literature.

Atlas of Stereochemistry (2 vols, W. Klyne and J. Buckingham, eds, Chapman and Hall, 2nd edn, 1978) gives absolute configurations of organic molecules with extensive use of stereochemical formulae. Volume 1 covers the literature to 1971, whilst Volume 2 covers the period to mid-1976. A supplement, edited by J. Buckingham and R.A. Hill, appeared in 1986 and extended the literature coverage by another seven years or so.

Organic Analysis (J. Mitchell, I.M. Kolthoff, E.S. Proskauer and A. Weissberger, eds, Interscience, Vol. 1, 1953; Vol. 4, 1960) is a series which aimed to present up-to-date information on quantitative chemical and physical methods of organic analysis. The first three volumes were concerned with functional group analysis, and the fourth gave surveys of newer instrumental techniques and includes a chapter on enzymic analytical reactions. Other still useful works are: *Organic Functional Group Analysis by Micro and Semimicro Methods* (N.D. Cheronis and T.S. Ma, Interscience, 1964); *Quantitative Organic Microanalysis* (A. Steyermark, 2nd edn, Academic Press, 1961); *Spot Tests in Organic Analysis* (F. Feigl, 7th edn, Elsevier, 1966).

Elementary Analysis Tables (G. Ege, Verlag Chemie/Wiley, 1966) are computer-prepared anti-composition tables giving a range of possible empirical formulae in the range C_1–C_{40} for any combination of elemental analysis figures for compounds containing carbon, hydrogen, nitrogen, oxygen or sulphur. *Tables of Percentage Composition of Organic Compounds* (H. Gysel, 2nd edn, Birkhauser, 1969) presents tables of elemental analyses and molecular weights for a wide range of empirical formulae containing C, H, N and O (S). Formulae C_1-C_{50} (hydrocarbons to C_{60}, and C, H, O to C_{52}) are included. *Computer Compilation of Molecular Weights and Percentage Composition for Organic Compounds* (M.J.S. Dewar and R. Jones, Pergamon Press, 1969) gives tables of elemental analyses, molecular weights and isomolecular weights (for use in mass spectrometry) for a wide range of empirical formulae containing C, H, and one or two heteroatoms (heteroatoms are Br, Cl, F, I, N, O, P). Formulae C_1-C_{30} (hydrocarbons to C_{40}) are included. The instructions are written in English, French and German in the above three works.

Melting Point Tables of Organic Compounds (W. Utermark and W. Schicke, 2nd edn, Vieweg, Braunschweig, 1963) gives the name, structure, empirical formula, molecular weight, boiling point, refractive index, *Beilstein* reference and some properties and reactions of a large number of organic compounds arranged in order of melting point from −189.9°C to 500°C. *Handbook of Tables for Organic Compounds*

Organic chemistry: other reference works 183

Identification (Z. Rappoport, 3rd edn, Chemical Rubber Co., 1967) is a supplement to the *Handbook of Chemistry and Physics* which includes the melting or boiling points and, in some cases, the refractive index and/or densities of over 8150 organic compounds, in addition to the melting points of some of their more common derivatives. (See also Chapter 7.)

Chromatographic and Electrophoretic Techniques (2 vols, I. Smith and J.W.T. Seakins, eds, 4th edn, Heinemann, 1976) has a bias very much towards the practical aspects of chromatography (Vol. 1) and electrophoresis (Vol. 2).

Bibliography of Paper Chromatography and Survey of Applications (I.M. Hais and K. Macek, eds, Czechoslovak Academy of Sciences/ Academic Press, Vol. 1, 1960; Vol. 2, 1962) are two volumes which list titles of original papers, authors and references to the journal, year and page. The entries are grouped by compounds separated, and within each group they are listed under authors in alphabetical order. Volume 1 covers the literature from 1943 to 1956; Volume 2 from 1957 to 1960. Volume 1 is published in Czech but a supplement is provided to enable English-speaking readers to use it. The title and coverage have been extended to become *Bibliography of Paper and Thin-Layer Chromatography* (K. Macek *et al.*, eds, Elsevier), published as supplementary volumes to the *Journal of Chromatography*: Suppl. Vol. 1 (1968) covers 1961–65; Suppl. Vol. 2 (1972) covers 1966–69; Suppl. Vol. 5 (1976) covers 1970–73. Other supplementary volumes in these series survey the following fields: Suppl. Vol. 3 (1973), Column chromatography, 1967–70; Suppl. Vol. 4 (1975), Electrophoresis, 1968–72; Suppl. Vol. 6 (1976), Liquid column chromatography, 1971–73. Further information is also published in the *Journal of Chromatography* series.

Chromatographic Methods (A. Braithwaite and F.J. Smith, 4th edn, Chapman and Hall, 1985) is a more modern text covering both gas chromatography and high-performance liquid chromatography (HPLC).

Gas Chromatography (A.B. Littlewood, 2nd edn, Academic Press, 1970) is a useful general text which is designed as a reference work for people using this important technique. The large theoretical section contains chapters on the theory of the design and operation of columns and on the principles of detection. The practical section describes, in general terms, the separation of compounds by classes. References are given to papers containing the practical details. *Modern Practice of Gas Chromatography* (R.L. Grob, ed., 2nd edn, Wiley, 1985) provides further coverage of the theory and applications.

Handbook of Chromatography (F. Zweig and J. Sharma, eds, Chemical Rubber Co., later CRC Press, 1972–) provides tabulated data. There are separate series of volumes dealing with particular types of compounds, e.g. drugs, hydrocarbons, steroids, etc.

Thin Layer Chromatography, a Laboratory Handbook (E. Stahl, ed.,

2nd edn, Allen and Unwin/Springer, 1969) is a still useful handbook dealing with instrumentation, techniques and applications of thin-layer chromatography to a wide range of organic compounds. Much tabulated data and a multitude of literature references are included.

Physical methods of structure determination

Physical methods have made such an enormous contribution to structural determination in organic chemistry in recent years that no account of organic chemistry would be complete without mentioning them. A large number of tables and books have appeared in the last few years to satisfy the demand for information in this rapidly expanding field. The following list is meant to supplement with interpretative works the section on spectroscopic data collections dealt with in Chapter 7, and includes examples from the most important fields, including coverage of the more recent developments in nuclear magnetic resonance. A useful general introduction to many methods may also be found in many organic chemistry textbooks, and in *Weissberger*, *Houben-Weyl*, etc.

General texts

The original work (1955) of *Determination of Organic Structures by Physical Methods* (F.C. Nachod *et al.*, eds, Academic Press, Vol. 1, 1955; Vol. 6, 1976) devoted a chapter to each physical method and technique; interpretation of results and applications were discussed. The methods covered include optical rotation, magnetic susceptibilities, dipole moments and X-ray diffraction, in addition to the more usual infrared (IR), mass spectrometry (MS), nuclear magnetic resonance (NMR), ultraviolet (UV) and visible (vis) spectroscopies. Later volumes contain expanded coverage of the techniques; Volume 4 is entirely devoted to the various techniques of NMR, including ^{15}N, ^{13}C and ^{31}P nuclei.

Organic Spectroscopy (D.W. Brown, A.J. Floyd and M. Sainsbury, Wiley, 1988) and *Spectroscopic Methods in Organic Chemistry* (D.H. Williams and I. Fleming, 4th edn, McGraw-Hill, 1987) are two very useful general texts which cover the basic techniques of UV, vis, IR and NMR spectroscopy and MS.

Physical Methods in Heterocyclic Chemistry (A.R. Katritzky, ed., Academic Press, 1963–74) is a comprehensive account of the application of physical methods to the wide field of heterocyclic chemistry. Volume 1 covered non-spectroscopic methods and Volume 2 covered spectroscopic methods; these both appeared in 1963. The series was updated later, Volumes 3 and 4 (1971) covering methods other than X-ray structural analysis, while Volume 5 (1972), sub-titled 'Handbook of molecular dimensions', comprised tabulated data for 1367 compounds.

Volume 6 (1974) was devoted to the more recently introduced techniques, including UV, photoelectron spectra and microwave spectroscopy. The emphasis of the series is mainly upon the interpretation of data.

Handbook of Organic Structural Analysis (Y. Yukawa, ed., Benjamin, 1965) is a collection comprising 20 000 entries covering a wide range of physical properties. Dissociation constants, IR, NMR, optical rotatory dispersion, redox potentials, thermochemical constants and UV are among the range of physical properties listed. Within each section active groups (e.g. chromophores for spectroscopy, acidic or basic groups for dissociation constants) are listed with the relevant physical data and literature references for a number of model compounds.

Tables of Spectral Data for Structure Determination of Organic Compounds (E. Pretsch, T. Clerc, J. Seibl and W. Simon, 2nd edn, Springer-Verlag, 1989) is an English translation of the third German edition, containing summary tables of spectroscopic data. The combination tables summarize the information for the various functional groups. These are followed by tables devoted to the major spectroscopic techniques (^{13}C-NMR, ^{1}H-NMR, IR, MS, UV/vis) each of which is arranged by the various functional groups. The book represents a very compact summary of reference material in a readily accessible format.

Infrared

It is Dr Bellamy's original volume *The Infra-red Spectra of Complex Molecules*, Vol. 1 (L.J. Bellamy, 3rd edn, Chapman and Hall, 1975); Vol. 2, *Advances in Infra-red Group Frequencies* (L.J. Bellamy, 2nd edn, Chapman and Hall, 1980) of 1950 (2nd edn., 1958, published by Methuen but available through Chapman and Hall) that is generally accepted as the standard text in this field. It presents a critical review of the data on which the usual empirical spectra/structure correlation tables are based. The book contains correlation tables, an empirical discussion of the characteristic bands for the more important classes of compounds and the factors which can influence their frequencies. Volume 1 essentially presents the data, while Volume 2 discusses the reasons for these known facts in chemical and physical terms.

Infra-red Characteristic Group Frequencies (G. Socrates, Wiley, 1980) provides further information for additional classes of compounds.

An Introduction to Practical Infra-red Spectroscopy (A.D. Cross and R.A. Jones, 3rd edn, Butterworths, 1969) is an excellent text containing correlation tables and short discussions on each of the more common functional groups. It is particularly suitable for undergraduates or for occasional users dealing with fairly straightforward compounds.

Laboratory Methods in Infrared Spectroscopy (R.G.J. Miller and B.C. Stace, eds, 2nd edn, Heyden, 1972) concentrates upon the practical problems involved in obtaining an infrared spectrum. It is very useful for the spectroscopist with a problem.

Mass spectrometry

Interpretation of Mass Spectra (F.W. McLafferty, 3rd edn, Oxford University Press/University Science Books, 1982) is a useful introductory text which contains chapters on the production of spectra and on the fragmentation patterns of compounds of different groups. Also included is a useful appendix comprising tables of fragmentation data.

Mass Spectrometry Principles and Applications (I. Howe, D.H. Williams and R.D. Bowen, 2nd edn, McGraw-Hill, 1982) describes how a mass spectrometer can help in studies of chemistry, biochemistry and medicine. It concentrates on the principles involved and provides selected examples and applications.

Three related volumes give fragmentation patterns for selected compounds and groups of compounds with explanations and specimen spectra: *Mass Spectrometry of Organic Compounds* (H. Budzikiewicz, C. Djerassi and D.H. Williams, Holden-Day, 1967); *Structural Elucidation of Natural Products by Mass Spectrometry* Vol. 1, *Alkaloids*; Vol. 2, *Steroids, Terpenoids and Sugars* (same authors, Holden-Day, 1964). A complementary work is *Biochemical Applications of Mass Spectrometry* (G.R. Walker, ed., Wiley, 1972), which discusses the fragmentation patterns of important biochemical molecules.

Mass and Abundance Tables for Use in Mass Spectrometry (J.H. Beynon and A.E. Williams, Elsevier, 1963), *Table of Meta-Stable Transitions for Use in Mass Spectrometry* (J.H. Beynon, R.A. Saunders and A.E. Williams, Elsevier, 1965) and *Table of Ion Energies for Metastable Transitions in Mass Spectrometry* (J.H. Beynon, R.M. Caprioli, A.W. Kunderd and R.B. Spencer, Elsevier, 1970) are sets of tables (see also Chapter 7) which are computer-calculated and printed with instructions in several languages. They assist in the extraction of data from mass spectra and are of considerable assistance in their interpretation. The third book is designed to aid in the interpretation of ion kinetic energy spectra.

Tables for Use in High Resolution Mass Spectrometry (R. Binks, J.S. Littler and R.L. Cleaver, Heyden/Sadtler, 1970) are tables useful to the operators of mass spectrometers for the technique of mass matching, as used to determine accurate masses and, hence, elemental compositions.

Ultraviolet

The Theory of the Electronic Spectra of Organic Molecules (J.N. Murrell, Chapman and Hall, 1971) is a reprint of the original volume published by Methuen in 1963 and presents a quantum-mechanical approach to the

interpretation of spectra, with a basic introduction to the necessary quantum mechanics.
Gillam and Stern's Introduction to Electronic Absorption Spectroscopy in Organic Chemistry (E.S. Stern and C.J. Timmons, 3rd edn, Arnold, 1970) is a revised edition of the original work, which appeared in 1958. It is a more empirical and practical book, containing an introduction to light absorption, the origin of spectra and some experimental techniques, in addition to a discussion of the absorption data for typical chromophores and an outline of some applications in analytical and structure determination work.

Nuclear magnetic resonance

General/proton
Volume 1 (1965) of the comprehensive treatise *High Resolution Nuclear Magnetic Resonance Spectroscopy* (2 vols, J.W. Emsley, J. Feeney and L.H. Sutcliffe, Pergamon Press, 1965–66) is concerned with basic theory and spectral analysis, while Volume 2 (1966) deals with structural applications. The complete work contains a considerable quantity of data for reference purposes. Despite the very significant advances that have been made in the field of NMR since the publication of these volumes, they are nevertheless still of great value. A less comprehensive text is *Applications of Nuclear Magnetic Resonance Spectroscopy in Organic Chemistry* (L.M. Jackman and S. Sternhell, 2nd edn, Pergamon Press, 1969), but it provides a thorough basic theoretical background, necessary for the interpretation of NMR spectra. It, too, contains much tabulated data and many original literature references. Also of value is *NMR Spectroscopy, an Introduction* (H. Günther, Wiley, 1980) which is a translation of an earlier work *NMR-Spektroskopie* published by Georg Thieme Verlag in 1973.

Modern NMR Spectroscopy (J.K.M. Sanders and B.K. Hunter, Oxford University Press, 1987) is a book providing a non-mathematical, descriptive approach to modern NMR spectroscopy, and describes the various pulse sequences now available as well as two-dimensional methods. The emphasis throughout is on the ability of NMR to solve chemical problems.

Interpretation of NMR Spectra (K.B. Wiberg and B.J. Nist, Benjamin, 1962) contains a collection of theoretically calculated second-order NMR spectra presented in a computer-type visual format. It is a very useful manual for the estimation of coupling constants and chemical shifts from complex spectra.

Other nuclei
Carbon-13 NMR Spectroscopy (J.B. Stothers, Vol. 24 of *Organic Chemistry*, a series of monographs; see p. 147) is a useful introductory

text to the rapidly expanding field of ^{13}C-NMR. *Carbon-13 NMR Spectroscopy* (H.-O. Kalinowski, S. Berger and S. Braun, Wiley, 1988) and *Carbon-13 NMR Spectroscopy* (E. Breitmaier and W. Voelter, 3rd edn, Wiley, 1988) are further texts which include much tabulated data and many literature references. They also include extensive coverage of the techniques of pulsed NMR and Fourier transformation.

Nitrogen NMR (M. Witanowski and G.A. Webb, eds, Plenum Press, 1973) deals with the theory, technique and applications of ^{14}N and ^{15}N NMR.

NMR and the Periodic Table (R.K. Harris and B.E. Mann, eds, Academic Press, 1978) presents a comprehensive survey of NMR studies of all the elements of the periodic table for which such information exists, and includes much tabular material. It is particularly valuable for metallorganic compounds.

Section C: Industrial Chemistry

CHAPTER ELEVEN

Patents

T.S. EISENSCHITZ

The nature of a patent

A patent is a bargain between an inventor and the state. The inventor receives a limited-time monopoly to exploit the invention. This is the most public and visible aspect of the patent grant. Because the main purpose of the system is to facilitate the spread of knowledge of technological advances, the quid pro quo of the monopoly is for the inventor to deposit for the public a written description of the idea. This description must be in sufficient detail for someone 'skilled in the art' to be able to carry out the invention.

The monopoly on a patent lasts for 16–20 years depending on national law, but the information continues to be available and useful indefinitely. In most countries specifications are published and available for sale or for consultation in specific libraries. Increasingly, patents are being searched online in bibliographic or full-text databases for both technical and commercial/marketing purposes.

Patents are instruments of national or regional law and one grant establishes the bargain over a country or a group of countries. Therefore, patents appear in families of similar documents relating to protection in a number of territories. They are held together and defined under the International (Paris) Convention signed in 1883. This identifies the date, country and number which relate to the very first application of the series (the basic patent) and are called collectively the 'priority data'. The defining characteristics of patentable inventions are that they should embody knowledge which is new and not obviously deducible from that already known. The priority date supplies the time limit for judging these characteristics (Eisenschitz, 1987). Inventions must also be useful, but this is not a strong constraint. Broadly speaking, any vendable product must be useful to the buyer, and this satisfies the criterion.

This chapter looks at the systems for obtaining a patent. The details of any one nation's law are best looked up when required. There are a number of compilations in existence (e.g. *Patents Throughout the World*, 1988).

How to apply for a patent

What systems exist? There are three major systems of patent administration. These are *deferred examination, traditional patents* and the *Comecon system.* There are, however, some general considerations which apply to all three systems. (For a survey of patent problems, see Phillips, 1985.)

Who can apply for a patent? Any inventor can apply on his or her own invention. In most countries (including the UK), where the inventor is an employee, the employing organization can apply in its own right. The major exception to this rule is the USA, where the applicant must always be the inventor. An employee may then assign the rights to his or her employer. Corporate patentees are then called 'assignees'.

What happens when there is more than one claim to an invention? When an application is lodged with a patent office, the date and time are recorded. In most countries the person who gets to the patent office first gets the patent if there is any dispute about priority. Exceptions are the USA and Canada where researchers keep detailed records of their work and conclusions. To establish the right to the patent, an inventor must show when he or she first had the idea and then that it was worked on steadily to bring to fruition. Proceedings concerned with priority are called 'interference proceedings'. However, the US authorities are considering changing to a first-to-file system in the interests of international harmonization (EPO, 1986*b*).

How is a patent granted? Patent grant procedure usually involves an examination of the contents of a written specification in comparison with the laws and practice of the country concerned in order to check that the invention satisfies all the requirements of patentability. A major consideration is the novelty of the invention and for this purpose a search of the prior art is carried out.

Deferred examination

The essence of deferred examination is that the prior-art search is carried out in advance of the examination. The applicant is given time to consider the results of the search and then requests substantive examination.

The deferred examination system is now in widespread use. Most European countries use it as do Australia, Japan and many others. It began in Holland in the mid-1960s as a way of cutting the backlog of applications by allowing withdrawal of a hopeless case without the need for a full examination. From an information point of view, a particularly important achievement is the standardization of early publication at 18 months from the priority date, so that not only is a publication guaranteed but members of the same family should all appear at around the same time.

Because of membership of international bodies, there are three routes to a UK patent: the national system (Patent Office, 1986a,b), the PCT (Patent Co-operation Treaty) application (WIPO, 1985a) or the European patent (EPO, 1986a). It is useful to start with a national application, but any patent likely to be important will be put into the European and PCT routes.

The UK national system
An application form and fee are submitted accompanied by a specification which sets out the invention. The Patent Office carries out a formal check that all essential components of the specification are present, for instance that an inventor is named and that there is at least one claim (description of the invention for legal purposes). The document is given a classification code and sent to the appropriate subject specialist for the prior-art search. In theory, the examiner searches all relevant literature previous to the priority date. In practice, a selection of patents and some technical literature known to be relevant to that field are searched. Material is selected which might indicate that the proposed invention has already been described or that it is obvious from previous knowledge. No judgements are made at this stage. The literature is assembled and a list is sent to the applicant. This search is completed within 18 months of the priority date.

At or soon after 18 months, the application is published as submitted with the addition of the search report and the Patent Office-assigned classification codes and document numbers. The applicant now has six months to decide whether to continue with the application or to withdraw. If the application is to proceed, a request for substantive examination must be lodged.

The examination consists of looking at all possible aspects of patentability, especially novelty and the inventive step. Also, the quality of the descriptions is considered, as is possibility of the invention falling into a forbidden category. Third parties are not allowed to participate in the examination process but the examiner has a duty to consider any comments sent in as a result of the published application. Competitors are more likely to know of obscure pieces of prior art or of any use of the invention before the priority date than the examiner.

The application could be accepted or rejected out of hand but often a request for amendment is made and there will be much argument and discussion to and fro. (If the two do not agree there is always the possibility of appeal at this or any other stage.) When agreement is reached and the application accepted, there will be a simultaneous second publication and grant of the patent. After grant the lifetime of the monopoly is 20 years from the date of application provided that annual renewal fees are paid. A patent, however, remains open to challenge during its entire lifetime. At any time an unsuspected piece of prior art could come to light. Normally a challenger would then go ahead and infringe the patent and leave it up to the patentee to decide whether or not to sue. This occurs with very successful inventions where no effort is spared to try to break the monopoly. No extensions to the patent lifetime are allowed. This hits particularly hard at pharmaceutical patents, where safety testing can take so many years that little time is left for exploitation. However, an increasing number of countries do allow extensions.

The PCT (Patent Co-operation Treaty)
The PCT is a system of centralized searching. Member states take the results of this search and then decide by their own procedures whether or not to grant the patent. As an option, a preliminary examination and opinion as to grant can also be given. This treaty is best seen as an extension of the Paris Convention. It does not create yet another type of patent. It does, however, create another type of document — the PCT published application. Member states come from all regions of the world. There is a very comprehensive search. Minimum documentation is defined to be the patents from 1920 onwards of ten of the largest and most active patent offices and also articles from a selection of key journals in various fields over the last five years. The ten offices are those of the USA, USSR, Germany up to 1945, then the FRG and now Germany again, France, the UK, Switzerland, Japan, OAPI (French-speaking African federation), EPO and PCT. Apart from this material, each authority searches its own specialized material. Only offices with a large searchable collection can be authorities for the international search. The application is published complete with search report 18 months after the priority date and then enters the national phase in each designated state. Designated states are those in which the application requests protection and they have to be listed when the application is submitted. It is permissible to reduce the list but not to increase it as this would extend protection from that originally notified.

Another advantage is that of language. English, French, German, Japanese or Russian is accepted for the published application. English is the preferred language; the abstract and search report are always translated where the specification is in a language other than English.

The European patent
By a Convention signed in 1973, 11 countries, and growing, formed a region in which one centrally-granted patent would become a bundle of national patents in the designated states. This leads to a substantial cut in the number of documents associated with the patent family. Documents are in one of three languages only (English, French and German), with claims translated into all three. The system is one of deferred examination and works exactly as the UK version. The system led to some harmonization of laws. The members are all the EC countries plus Austria, Sweden, Switzerland and Liechtenstein. All of the EC countries were intended to form one country under the EPC by way of a Community Patent Convention which would unify the contents of their national patent laws. Much of the groundwork has been done by now (EPO, 1986b), and the Single Market agreement for 1992 is acting as a spur to progress.

Traditional patents

These used to be the predominant type of patent until superseded by deferred examination. When the examiner receives a patent he or she embarks on the full examination and carries out the prior-art search as part of this process. Only one document is published and this is after the specification has been accepted by the examiner. This may be simultaneous with grant but is more likely to be followed by an opposition period as this is the first opportunity for public comment. Grant then follows after any objections have been dealt with or at the end of the fixed period.

The UK Patents Act 1949 was of the traditional type. Quite a number of recently independent countries inherited such systems from their colonial powers and a number still use them. The US system is of this type as well.

The Comecon system

This system was used by the USSR and its Comecon partners (not including China, which has a full deferred examination system), though with the virtual collapse of Comecon in 1989–90, the future is hard to predict. It is a dual arrangement. Foreign applicants receive an ordinary patent of the traditional type. National inventors are given an inventor's certificate which confers recognition and an appropriate reward. Exploitation is left in the hands of the state, though this may well change in post-communist societies.

Within the current negotiations for revision of the Paris Convention there is pressure to include inventor's certificates in its scope. Changing the patent system is not a priority for the countries of Eastern Europe, but it seems likely that as new economic and political structures are introduced, inventor's certificates will be phased out and either deferred examination or a traditional type of patent system will be available to all comers. Some of the Eastern European countries may join the European Convention.

Contents of the specification

Patent documents are highly standardized. This is a great help to readers as the basic contents and layout are very similar in all cases. The three main divisions are into the *front page*, the *body of the specification* and the *claims*.

The front page

The front page of a modern UK patent is shown in Figure 11.1. It contains the title, abstract and drawing and bibliographic details mainly added by the Patent Office. Each item is labelled with a number on the left-hand side. Thus the application number is 21. These are INID numbers allocated by the World Intellectual Property Organization (WIPO), which concerns itself with international standardization. They are the same in all countries and are intended as field labels for computer processing. They indicate to a searcher which names and numbers are which if the specification is in an unfamiliar language. The title alone is often too brief to be helpful but it is informative if combined with the abstract. Provision of a drawing is optional and depends on the nature of the invention.

Two classifications are given, the domestic and the international. The field of search is also given, in the domestic classification. The international classification is administered by WIPO and all adherents put its symbols on patents to aid multi-country searching (WIPO, 1986). Many countries use only the international classification and have no separate domestic system. The UK and US classifications are two of the major domestic schemes left. Documents found in the search are listed but under the EPO and PCT systems these are at the back of the specification and give details of degree of relevance and which claims are affected.

Other entries include details of the applicants, the inventors and their agents. Application date and number are given and also the priority date, number and country. Priority data are valuable beyond the one patent as they determine the members of the family and the time available for convention applications.

The body of the specification

The body of the specification contains the disclosure to the public. It

Figure 11.1. Front page of UK patent.

(12) UK Patent Application (19) GB (11) 2 182 658 (13) A

(43) Application published 20 May 1987

(21) Application No 8624630

(22) Date of filing 14 Oct 1986

(30) Priority data
(31) 8525321 (32) 15 Oct 1985 (33) GB

(71) Applicant
Glaxo Group Limited,
(Incorporated in United Kingdom),
Clarges House, 6/12 Clarges Street, London W1Y 8DH

(72) Inventors
Ian Frederick Skidmore,
Lawrence Henry Charles Lunts,
Harry Finch,
Alan Naylor,
Ian Baxter Campbell

(74) Agent and/or Address for Service
Elkington and Fife, High Holborn House, 52/54 High Holborn, London WC1V 6SH

(51) INT CL⁴
C07C 91/40 A61K 31/16 31/135 31/535 C07C 101/48 103/00 143/74 C07D 295/00

(52) Domestic classification (Edition I)
C2C 220 226 227 22Y 280 281 282 29X 29Y 304 313 31Y 321 322 323 32Y 338 342 34Y 360 361 362 364 366 367 368 36Y 385 455 456 45Y 490 492 493 509 50Y 510 51X 530 573 583 591 623 624 628 62X 630 634 640 643 650 652 65X 660 662 668 680 682 688 699 802 80Y AA KH LE LG LS SJ
U1S 1321 2414 2415 2416 2418 C2C

(56) Documents cited
GB A 2165542

(58) Field of search
C2C

(54) Dichloroaniline derivatives

(57) Compounds of the formula (I)

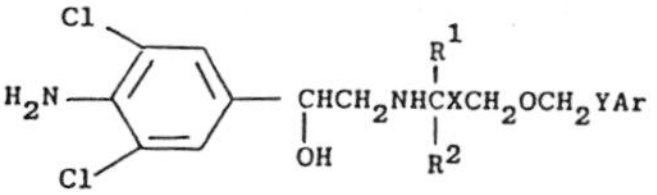

$$H_2N-\underset{Cl}{\underset{|}{\overset{Cl}{\overset{|}{C_6H_3}}}}-CHCH_2NHCXCH_2OCH_2YAr \quad (I)$$
$$\qquad\qquad\quad \underset{OH}{|} \quad \underset{R^2}{\overset{R^1}{|}}$$

wherein
X represents a bond, C_{1-6} alkylene, C_{2-6} alkenylene or C_{2-6} alkynylene and
Y represents a bond, C_{1-4} alkylene, C_{2-4} alkenylene or C_{2-4} alkynylene, the sum total of carbon atoms in X and Y being not more than 8;
Ar represents a phenyl group substituted by one or more substituents selected from nitro, $-(CH_2)_qR$ (where R is C_{1-3} alkoxy, $-NR^3R^4$ (where R^3 and R^4 each represent a hydrogen atom, or a C_{1-4} alkyl group, or $-NR^3R^4$ forms a saturated heterocyclic amino group which has 5-7 ring members), $-NR^5COR^6$ (where R^5 represents a hydrogen atom or a C_{1-4} alkyl group, and R^6 represents a hydrogen atom or a C_{1-4} alkyl, C_{1-4} alkoxy or $-NR^3R^4$ group), and q is 1 to 3], $-(CH_2)_rR^7$ [where R^7 represents $-NR^5SO_2R^8$ (where R^8 represents a C_{1-4} alkyl, phenyl or $-NR^3R^4$ group), $-NR^5COCH_2N(R^5)_2$ (where each of the groups R^5 represents a hydrogen atom or a C_{1-4} alkyl group), $-COR^9$ (where R^9 represents hydroxy, C_{1-4} alkoxy or NR^3R^4), $-SR^{10}$, $-SO_2R^{10}$, $-CN$, or $-NR^{11}R^{12}$ (where R^{11} and R^{12} represent a hydrogen atom or a C_{1-4} alkyl group, at least one of which is C_{2-4} alkyl substituted by a hydroxy, C_{1-4} alkoxy or NR^3R^4 group), and r is 0 to 3], $-O(CH_2)_qCOR^9$, or $-O(CH_2)_tR^{13}$ [where R^{13} represents hydroxy, NR^3R^4, $NR^{11}R^{12}$ or a C_{1-4}alkoxy group optionally substituted by hydroxy, C_{1-4} alkoxy or NR^3R^4, and t is 2 or 3].
R^1 and R^2 each represents a hydrogen atom or a C_{1-3} alkyl group, the sum total of carbon atoms in R^1 and R^2 being not more than 4;
and physiologically acceptable salts and solvates (e.g. hydrates) thereof, have a *stimulant action* at β_2-*adreno-receptors* and may be used in the treatment of diseases associated with reversible airways obstruction such as asthma and chronic bronchitis.

Formulae in the printed specification were reproduced from drawings submitted after the date of filing, in accordance with Rule 20(14) of the Patents Rules 1982.

starts with an introduction which sets out the background and prior art of the invention. It is usually quite brief in case further prior art is discovered which changes the perspective of the invention. A few countries, notably the USA, require a complete disclosure of the prior art as known to the applicant. Even in this case disclosure can be made to the patent office on a form and need not be discussed on the specification. There follows the specific description of the invention. This will be firstly in general terms and then elaborated in examples. In chemical patents there may be a lot of examples, as reactions can be carried out in different ways and many compounds have analogous behaviours.

The claims

The claims set out the inventive aspects which are to be protected and stake the legal claim. Claims run in sequences starting with the most general formulation and then narrowing as specifics of the actual operation are included. There may be more than one sequence as a compound is synthesized and then used in some way or perhaps requires a delivery system. The applicant wishes to claim widely so that an infringer cannot make an insignificant change and circumvent the patent. The examiner is concerned to ensure that no part of the prior art is included in the claims on the one hand and that they do not wander off into flights of fancy on the other. A great many claims are submitted so that a lesser number may survive, since it is not permissible to widen the scope of claims once entered; they can only be maintained or narrowed.

Specifically chemical inventions

What kinds of inventions can be protected in the chemical area? Major categories are as follows:

A new compound will be protected however made or used.

A known compound will be protected when used in a novel process or made in a new way.

A compound may need to be combined with others for storage or for delivery to its site of action. These factors can also be protected.

Patent systems are particularly heavily used by the pharmaceutical, pesticides and related industries. This is because the patent protects the R&D process rather than exploitation of a final product. Chemical patents appear in many other areas. For instance, new alloys and polymers are required for many purposes. The patent will appear as a chemical patent if the actual invention is the development of the chemical.

This is one reason why there are so many chemical patents. The difficulty of relating the technology of an invention to that of its application is one reason for the rather low use of patent information made by industrial researchers.

Patent life extension

A number of countries have recently allowed extension of life for pharmaceutical patents to allow for the long testing procedures. The USA and Japan allow up to five years to give up to 14 or 15 years respectively of life from the date of approval. The EPO has approved in principle up to five years, for 15 years life from approval. However, France has already chosen up to seven years for 17 years life for its national system.

Markush formulae and selection patents

All patents are required to describe novel products or processes. They need not have been carried out in practice (there are many paper inventions), but they should have been thought through.

In chemistry, this novelty requirement is modified. By reasoning on the basis of homologous series, it is possible to claim an invention for a whole host of compounds from just one or two compounds made and tested in the laboratory. These compound representations are called 'Markush formulae' after the US chemist who was the first to file a patent application involving such generic formulae in its claims (see also Chapter 5). Their use is now widespread and has led to the evolution of a special kind of patent, the 'selection patent', to deal with unexpected discoveries made concerning speculative compounds.

At the turn of the century, during work with dyestuffs, it was realized that activity stemmed from a small part of a molecule and changed in a regular fashion in homologues. Markush had to defend and justify the practice of taking this seriality for granted in patent claims. Thereafter it became common practice and seemed to be fully justified. The practice was then transferred to the newly emergent pharmaceutical industry and also to pesticides and veterinary products. Here the justification is much less clear because the compounds must interact with the metabolism of a living organism. This leads to unpredictability; related compounds do not necessarily have smoothly varying activity. Nevertheless, by the force of tradition, patents in these fields are submitted and accepted with Markush claims. The tradition is justified by arguing that the main claim is for a new compound as such and activity in an organism is a secondary issue. However, since the patent would not have been applied for without significant activity, this reasoning is unsound. An example of a Markush formula is given in Figure 11.1. It involves the following symbols: X, Y, Ar, R, R^1 and R^2. Each of these symbols embodies a number of choices so that their combination gives rise to a very large number of choices indeed. In the patent of Figure 11.1, 18 examples are given.

They are said only to illustrate the invention, not to exclude other forms. These will have been all the ones made, as all known possibilities must be disclosed. Applicants are keen to disclose all variations because if there is a challenge to the patent, only the examples given will be accepted as being truly invented and worthy of protection. It must be emphasized that the number of individual compounds in these cases runs typically into the millions, and may be infinite. Goehring and Sibley (1989) discussed the difficulties caused for information systems when very large numbers of compounds are specified.

To deal with discoveries involving compounds claimed but not made, the concept of the selection patent has been developed. This scheme is virtually unique to chemical patents. Selection patents are issued for one or a sub-group of the substances within the group already patented. The sub-group must be clearly distinguishable and each member of it must possess an unexpected property not found in members of the rest of the group. A particular problem is raised by the word 'unexpected'; the property must involve some inventiveness of its own. Then a second patent is granted because the compounds were never truly invented in the first place. An example of a selection patent is given by the case of amoxycillin. Beechams were researching synthetic penicillins. They made and patented a whole group of compounds called ampicillins which included amoxycillin, none of which seemed to have any particular advantage. In subsequent testing it was discovered that pure amoxycillin had a particularly high absorbance rate in the stomach. Therefore a separate patent for amoxycillin was applied for. It was decided by the Court of Appeal that there was sufficient inventiveness for a patent to be granted. In the case of amoxycillin, the same company owned the wider patent as applied for the selection patent, but this need not be so.

Outside the chemical area, lists of alternatives are much shorter and it is usually felt that technical reasoning will cover all eventualities. Therefore selection patents are much harder to obtain outside chemistry.

These discussions illustrate a major disadvantage of patents as sources of technical information. Facts are often obscured in lists of possibilities. The workable alternatives and the constraints which determine them have often not been elucidated when the patent is applied for. Therefore, much routine work is left to be done, and many researchers feel that their time is better spent working from first principles and ignoring patents (Jalloq, 1982).

A problem of obviousness

The patents described above deal with new compounds, but effective and economic methods of synthesizing them on a commercial scale are also required. Computerized synthesis-prediction systems are being developed and refined (see Chapter 5). If these ever become a routine procedure, it could be argued that the resulting compounds were not produced by means of human inventiveness, and patents would no longer be available for making them, as the syntheses would be judged to be obvious (Blick, 1979).

The availability of patent literature

Patents tend to be held separately from other forms of literature in libraries. Because of this they are often overlooked by researchers who are conditioned from their academic training to think almost entirely in terms of the journal literature.

In chemistry, patents tend to be quite well used compared with other subjects. This is due largely to the policy of *Chemical Abstracts* to integrate patents with other literature. Those covered are the patents describing newly synthesized chemicals. Known chemicals applied in various novel ways are usually not covered. The policy is to cover material of interest to chemists, and most process and other patents are deemed to be of interest to the application industry rather than to chemists. In the central domain of new chemicals, the coverage is almost perfect, at least for countries where the documents are readily accessible and the language is not too esoteric (Oppenheim, 1974). The abstracts of *Chemical Abstracts* give details of the compound, its synthesis and information about basic properties (see Chapter 3). In a study of citations to patents found in the *Science Citation Index*, Tewnion (1982) found that in chemistry the majority cited the abstract number from *Chemical Abstracts*. Information referred to was available in the abstract and it seems likely that the patent specification was not consulted at all in many cases.

Apart from *Chemical Abstracts*, patents are abstracted in the journals of many smaller specialist organizations. A major characteristic to watch out for is the quality of the indexes. Which aspects of the patent are searchable? For instance, priority details are important and so is the patent number (Mann and Hellyer, 1980). Examples of high-quality specialist patent services in a restricted subject area are the American Petroleum Institute's file, APIPAT, and the Institute of Paper Chemistry's PAPERCHEM. Like *Chemical Abstracts,* these embrace wide multi-country coverage in their chosen subject areas. Priority details are important from the *legal* point of view but not really important re patents as an information source.

Ultimately, the best service is given by those organizations that specialize in patent information and give complete coverage in one or many countries. Derwent Publications Ltd's World Patent Index (WPI) and Chemical Patent Index (CPI) (see below) need to be mentioned here. From Derwent's database a variety of indexing and abstracting bulletins are printed to meet the needs of different groups of users. An increasing number of these printed bulletins are being produced for single or very small groups of customers (Shenton, 1987).

An argument often heard against using patent literature is that any worthwhile information will be reproduced in more accessible form in journal articles. The overlap is highest in chemistry at around 20 per cent but this is a low figure considering the importance of chemical inventions (OTAF, 1977). However, it seems that when entire research projects are considered as delineated by groups of patents, then the overlap with articles is much higher (Eisenschitz et al., 1986). Articles can cover any aspect of the project and the overlap with patents is not one-to-one. This reflects the fact that articles and patents are written for different purposes and therefore at different stages of the lifetime of a project. Results from a group of UK pharmaceutical patents gave an overlap of 60 per cent, much higher than any previous studies. Within that total, a subset of patents of high scientific interest yielded a 100 per cent overlap, while another subset of more commercially-oriented variations produced a much smaller overlap. Other subject areas could be expected to generate lower overlaps, as the scientific interest of patent contents is very variable. In a food-processing study the total overlap was only 25 per cent. This suggests that much patent literature is of little scientific interest, although in an unknown number of cases the main reason for lack of journal publication will be commercial secrecy, irrespective of scientific concerns. The encouraging point is that in areas of great interest the overlap is high. A problem is that patents are rarely cited in journal references so that it is difficult for a researcher to locate them. It remains true that many of the operational details are given in patents and nowhere else.

Searching

Patents can be searched using general literature services or using those specific to patents. General services are useful in that the patent search is integrated with the searches for all other forms of literature. Nevertheless, for completeness of subject coverage and width of country coverage, the specialist services are required. As most services are now available online, they can be accessed as and when required. Many data-elements are searchable online, so one is not restricted to the printed indexes.

Specialist services

Each country maintains its own integral collection of national patents and appropriate indexes. Most countries also collect the patents of selected other countries and hold these as single-country collections too, i.e. each national collection is distinct and must be searched separately. An exception is the Dutch national patent collection, which now forms the basis of the searching branch of the European Patent Office. In this case all the patents are classified in the same scheme and searchable as one file.

With online systems it is much easier to have a single integrated collection, and two will be described. There are also a number of single-country services which, being online, are more usable than the printed versions. In the online services, what is available is bibliographic information and sometimes an abstract. Drawings, claims and the body of the specification have to be consulted in hard copy (WPI, 1985), though Derwent's abstracts contain the main claims. *Currentscan* is a new service from the British Library's Science Reference and Information Service, which alerts subscribers to new UK, EPC or PCT patent applications which fit their profiles within 48 hours of publication.

Single-country services
For many years, only US patents were available as a database of their own. More recently, the French, European and PCT databases have been made available.

US patents have been printed using computer technology since 1971 and full-text tapes from that date have been leased to any organization prepared to pay for them. These are only complete from 1975 onwards, as in the years 1971–74 there were still some unsolved problems relating to diagrams. The Orbit online service offers a separate US database, called USP, of front-page information in a file set up by Derwent Publications. The full text is not provided but all claims are given. Missing pre-1975 documents have been added. Another US producer, IFI Plenum, has US patent bibliographic details mounted in a file called CLAIMS. This file extends back to 1950 for chemical patents and thus is unique in its extent of backfile. It is indexed with a chemical indexing system derived from that of the DuPont Company. Non-chemical patents extend back to 1963. It is mounted on Dialog, Orbit and STN International. The only US full-text file available at present is LEXPAT. It goes back to 1975 and is mounted by Mead Data Central, who own the legal databases LEXIS and NEXIS and are therefore experienced in handling full text.

The French Patent Office has created a set of databases of French patents since 1969 and also of all European patents. These files are called FPAT and EPAT respectively and are mounted on Questel.

UK patents will soon be available as a separate database from the Patent Office. This will include European patents. However, it will not be accessible via the telecommunications networks, only by use on the Patent Office's own machines.

CHINAPATS and JAPIO covering Chinese and Japanese patents, respectively, are both available via the Orbit host. To ascertain the latest state of patent databases (since new ones are likely to start up at any time), it is advisable to consult the available directories and lists which are published regularly. In particular, the Derwent journal *World Patent Information (WPI)* has an updating feature on database availability.

CD-ROM services are ideal for patents. The EPO offers a series called ESPACE covering European and PCT specifications and national patents of a few countries. These are facsimiles and can be searched on front-page information only.

US patents on CD-ROM are full-text searchable from a number of vendors, for example Micro-Patent of Cambridge. CD-ROM products are popular and new ones are appearing quite steadily. An up-to-date directory, e.g. Sibley (1991), is required for an overview.

Multi-country services
There are two sets of services. Inpadoc covers 55 single-country sources of patents (Piltch and Wratschko, 1978) and Derwent Publications covers 33 (Dixon and Oppenheim, 1982). Inpadoc's strength is its very wide coverage but it gives bibliographic information only, without abstracts, and reproduces what the various national patent offices provide. Derwent covers fewer countries but nearly all the material is produced by their own staff and contractors so that the product is closely tailored to its users. Thus the two services are complementary.

Inpadoc — The International Patent Documentation Centre
Inpadoc was a private company set up by the Austrian government in association with WIPO. Any patent-issuing authority can join the currently 56 members. They send their weekly or monthly records to the Centre in Vienna where they are entered without modification into the computer database. The output is then the INPADOC database plus a series of indexes, on computer-output microfiche. Patent offices of member states have access to all the output. There are also obligations on the offices to provide free access to most of the indexes to fulfil WIPO's aims of technological development through access to information and products. Other organizations can subscribe to the output. Most files go back to 1968 but one must check the starting date for each country. All words, numbers and dates are searchable. The database is available by direct line from Vienna. It is also hosted in Europe and in the USA on the Orbit computer, and by other agreements elsewhere. It is updated weekly. On Orbit, Derwent accession numbers have been added to all records to facilitate cross-file searching. This is particularly useful as it allows an abstract to be consulted.

The computer-output microfiche indexes arrange the patents as follows: in International Patent Classification (IPC) order; in applicant/patentee order; in inventor order; in numerical order for each country; and giving all previous family members. The weekly collection is published together as a *Gazette;* quarterly, annual and five-yearly cumulations of each index are published separately.

INPADOC is only of moderate use for chemical searching, as only the IPC is available as a subject guide. However, for comprehensive family searching there is no rival. Families are sought by priority data and there is a recursive algorithm so that multiple priorities may be followed through. Non-convention equivalents cannot be found in this system, however another valuable command is the GET command. Once a set of references has been created it enables time series and ranked lists to be produced. These are valuable for commercial intelligence as similar work and active firms or inventors can be identified. Uses of this information are discussed later.

Inpadoc was taken over by the EPO in 1991 and in September 1991 changed its name to EPIDOS, which covers all the EPO information services. Its Directorate is based in Vienna and its newsletter *EPIDOS News* should keep users up to date.

Derwent Publications Ltd

Derwent covers 33 sources of inventive information. There are 29 single-country sources, the EPO and WIPO-PCT, and also two defensive-publication journals (*Research Disclosures* and *International Technological Disclosures*). These journals publish brief accounts of inventions for the purpose of establishing a publication so that no one can apply for a patent later.

All patents are entered on to the Derwent database. If a new document is an equivalent it is noted and the patent number is added to the database. Additional information is added if appropriate. This includes designated states, examiners' search reports and additional assignees. A new abstract is written for some of the more important countries. If the document is a basic, then it is given an informative title, abstract and indexing terms by Derwent staff with the aim of bringing out the information of use to industrial searchers as well as the inventive information useful to patent offices and agents. The product consists of bibliographic details and abstracts. Various printed products are produced from the database, containing the information repackaged in a variety of ways for different user groups. The database on pharmaceutical patents goes back to 1963 and other chemical areas started at various dates up to 1970. All other subjects started in 1974.

The chemical file called CHEMICAL PATENTS INDEX (CPI) has deep chemical coding designed for substructure searching. ELECTRICAL PATENT INDEX started in 1981 also has some deep indexing to open a new market. All other matter has only a rather simple classification applied to it and the file is called WORLD PATENT INDEX (WPI). Anyone can access it online, but only subscribers to the hard copy are allowed to use the deep indexing.

National classifications are not available but one has the general

Derwent classes and also the International Patent Classification terms as applied by the national offices. Equivalents searching is by Derwent accession number which means that only members of the immediate family can be traced. Also, there are far fewer countries than in the INPADOC service. But related priorities are listed and one can search directly on priorities to get the extended family. The WPI file was closed in 1980 and a second file WPIL opened for the latest input. WPI remains open for subsequent family members as these have to be checked and assigned the same accession number, much more cumbersome than using the priorities. The one item of family searching in which Derwent does excel is that of non-convention priorities. Non-convention applications apply particularly to the pharmaceutical area where developments can be rapid over a long enough time scale to open up possible markets after the time limit has expired.

The statistical package Patstat-plus

Derwent has organized its statistical functions into a package to be run on a microcomputer (Oppenheim, 1983). It is called Patstat-plus in its present version which has been developed and refined well beyond the stage described by Oppenheim. It is very powerful, allowing for a number of cross-correlations as well as basic ranking and time series. In Orbit, a hit set is created and then statistical operations are carried out using the basic statistical command GET on the mainframe computer. It ties up a lot of central computer unit time and has to be done at once following the search (Terragno, 1984). With Patstat-plus, the created set is downloaded into the microcomputer and then statistical operations can be undertaken at leisure. Although only selected required lines need to be downloaded, this can take a very considerable time and is a disadvantage of the system. Derwent databases are available on the Orbit, Questel and Dialog hosts. Questel has an online statistical command called Memsort, and Orbit has the GET command, so the Derwent package can be bypassed for quick analyses on two of the three hosts. It is unclear at present which approach is preferred by users. As with most of Derwent's services, Patstat-plus seems to be most appropriate for big and frequent users. Occasional and small-scale users may be better off with the online software where there is no additional outlay and one pays only for what one uses. For a large user, Patstat-plus is much more powerful and versatile and analyses can be done at leisure. Patstat-plus was developed for use with Derwent's WPI files but has recently been extended to accept data from its US Patent Office files as well. The results of different analyses can be merged and edited.

Use of the information

The main aim of the patent system has always been to transmit technical information. Apart from this, some commercial information is also transmitted by way of statistical analysis of numbers of documents. The two types of information have different users, and therefore patent information is now used as much if not more in strategic planning and marketing as in R & D and other technical departments.

Technical information

For a chemist, technical information could be such things as:

structure and properties of new compounds

how they are made

systems of storage on the shelf or of delivery to their site of operation, e.g. pharmaceutical formulations, spreading agents for paints, etc.

means of preservation of a substance.

Many variations on these themes are also possible.

Apart from purely chemical patents there will also be chemical aspects to inventions in other technologies, such as metallurgy or laser crystals.

A patent can be very precise; it could give just what is wanted or else miss the mark entirely. But often what is useful is unlikely to be a precise answer; problems do not usually recur in identical form. An idea of the types of problems and solutions encountered through a range of patents is usually better and then the enquirer can use these as adapted to the particular circumstances. This account, of course, excludes the circumstances where the aim of the work is to reproduce an invention for legal and competitive purposes. The purely informational aspects are of much greater and more general importance as they are not limited by time or nationality.

Technocommercial information

Technocommercial information is derived from by the type of statistical analyses described in the previous section. Either a package is available or software is provided by the online host computer. Companies and individuals active in a particular subject field can be identified. Those working on topics most similar to your own are of particular interest. The most active countries, branches of a firm or individual inventors will be revealed and the most productive technologies in a country or worldwide can be identified. Clearly, these types of analyses are normally of more commercial than technical interest. Their use in commercial forecasting opened a whole new market of patent users. Recognition of this wide interest in patent statistics is evidenced by the recent development by the Battelle Institute of a package called Patents-PC which looks similar to Patstat-plus (on a brief assessment) and analyses the output of an online search of any patent database (Battelle Institute, 1987).

Online searching of chemical structures

We have so far concentrated on searching for patent bibliographic information. But because chemical patents act as a kind of register of newly-synthesized chemicals and also because of the prevalence of Markush-type claims (see Chapters 5 and 6), patent searchers need to be able to carry out complex structural searches.

There are two problems here (Barnard, 1984):

to represent three-dimensional chemical structures in a form suitable to enter on to a computer (see Chapter 6), and

to represent entire Markush groups in one formula rather than the individual members.

Until recently, the only commercially available system which allows both input and retrieval of Markush formulae has been Derwent Publications Ltd's fragmentation code (Jackson, 1984). Very simply, fragments are coded separately and are then linked together. A series of compounds is handled by means of ranging from... to... on the computer. A front-end system called Topfrag has just been introduced for use with these codes. The user inputs a diagram and the program works out all the codes. At present only one structure at a time can be input. The system will be enhanced to include at least limited generic searching. However, various systems are being developed which are based on new ideas about topological representations and which allow the direct use of structural diagrams, the way in which most chemists think about a problem. If they can input a diagram, users would not need elaborate training as is at present required to understand the codes. The three schemes we will consider are (Barnard, 1987):

1. A development of DARC to include generics called Markush DARC.

2. The American Chemical Society's system (MARPAT) for use with the Chemical Abstracts Service's files (see Chapter 5).
3. The work at Sheffield University.

Markush DARC

Markush DARC is being developed from the existing generic DARC system by its French suppliers, Télésystèmes, in agreement with the French Patent Office and Derwent Publications Ltd. The first version was launched commercially in January 1989 but only runs on post-1986 patents. Derwent will allow the new file, WPID, to run in parallel with Topfrag. There will be a series of upgrades and the older system will be relegated for use with the back files only. There is no intention, at present, on the part of Télésystèmes, to convert older files. However, the French Patent Office does intend to convert starting with its oldest French patents online from the 1960s.

The Markush structures are represented in the computer by a single connection table concatenating all the partial structures for the constant part. Values are given for the variables and link descriptors to show the way they are connected. Generic groups are represented by a set of 21 'superatoms' which can be qualified by various attributes. Examples of superatoms are 'halogens', 'carbocyclic aromatic' and 'polymer end group'. They seem to be as general as the claims in patent documents. The search will involve a fragment screening stage followed by an exhaustive atom-by-atom match.

The Chemical Abstracts Service system

As with DARC, this system is based on connection tables. Variables are defined by means of nomenclature as well as by reference to components of the structure diagram. Nomenclature includes such terms as 'alkyl'. With the diagrams, a hierarchy of generic group types may be used and they are qualified by attributes such as element counts, ring sizes and number of rings. This system is protected by a software patent (American Chemical Society, 1987). A commercial version called MARPAT has been launched and development work is continuing (see also Chapter 5).

The Sheffield project — GENSAL

GENSAL is still at the development stage at Sheffield University Department of Information Studies. It is a much more radical system than the other two. GENSAL is a formal language similar to a programming language. It represents the structure by its topography. A test database has been created and commercial interest in this approach is high.

Acknowledgements

I thank Deborah Bumstead for helping me with last-minute updating. I am grateful to the Controller of Her Majesty's Stationery Office for permission to reproduce the front page of British Application no. 2 182 658 A.

References

American Chemical Society (1987), *Storage and Retrieval of Generic Chemical Structure Representations*, US Patent 4 642 762.
Barnard, J.M., ed., (1984), *Computer Handling of Generic Chemical Structures, University of Sheffield Conference*. London: Gower.
Barnard, J.M. (1987), Online graphical searching of Markush structures in patents. *Database*, 10(3), 27–34.
Blick, A.R. (1979), Computer assisted chemical synthesis packages. *Journal of Information Science*, 5, 227–229.
Battelle Institute (1987), *Patents-PC — A Method and Personal Computer Based Software for Analyzing Published Patent Data*. Product brochure.
Dixon, M.D. and Oppenheim, C. (1982), Derwent Publications patent information services. *World Patent Information*, 4(2), 60–65.
Eisenschitz, T.S., Lazard, A.M. and Willey, C.J. (1986), Patent groups and their relationships with journal literature. *Journal of Information Science* 12(1), 35–46.
Eisenschitz, T.S. (1987), *Patents, Trade Marks and Designs in Information Work*. Beckenham: Croom Helm.
EPIDOS News, available from Schottenfeldgasse 29, A-1072 Wien, Austria.
EPO (1986a), *How to Get a European Patent*. Munich: European Patent Office.
EPO (1986b), *Annual Report 1986*. Munich: European Patent Office.
Goehring, K.E.H. and Sibley, J.F. (1989), A giant step for mankind? *World Patent Information*, 11(2), 5–10.
Jackson, S.E. (1984), Experiences of a patent searcher. In J.M. Barnard, ed., *Computer Handling of Generic Chemical Structures, University of Sheffield Conference*, pp. 30–37. London: Gower.
Jalloq, M.C. (1982), *Use of Patent Literature by Academics*. British Library R & D Report no. 5770.
Mann, H. and Hellyer, A. (1980), Coverage of UK patent specifications by abstracting journals. *World Patent Information*, 2(1), 27–28.
Oppenheim, C. (1974), Patents coverage by *Chemical Abstracts*. *Information Scientist*, 8, 133–138.
Oppenheim, C. (1983), A microcomputer program for the statistical analysis of patent databases. *World Patent Information*, 5(4), 209–212.

OTAF (1977), *Patents as a Technological Resource*. 8th OTAF Report, pp. 23–27. Washington, DC: US Department of Commerce.
Patent Office (1986a), *Patents as a Source of Information*. London: Department of Trade and Industry.
Patent Office (1986b), *Introducing Patents*. London: Department of Trade and Industry.
Patent Office (1986c), *Annual Report of the Comptroller*. London: Department of Trade and Industry.
Patents Throughout the World (1988). New York: Trade Activities Inc.
Phillips, J., (ed.), (1985), *Patents in Perspective*. Oxford: ESC.
Piltch, W. and Wratschko, W. (1978), INPADOC — a computerised patent documentation system. *Journal of Chemical Information and Computer Science*, **18**(2), 69–75.
Shenton, K. (1987), Personal communication.
Sibley, J.F. (1991), *Online Patents, Trade Marks and Service Marks Databases 1992*. London: Aslib.
Terragno, P.J. (1984), The GET command. *World Patent Information* **6**(2), 69–73.
Tewnion, L. (1982), A Study of the *Patent Citation Index*. MSc Thesis, The City University.
WIPO (1985a), *PCT Applicant's Guide*. Geneva: World Intellectual Property Organization.
WIPO (1985b), *International Patent Classification Schedules*, 4th edn. Geneva: World Intellectual Property Organization.
WPI (1985), Online patent information. *World Patent Information*, **17**(1–2).

CHAPTER TWELVE

Technocommercial information

P. CARTER

With an estimated turnover of $1 215 billion in 1990 worldwide and over 100 000 products on the market, the chemical industry is diverse and complex. Definitions of the industry and its sectors differ because of the nature of its products. This may complicate the task of searching for information, as terminology used in the industry may differ from, for example, IUPAC nomenclature or statistical classification schemes. The industry often uses terms based on product derivation to describe its sectors — e.g. oleochemicals and petrochemicals. Other terms like fine chemicals, speciality chemicals, performance chemicals and effect chemicals, are also in common use. Even within these terms, definitions may differ: some would include polyolefins under the heading of petrochemicals, others under plastics. The trade press uses the language which industry uses, so a search for information should also include alternative terminology.

Although its products and processes differ from other industries, many sources of information are the same. Directories such as *Kelly's Business Directory* (Kelly's Business Directories, East Grinstead, UK, annual) or *Kompass* (Kompass Directories, East Grinstead, UK, annual) will include chemical companies. Ownership of a chemical company may be found from *Who Owns Whom* (Dun and Bradstreet, London, annual), and senior executives in the industry will probably have biographies in *Who's Who* (A. &C. Black, London, annual). Financial information can be obtained from Companies House records or databases like DISCLOSURE (Disclosure, London) online via Data-Star or Dialog. Daily newspapers and news services like Reuters provide very current sources of information on the chemical industry and issues which may affect it. This chapter will try to identify major sources of information which are specific to the chemical and allied industries.

Basic reference works

A chemical dictionary is a prerequisite for anyone concerned with technocommercial information. One of the most useful is *Hawley's Condensed Chemical Dictionary**. Chemicals, abbreviations, processes, trade names, equipment and terminology are listed alphabetically. For chemicals and raw materials, chemical names and synonyms are followed by chemical formula, physical properties, derivation, grades available, brief hazard data, uses and shipping regulations. The *Concise Chemical and Technical Dictionary** restricts entries for each chemical to synonyms, formula and brief physical property data. Trade names are included, although the owners are not identified. More comprehensive technical and commercial information can be found in *Kirk-Othmer's Encyclopedia of Chemical Technology** and *Ullmann's Encyclopedia of Industrial Chemistry**. *Kirk-Othmer* contains nearly 1300 monographs prepared by expert reviewers. About half of the monographs deal with chemical substances such as sulphuric acid, or groups like aluminium compounds. Other monographs cover topics such as industrial processes, end-use sectors like cosmetics or adhesives, fundamental subjects like thermodynamics, and scientific and technical themes such as catalysts. The supplement volume has topical entries such as environmentally degradable plastics and process energy conservation. Product monographs usually have some historical data, physical, chemical and toxicological properties and safety data, commercial production methods and uses, as well as information on derivatives and related compounds. Some entries have brief statistical or economic data (production, trade, prices), usually for the USA, although much of this information is out of date, particularly in the earlier volumes. Each monograph is followed by a list of references. *Kirk-Othmer* is also available online as its full text, including tables, through STN. Like *Kirk-Othmer, Ullmann* is also published sequentially — the first volume of the fifth edition was produced in 1985. A total 36 volumes will be published containing 1000 monographs. Each monograph, again compiled by experts, is preceded by a table of contents, and followed by a list of references. The information in the monographs is similar to that in *Kirk-Othmer* and includes economic data.

Although principally a buyers' guide, the *Chemical Industry Directory* (Benn Business Information Services, Tonbridge, UK) has other information. The first section has buyers' guides for chemicals, plant and equipment, and laboratory apparatus and scientific equipment, and also a list of trade names. This is followed by details of chemical and oil-storage depots worldwide, and a list of professional and trade associations. *Who's Who in World Petrochemicals and Plastics* (Who's Who Information Services, Houston, TX) does not have biographical details, but lists over 2700 corporate offices with addresses, telephone,

---

*Further information on the titles asterisked is given in Chapter 4, where details of other reference works can also be found.

telex and fax numbers, together with names and job titles of senior executives, especially in sales, marketing or purchasing. In addition to petrochemical and plastics manufacturing or trading companies, trade associations and other organizations are also included. A second section has an alphabetical list of 7200 executives, with their corporate affiliation.

Periodicals and secondary sources

Trade journals for technocommercial information are of two main types: (i) those which cover the whole industry, and (ii) those which concentrate on specific sectors. Coverage may be international, by broad geographic area or by country. Titles, frequency and publisher details for chemical industry periodicals can be found in *Ulrich's International Periodicals* (see Chapter 3). For the UK, *British Rate and Data* (Maclean Hunter Ltd.) lists all periodicals containing advertising; the US equivalent is *Standard Rate and Data* (Standard Rate and Data Service). Relevant titles can also be found in the source lists of abstracting services such as *Chemical Industry Notes* (Chemical Abstracts Service.)

The two best known weekly English-language trade journals for the whole industry internationally are *European Chemical News (ECN)* and *Chemical Week International (CWI)*. Both journals report major news on the chemical industry: company financial results, new projects, market and price reports, new technologies, environmental issues, and legislation, as well as features in greater depth or special reports on topical issues. *ECN* (Reed Business Publishing, 1962–) also includes appointments of senior personnel in the industry and announcements of forthcoming conferences, meetings and symposia. *ECN* is available online in full text, through Data-Star. There are two files: CNEW, the historical file from 1984 to date, and CNEX, which contains the most recent issue. CNEX is available on each Friday before the journal is published on Monday. *ECN* also has seven *Chemscope* supplements each year which include reviews on the European chemical industry, the environment, new projects (including fertilizers), new processes and independent tank-storage facilities worldwide. The *Chemscopes* are not included in the online files. *CWI* (Chemical Week Associates, New York) is the international version of the US *Chemical Week*. When first published, the editorial content of *CWI* differed from that of *Chemical Week*: this has now been harmonized. Coverage is similar to that of *ECN*.

Non-English-language periodicals for the chemical industry include *Europa Chemie* (36 issues a year) and *Chemische Industrie* (monthly), which cover the German and European industries and are both published in German by Handelsblatt, Düsseldorf, in conjunction with the Verband der chemischen Industrie (VCI), the German chemical industry trade association. *Chemische Industrie* has short abstracts and contents in English. Also monthly, with short English abstracts, is the French *Informations Chimie* (Société d'Expansion Technique et Economique, Paris), which covers the chemical industry internationally. The Chemical Daily Company, Tokyo, produces a daily bulletin in Japanese with news on the Japanese chemical industries: the main stories are reproduced in English in the weekly tabloid *Japan Chemical Week*.

Also in distinctive tabloid format is the weekly *Chemical Marketing Reporter*, formerly *Oil, Paint and Drug Reporter* (Schnell Publishing, New York). This periodical concentrates mainly on US news and is well known for its list of chemical prices. It is particularly strong on new legislation, and also has details of chemical imports into the ports of New York and New Jersey, listing consignee, quantity, ship's name, port of origin and date of arrival into the USA. *Chemical Marketing Reporter* is also available in full text online in the PREDICASTS PROMPT database: the price and import information is not included. Schnell also publishes *Chemical Business* (monthly) which mainly comprises reviews or features on sectors or companies in the US chemical industry.

In the UK, the Society of Chemical Industry (London) publishes the semi-monthly *Chemistry and Industry*. This has the main news — particularly for the UK and Europe — and features or articles which are often of a semi-technical nature. *Chemistry in Britain* (monthly, Royal Society of Chemistry) again has a news review and technical or semi-technical articles, book reviews, and RSC news, and is intended to keep the individual chemist up to date on economic, political and social factors affecting the scientific community.

Periodicals for specific sectors of the chemical industry usually have short news items, with more emphasis on new products and features on relevant companies or product groups, often of a semi-technical nature. Although it contains news about the chemical industry generally, *Manufacturing Chemist* (Morgan-Grampian) tends to emphasize news and features on the cosmetics and toiletries, pharmaceutical and aerosol sectors of the industry. It also has details of relevant patent applications made to the European Patent Office, giving the title, applicant's name and number of the application. *Speciality Chemicals* (FMJ Publications, Redhill, UK) and *Performance Chemicals* (Reed Business Publishing) cover the sectors of high-value, low-volume chemicals and products such as biocides and surfactants, which are chosen for their effects or performance rather than their chemical composition. *Performance Chemicals*, published every two months, has regular listings of the manufacturing capabilities of contract and special organic chemical manufacturers, giving details of equipment, expertise in particular reactions and handling of 'difficult' chemicals, and also listings of biocide and surfactant manufacturers. The rubber and plastics industries are covered in Europe by periodicals such as *Plastics and Rubber Weekly* and

European Plastics News (both EMAP-Maclaren, Croydon, UK), and in the USA by *Modern Plastics* (McGraw-Hill) and *Plastics World* (Cahners Publishing, Newton, MA). Trade and technical information on the surface coatings industries may be found in journals such as *Polymer, Paint, Colour Journal* (FMJ Publications) and *Farbe und Lack* (Gesellschaft Deutscher Chemiker, Hannover). The British Sulphur Corporation produces *Sulphur, Nitrogen* and *Phosphorus, Potassium* covering fertilizers containing those elements; *Farm Chemicals* (Meister Publishing) has information not only on fertilizers but also on other chemicals used in the agricultural industries like herbicides, fungicides and other pesticides (see also Chapter 15). As chemicals are used by many industries, it may also be necessary to seek information from sources for those industries, for example articles on food additives may be found in periodicals like *Food Flavourings, Ingredients, Packaging and Processing* (MBC Trade Publications).

Information on the chemical industries of Eastern Europe and the former Soviet Union can be found in the monthly journal *Eastern Bloc Chemicals* (Eastern Bloc Research, Newton Kyme, Yorkshire, UK). This has news on each of the countries in Eastern Europe: the section on the former USSR is divided into production, raw materials, aromatics, pharmaceuticals and other sectors, engineering, pollution and trade. There is also a section on trade between the Eastern European countries and the rest of the world: each issue contains much statistical data. Also covering an increasingly important geographic region is the quarterly magazine *Asia-Pacific Chemicals* (Reed Business Publishing). In addition to company, project and other news, details of forthcoming conferences are given, followed by features on sectors such as petrochemicals and man-made fibres.

In addition to the trade press and learned society publications, there are also a number of newsletters on the chemical industries. The longest established in the UK is *Chemical Insight* (Reed Business Publishing). As its title suggest, *Chemical Insight* presents a very personal view of the industry by its editor. Published twice a month, each issue is on a topical theme and often gives an 'insight' into the thoughts and goals of senior executives from major chemical companies. A popular feature is *Chemical Insight*'s annual league tables of the top chemical and pharmaceutical companies worldwide, ranked by turnover and other criteria. Also published twice monthly *Chemical Matters* (Chemical Matters, London), originally *World Petrochemicals Analysis*, contains brief news items — company results, mergers, lawsuits, new projects and also topical features. *Chemical Matters* is edited by Hilfra Tandy in a very distinctive, forthright style. Special reports are also produced separately: a recent review covers the international olefins industry giving production, trade and other economic data, present and future plant capacities and other details. *Focus on Chemicals* (Stuart Walmsley, Burwash, Sussex,

UK) written by a former City analyst, concentrates on single topics, for example the titanium dioxide industry. The most recent UK newsletter is *Chemical Outlook International* (Business International, London), which is published semi-monthly, each issue containing about six features or special reports.

Most major chemical companies produce their own house journal or newsletter, to inform their employees of news and developments within the company. These journals are useful sources of information — many companies will make them available on request to non-employees. ICI has recently made available online through PROFILE the full text of some of its house journals and its press releases. Examples of house journals are BP Chemicals' *Double Bond*, ICI's *The Roundel* and Shell's *Shell Petrochemicals*.

As with periodicals, secondary sources of information may cover the whole of the chemical industry or specific sectors. For the chemical industry as a whole, there are two major sources. *Chemical Industry Notes (CIN)* is produced by Chemical Abstracts Services and is available in weekly hard-copy form, or as an online database on Data-Star, Dialog and other hosts. *CIN* abstracts information from about 100 periodicals which include the major trade journals, newspapers, and economic and business journals worldwide, but with a US emphasis. All areas of the chemical industry are covered from prices and production to sales and products. The hard-copy version has short abstracts and permuted keyword and corporate indexes. The online version includes CAS Registry Numbers and an online thesaurus of geographic terms. The Royal Society of Chemistry produces the online *Chemical Business NewsBase (CBNB)*; hosts include Dialog, Data-Star, and Pergamon Financial Data Services. *CBNB* contains abstracts and brief factual items covering all aspects of the chemical and allied industries and their end-use sectors. In addition to the main periodicals, sources include information from chemical and pharmaceutical companies such as house journals, and also stockbroker and market research reports, directories and handbooks. Launched in 1985, *CBNB* is very much European oriented but includes major sources from Japan and the USA. Descriptors are added to the abstracts to enable users to search by geographic location, CAS Registry Numbers, industry codes and economic and business terms. Recent developments include the addition of more environmental information from journals such as *Europe Environment* (a fortnightly briefing service from Europe Information Services, Brussels), and also by downloading from databases such as SCAD and POLIS to provide regulatory information on the environment. In addition to its online database, the RSC also provides weekly printed current-awareness services, *Chemical Business Bulletins*, which are extracts from *CBNB*. There are 11 titles including *Agrochemicals*, *Fertilisers*, and *Speciality Chemicals*. The online *East Europe Chemical Monitor* provides a comprehensive source of business information, including company news, market information, new plant capacities and details on new products. Agrochemicals, petrochemicals, plastics, pharmaceuticals and synthetic fibres are covered. Information is obtained by Business International, Vienna, from about 100 commercial and trade press sources from Eastern Europe: the online host is Data-Star.

More general technocommercial databases, although not specifically for the chemical industry, should not be forgotten in any search for information. For example, *Predicasts Prompt* (Predicasts Inc.), available online on various hosts and also in hard copy, has a very wide coverage of sources and is particularly useful for market-type information. Reuter's TEXTLINE is a good starting point for information on companies. Examples of secondary sources for sectors of the chemical industry include *World Surface Coatings Abstracts* (Paint Research Association, London) for the paint and surface coatings industries, and *Rapra Abstracts* for the rubber and plastics industries (Rapra Technology, Shawbury, Shropshire, UK): both of these are available in hard copy monthly and online through Pergamon.

Catalogues and buyers' guides

Catalogues and buyers' guides have several uses. They can be used to locate suppliers of chemicals for manufacturing or research purposes or for market intelligence to discover which companies are manufacturing or supplying particular chemicals. They are also useful sources of addresses and telephone numbers of companies.

For laboratory or research purposes, a collection of catalogues can be organized from suppliers such as Aldrich, BDH and Kodak. The Aldrich catalogue is also available on CD-ROM, which lists some 19 000 chemicals and is searchable by chemical name, CAS Registry Number or molecular formula. Also on CD-ROM, Aldrich publish ALDRICHEM DATA SEARCH, which includes chemicals from their general catalogue as well as rarer chemicals. Information on 50 000 chemicals can be searched by chemical name, CAS Registry Number, physical property data and substructure or structural class.

An alternative to collecting individual catalogues is Molecular Design's *Fine Chemicals Directory (FCD)*, a compilation of about 80–100 laboratory chemical catalogues, which can be supplied to customers in machine-readable format. *FCD* was originally produced by Fraser-Williams Scientific Services (Poynton, Cheshire, UK), who still maintain and update the directory. An online version, CHEMQUEST, can be accessed through Pergamon Financial Data Services, and contains information on 60 000 chemicals, their prices and suppliers. Searching is by chemical name, CAS Registry Number, molecular formula or Wiswesser Line Notation. CHEMQUEST will also support graphics searching from several types of graphics terminals.

Chemical buyers' guides for commercial chemicals usually consist of an alphabetical list of chemicals, often with a code to indicate the supplier, and a company directory section giving company addresses, telephone, telex and fax numbers. Trade names are often included. Directories Publishing Company (Clemson, SC) produce *Chem-Sources International* (biennial, the latest edition was published in 1990) and *Chem-Sources USA* (annual), which contain respectively 70000 and 100000 product listings of both laboratory and commercial chemicals from over 800 companies. Indications of whether the products are available in bulk or high purity are included, along with CAS Registry Numbers and trade names. These directories are also available online through STN in two files: CSCHEM contains information on the chemicals and their suppliers, while CSCORP has fuller details on the companies supplying the products. The *Directory of World Chemical Producers*, 1992/93 (Chemical Information Services, Dallas, TX, 1991) lists almost 50000, mainly commercial, chemicals from 5000 manufacturers worldwide. There is extensive cross-referencing between chemical synonyms, and a useful feature is the grouping of esters and salts of organic acids under the acid (for example, vanillic acid, methyl ester, rather than methyl vanillate), thus bringing chemically-related compounds together. The *Directory of World Chemical Producers* is also available on diskette in a database format for use on any IBM-compatible PC.

Some publishers of trade journals also produce annual buyers' guides: these include the *Chemical Week Buyers Guide* (Chemical Week, annual), and the *OPD Chemical Buyers' Directory* (Schnell Publishing, annual): these directories are sent automatically to subscribers to *Chemical Week* and *Chemical Marketing Reporter*, respectively.

Trade associations often produce directories of their member companies' products. The French Union des Industries Chimiques *Annuaire UIC* is available both in hard-copy form and on diskette, which can be searched using chemical names and CAS Registry Numbers or customs tariff numbers. Many trade associations have now collaborated with publishers to produce directories. Examples are *Where to Buy Chemicals Plant and Services*, formerly *Where to Buy Chemicals and Chemical Plant* (Where to Buy, Redhill, Surrey, UK, annual), which is the official directory of the British Chemical Distributors and Traders Association, and the *Buyers' Guide, Plastics, Rubber and Chemical Products* (Kompass Publishers, East Grinstead, UK, annual), which is published in association with the British Plastics Federation. The UK Chemical Industry Association's (CIA) *Directory of Products and Buyer's Guide 1992* (Hanslet Information Services, Cambridge, UK, 1991) lists over 10000 products and 4500 trade names from more than 1000 UK suppliers. In addition, there is a classified section for contract processing services. The *Firmenhandbuch der Chemische Industrie* (13th edn, Econ Verlag, Düsseldorf, 1988) is supported by the VCI and is probably one of the longest established chemical directories. Otto Wenzel, the editor of the first edition, published in 1888 under the title *Addressbuch und Warenverzeichnis der chemischen Industrie des Deutschen Reiches* (Directory and Register of Goods of the Chemical Industry of the German Empire) faced the same difficulties that publishers of chemical directories face today. He was able to compile a comprehensive list of sites manufacturing chemicals from statutory information provided to insurance companies, but had great difficulty with the listing of chemical products. He found that the manufacturers who provided the information listed the same products 'under completely different, locally usual trade names'. Chemical directories require extensive cross-referencing between chemical synonyms, particularly as non-systematic names are still in common use in the chemical industry. If a directory is not well cross-referenced, it should be used in conjunction with a chemical dictionary to provide alternative names.

In addition to the directories mentioned above, there are many other directories for specific sectors of the chemical industry. It would be impractical to indicate more than a few examples of these; trade associations should be able to advise on directories relevant to their sectors of the industry. The CBD (Beckenham, UK) *Directory of British Associations* (G.P. and S.P.A. Henderson, eds, 10th edn, 1990) and *Directory of European Industrial and Trade Associations* (R.W. Adams, ed., 5th edn, 1991) are good sources for locating trade associations. Sources of supply for chemical intermediates can be found in *Fine Chemicals and Intermediates from British Manufacturers* (Special Chemical Statistics, London, 1989) which lists 2500 chemicals from 75 suppliers. This directory is published in conjunction with the Specialised Organics Information Service (SORIS) of Macclesfield, Cheshire. SORIS can advise enquirers of companies which can undertake custom synthesis of chemicals not commercially available. The Japanese *Speciality Chemicals Handbook* (by the editors of *Japan Chemical Week*, 3rd edn, Chemical Daily Co., 1989) contains information on aliphatic, aromatic and heterocyclic organic intermediates. In addition to chemical formula, chemical property, and uses for each chemical, the class number for the MITI (Ministry of International Trade and Industry) Existing Chemical Substances Inventory is given, along with the status of the chemical under the US Toxic Substances Control Act (TSCA). FMJ International Publications (Redhill, UK) produce the annual *Polymers, Paint, Colour Yearbook*, containing information on suppliers of materials and equipment used in the surface-coating industries, and also the British Adhesives and Sealants Association's annual *Adhesives and Sealants Yearbook and Directory*. Suppliers of pesticides can be found in the *Pesticide Index* (H. Kidd and D. Hartley, eds, Royal

Society of Chemistry, Cambridge, UK, 1988), which is a quick reference guide listing pesticides under approved, chemical, common and trade names on a worldwide basis. *Surfactants Europa* (G.L. Hollis, 2nd edn, Tergo Data, Darlington, UK, 1989) is a directory of surface-active agents available in Europe. Arranged by surfactant type (anionic, cationic, ethoxylates, etc.), it gives details on properties and applications, along with trade names and contact details of suppliers.

General business directories, such as *Kelly's Business Directory*, can also prove useful in tracing sources of supply for materials, particularly products like minerals, which fall outside the mainstream chemical industries. Trade associations are often able to advise on potential sources of supply of products from their member companies. The British Plastics Federation and the British Rubber Manufacturers Association jointly operate the Plastics and Rubber Advisory Service. Information on suppliers of chemicals, machinery and equipment which are used in the plastics and rubber industries is held on an in-house computer database. Enquirers are provided with supplier details, and suppliers are sent details of enquiries for their products.

Trade names

Tradename and trademark information is important to those who need to source a product as well as those wishing to check whether a particular trade name has been registered. However, tracing trade names for chemical products can be quite difficult. Most of the directories and buyers' guides listed in the previous section contain tradename listings. There are also directories of trade names for chemicals: these include *Chemical Synonyms and Trade Names* (W. Gardner, 9th edn, Technical Press, 1987) which lists over 20 000 trade names. This edition of *Gardner's* has been extensively revised; however, certain trade names without a manufacturer's affiliation have been retained from previous editions. *Thesaurus of Chemical Products* (M. and I. Ash, eds, Edward Arnold, 1986) contains 50 000 trade names from 900 companies: cross-references are provided between chemical and trade names. The same publisher also produces the *Chemical Products Desk Reference* (M. and I. Ash, eds, 1990). This has about 32 000 trade names, with indexes by function (for example, absorbants, pesticides, oxidizing agents) and also by chemical name. A list of manufacturers with addresses is also included. Although not specifically a directory of chemical trade names, *UK Trade Names* (10th edn, Kompass Publishers, 1988) can assist in tracing the owner of a trade name; however, pharmaceuticals are not included. American trade names may be traced through the *Chemical Week Buyers' Guide*, *OPD Chemical Buyers' Directory* or *Chem-Sources USA*.

Online sources include BRITISH TRADE MARKS (Pergamon Financial

Data Services). Data for this file are supplied by the UK Trade Marks Registry. The database has information on UK trade and service marks, including those for which application for registration has been made. Trade marks which have lapsed since 1976 are also included. One point to note: companies do not always inform the Trade Marks Office when a trade mark is sold or transferred to another company. Ownership should be confirmed with the registered owner. RAPRA TRADE NAMES, produced by Rapra Technology (Shawbury, Shropshire, UK), and also available online via Pergamon, has in excess of 22 000 trade names relevant to the plastics and rubber industries, including adhesives and chemicals. A hard-copy version, *Rapra New Trade Names*, containing about 5000 trade names, is published annually.

Trade names can be informative and it is sometimes possible to trace the owner and/or type of product from the trade name. For example, Bayhibit is a trade name for a series of corrosion inhibitors manufactured by Bayer, and a number of Hoechst's trade names begin with Hosta... . Some trade names have become generic in common use, for example DuPont's Nylon and Freon and Union Carbide's Cellosolve solvents, despite companies trying to resist such generic use.

Chemical prices

Prices for laboratory chemicals are best obtained from the current catalogues of suppliers. For commercial chemicals, particularly major products, several journals list prices on a regular basis. *European Chemical News (ECN)* publishes prices on a weekly basis for the major olefins and aromatics, and also methanol, ammonia and naphtha. The price information is obtained from European producers, consumers and merchants each Wednesday prior to publication of the journal on Monday. *ECN* also offers its price information as a fax service, so that information may be obtained every Thursday prior to publication. European contract and spot prices are quoted, along with the US price range. In addition, *ECN* has a monthly plastics market report, giving prices for polyethylene, polypropylene, polystyrene and poly(vinyl chloride). *Chemical Marketing Reporter* publishes prices on a weekly basis for some 3000 chemicals on the US market. *Chemical Week International* also publishes a weekly list of prices in the European and US markets for the major petrochemicals, some detergent intermediates and solvents. ICIS-LOR produces a series of weekly reports giving price and other market information for chemicals, crude oil, petroleum products such as gas oil and kerosene, and liquefied propane gas. Chemicals covered by this service include the major petrochemicals, and also chlorine and sodium hydroxide, sulphur and phosphate fertilizers. Average monthly European and US prices for petrochemicals, solvents, polymers and feedstocks (naphtha, propane and butanes) are quoted in *Platts International*

Petrochemical Report (McGraw-Hill), along with shipping rates. *CPI Purchasing* (Cahners Publishing) gives monthly contract and spot prices for over 40 major organic and inorganic chemicals and the major polymers, and for some of these publishes six-month price forecasts. A more detailed price outlook for some selected chemicals is also included.

Although it is relatively easy to obtain price information for the major chemicals, especially petrochemicals, published prices for fine chemicals are more difficult to find. Direct enquiries to traders or producers can be tried, and the Japanese *Speciality Chemicals Handbook* gives prices from Japanese producers for some organic intermediates. Another possible source is trade statistics, which usually contain both quantity and price information for imports and exports. The monthly trade statistics compiled by the UK Customs and Excise give prices for exports on a fob (free on board) basis, that is the cost to the purchaser abroad, including packaging, insurance and transport to the UK port whence consigned. Prices for imports are quoted for duty-free goods on a cif basis (cost, insurance, freight), that is the cost to the importer to the point of importation into the UK. Unit costs can be calculated by dividing the cost (in pounds sterling) by the quantity (which is expressed in kilograms).

Statistical and economic information

Statistical and economic information is collected and published by many different organizations — government, international bodies, banks, trade associations and others. Statistics compiled by governments and international bodies such as the United Nations are usually referred to as official statistics and those compiled by other organizations as non-official statistics. In some instances, statistics collected by non-governmental organizations may be used to produce official statistics. For example, in the UK the Fertiliser Manufacturers Association and the National Sulphuric Acid Association collect data from their member companies, which form the basis of statistics appearing in official UK sources.

A number of guides to sources of statistical information are published: CBD (Beckenham, UK) produces *Statistics Europe* (J.M. Harvey, 5th edn, 1987) which for both Western and Eastern European countries has details of the official central statistical offices, other major organizations collecting and publishing statistics, and libraries holding collections of statistical material. Euromonitor (London) produces the *International Directory of Non-official Statistical Sources* (1990) and the *European Directory of Non-official Statistical Sources* (D. Mort, ed., 1988), covering a variety of organizations — business schools, trade associations, banks, trade journals, etc. *Sources of Unofficial UK Statistics* (D. Mort and J. Siddall, Gower Publishing, 1990) has an alphabetical listing of organizations producing statistics, together with a subject index, which allow a researcher to locate sources of, for instance, statistics on the salaries of chemists, chemical engineers and other employees in the chemical industry.

Various classification schemes are used in the compilation of official statistics. Recent attempts, particularly by the European Commission (CEC) and EC member countries, have been made to harmonize the different schemes. With the advent of the Single European Market in 1992, this is necessary to ensure compatibility of statistical data. In the UK, production statistics are compiled using the *Standard Industrial Classification Revised 1980, SIC(80)*. This is an attempt to align the UK scheme with the EC's NACE classification. *SIC(80)* is based on activity rather than on commodities. A full description is given in *Standard Industrial Classification Revised 1980* (Central Statistical Office, HMSO, 1979). Trade statistics are often classified by commodity-based schemes. The system used in the UK, often referred to as 'The Tariff', is a single list of descriptions applicable to both imports and exports. This is based on the Harmonised Commodity Description and Coding System (usually called the Harmonised System or HS), and incorporates the EC's Combined Nomenclature (CN). Each commodity description is uniquely described by a nine-digit number (eight digits in the EC scheme): the first six digits indicate the chapter and main heading within the HS, the seventh and eight digits indicate the heading of the CN to which it correlates (the ninth digit adds an extra level of detail). For example, the tariff classification under the CN for acetylene is 2901 29 90.

An invaluable guide to the tariff classification of chemicals under the CN is the *European Customs Inventory of Chemicals* (Commission of the European Communities, Luxembourg, 1991) published in nine languages. Some 26 000 'preferred names' are listed together with 6000 synonyms. Preferred names are in accordance with IUPAC nomenclature, and priority is given to ISO (International Organization for Standarization) names for pesticides and plant growth regulators, and to WHO (World Health Organization) INN and INNM names (International Non-proprietary Names and Modifications) for pharmaceuticals and their derivatives such as salts and esters. The book is divided into four parts. Part I, an alphabetical list of chemical products, has for each product the tariff classification and the 'CUS' number (an identification number allocated by the EC's Directorate for Customs Union and Statistics, DG XXI). Part II is indexed by the CUS number; parts III and IV are cross-indexes between CAS and CUS numbers. Another commonly-used classification for trade statistics is the United Nations' *Standard International Trade Classification*, currently in its third revision (SITC Rev.3). The headings of this scheme correlate directly with those of the CN, although the order in which commodities are grouped differs between the two schemes.

In 1992 and beyond, there are likely to be significant changes in the tariff codes. The CEC's statistical office would like to incorporate the current ninth digit within the eight-digit CN of the HS. This would lead to some loss of detail in EC trade statistics. A common classification scheme for both production and trade statistics will be sought, although any new codes will be within the framework of the present CN system. Rules on the suppression of data (used to ensure commercial confidentiality) are also under review.

It is important for anyone using statistical information to be aware of the different classification schemes and their revisions, particularly when using data from different countries over a period of time, to ensure that the information is compatible. As a simple example, some countries (e.g. Italy) include man-made fibres in their definition of the chemical industry while others (e.g. the UK) do not. When classification schemes are changed, correlation tables may be published. Another factor for the user of statistics to remember is that data, especially when cumulated, take time to compile. The latest published data for a particular topic may therefore be several years old.

For official statistical and economic data in the UK, an essential reference for researchers is the *Guide to Official Statistics, No. 5* (Central Statistical Office, HMSO, 1990), which has details on all official (and some non-official) sources of information in the UK. The *Index of Commodities, PO 1000* (Business Statistics Office, HMSO, 1990) has an alphabetical list of products, and identifies the source in which the data are published. A list of the addresses of the manufacturing units which submit information used to compile the Business Monitor series is available in the *UK Directory of Manufacturing Businesses, PO 1007* (Business Statistics Office, HMSO, 1991), although manufacturers may request that their details are not included.

Production and sales information for the UK chemical industry can be found in the Central Statistical Office's *Business Monitor* series of publications (HMSO). The most detailed information available appears in 15 quarterly publications in the *PQ25xx* series (e.g. *PQ2511, Inorganic Chemicals; PQ2512, Basic Organic Chemicals*). These are cumulated into an annual series *PA25x*, for instance *PA251, Basic Industrial Chemicals*. Owing to disclosure rules under the Statistics of Trade Act, data on the output of individual manufacturers are protected. This means that if there are only one or two producers of a particular product, the information is aggregated into a more general heading: for example, UK sales of toluene, cyclohexane, xylenes, propan-2-ol, and other products are published under a single heading in *PQ2512*. Information in the quarterly monitors includes sales in £m, and for some activity headings, tonnage figures, with annual data for the two most recent years. Some trade statistics are also included. For fertilizers, synthetic rubber, sulphuric acid and ethanol, monthly production data are available in the *Monthly Digest of Statistics* (HMSO). In its drive to reduce the administrative burden on industry, the UK government took the decision to abolish 10 of the 15 *Business Monitors* in the *PQ25xx* series at the end of 1989. These data will now only be available in a new annual series *PAS25xx*. The quarterly series no longer available include those for dyestuffs, pesticides, resins and plastics, soaps and detergents and others.

Trade data for the UK are considerably more detailed than production data. However, information may be suppressed, or rolled up into a more general heading, if there are only a small number of importers or exporters of a particular commodity. There are, at present, three main sources for UK trade statistics, all of which are available monthly. The *Overseas Trade Statistics of the United Kingdom, MM20* (HMSO) contains full details of total product trade, including chemicals, but only partial information at the commodity level. Data are published under the headings of SITC Rev.3. The *Guide to the Classification for Overseas Trade Statistics — Correlation Tables* (HM Customs & Excise, HMSO, 1988) gives correlations between the second and third revisions of the SITC. More detail is available in data compiled by HM Customs & Excise and now marketed by five appointed agents (a list of these agents is published in *Business Monitor MM20*). For each tariff heading, full country of origin or destination details are given: the information is also cumulated so that the December statistics also include the year-end total. Most of the agents can supply data in various formats — microfiche, paper reports, by CN at the eight-digit level or by SITC, and customers can specify to which headings they wish to subscribe. Data are also available according to port of entry to, or exit from, the UK. It is also possible to obtain the identity of UK importers by commodity, although this may not be at the eight-digit level. The most detail, for imports only, is found in the monthly *Special Chemical Import Statistics* (Special Chemical Statistics Ltd, Millbank, London), which disclose information on certain individual products at the nine-digit level. Information is given on volume, value and country whence consigned. Subscribers may have the information provided on diskette. The *Specials* originate from the *Key Import Duty* statistics. Import duties on certain chemical products, particularly fine organic chemicals, were imposed from the early 1920s, to protect UK chemical manufacturers from what was perceived as unfair competition, particularly from the German chemical industry. The future of the *Specials* is in some doubt because of the EC's wish to abolish reporting at the nine-digit level.

Apart from statistics on products, other data on the UK chemical industry are also published. Employment statistics can be found in the monthly *Employment Gazette* (HMSO), which lists numbers of employees in the various sectors of the industry. Details of employee earnings are published in the *Employment Gazette's* New Earnings Survey, and also in the semi-monthly Incomes Data Services (London) *IDS Report*. The RSC and the Institution of Chemical Engineers both publish the results of surveys on the salaries of their members. The ACS publishes annual surveys each August, *Salaries 199x*, and each November, *Starting Salaries of Chemists and Chemical Engineers 199x*.

The principal UK official statistics relevant to the UK chemical industry were summarized in the monthly journal *British Business* (HMSO), produced by the Department of Trade and Industry. *British Business* also published a quarterly article on the UK chemical industry, which provided a compendium of up-to-date data by sector. Some of the information was not published elsewhere. Unfortunately, *British Business* ceased publication in 1989, but it is still a useful source for historic data. In the UK, the Chemicals EDC of the National Economic Development Office used to undertake long-term forecasting, and published its results. *Chemicals: A Positive Future* (NEDO, London 1988), is a detailed study of the chemical industry, assessing strengths and weaknesses and future prospects. The Chemical EDC has now been closed, but the role of long-term forecasting for the industry has been taken up by the Chemical Industries Association (CIA), which has published its study *Chemical Industry Main Markets, an Assessment by Area and Product Group 1989–2000* (CIA, London, 1990).

International economic and statistical data for the chemical industry are published by the United Nations' Economic Commission for Europe (UNECE) in its *Annual Review of the Chemical Industry*. This contains brief comments on the general economic situation, and recent developments in production, consumption, prices and international trade in the chemical industries of individual countries are followed by general statistical data, on production, employment and trade. For selected major products, production and trade statistics for individual countries are listed. The UNECE also produces an *Annual Bulletin of Trade in Chemical Products*, giving data for trade in terms of both value and quantity, by main sectors of the industry. Economic data and some forecasting are also presented by the UNECE in *Market Trends for Selected Chemical Products 1960–1985 and Prospects to 1989*. The survey covers major industrial chemicals, and includes, for selected products, consumption and price data.

In a series of publications *Basic International Chemical Industry Statistics 1963–19xx*, the UK CIA publishes data for Western European countries, Japan and the USA. The statistics include turnover (sales), home sales and investment (in US dollars and local currencies), as well as employment figures and various ratios (e.g. sales/employee). The data are also available on diskette. Information is obtained from official sources and unofficial statistics are obtained from other national trade associations. Similar data are included in the ESCIMO *(European Statistics Chemical Industry Monitor)* database, coordinated by CEFIC.

CEFIC, the European Council of Chemical Industry Federations, is a pan-European organization, whose members comprise national chemical industry associations and some of the major chemical companies operating in Europe. ESCIMO is a series of linked online databases providing statistical information on the European chemical industry. Each country's database is operated and maintained by the national federation (for example, the CIA, VCI and UIC) which provides and updates national data, often from official sources. The online host for ESCIMO is Wharton Economic Forecasting Associates (WEFA), which developed the system software. Access is currently restricted to the national federations and their members. CEFIC also compiles its own statistics and publishes annual surveys on olefins and aromatics (production, consumption and capacity data for Western Europe) and energy consumption by the chemical industry.

A major online source of trade statistics on chemicals is TRADSTAT. This was originally developed by the European Petrochemical Association and Unilever Computing Services, and is now marketed exclusively by Data-Star. TRADSTAT contains monthly import/export statistics, in terms of value and quantity, of all chemical products and commodities reported by the national customs or statistics authorities of some 20 countries. Using a database like TRADSTAT it is possible to estimate detailed trade data for chemicals for countries which do not normally publish such data, by using data from their main trading partners.

Statistical data on the EC is published by the Statistical Office of the European Communities — Eurostat. Information is published in hard copy and many of the series are also available online (via the hosts GEISCO, Datacentralen, WEFA and Reuters) and also on diskette or magnetic tape. Production, trade, employment and other data relevant to the chemical industry such as energy prices are available, as well as the various statistical classifications and their correlations used by Eurostat. The *Eurostat Catalogue,* available from the EC's Publications Office in Luxembourg and its agents (HMSO in the UK), describes the series that are available and in what formats they are disseminated.

Chemical manufacturing plants

In addition to reports of plant construction, expansions, whether planned or underway, and completed projects, which appear in the trade press, there are services specifically providing this type of information. The CHEMICAL AGE PROJECT FILE, CAPF (Chemical Intelligence Services, London), online exclusively through Pergamon Financial Data Services, has information on chemical plants worldwide from 1980 to date, with some 18 000 records. Details include plants currently under construction, planned or at the feasibility stage, and those underway but presently suspended. In addition to the main sectors of the chemical industry, CAPF includes mineral, metal processing and textile plants and pulp mills. Information includes plant ownership and location, contractor and process licensor details, capacity, feedstocks, cost and status. CAS Registry Numbers are included. Sources for the database include periodicals and information provided by plant owners or contractors.

Similar sources are used to compile the CHEM-INTELL chemical plant database (Chemical Intelligence Services, London, formerly Chemical Data Services), which has about 17 000 records. Chemical Intelligence Services also have a statistical database with production and trade data. However, both databases are confined to providing information on about 120 major chemicals worldwide; the plant data include all known plants operating, planned or under construction, and also plants which are idle or have been recently closed. Each record on the plant database is similar to CAPF in content: geographic and economic area descriptors (e.g. OECD, Asia) are added along with trade names, although CAS Registry Numbers are not included. Both databases are available online (Data-Star, Pergamon, Dialog) and also as hard-copy reports. Chemical Intelligence Services also publish a series of *Chemfacts* books: either on specific products — e.g. *Chemfacts Ammonia* (1989) which has plant and statistical data on ammonia worldwide — or by country, giving information for each of the 120 chemicals.

Stanford Research Institute publish a series of directories with information on chemical producers. The *SRI Directory of Chemical Producers, Western Europe*, produced annually since 1978, includes data on more than 3000 Western European chemical manufacturing companies — ownership, addresses, plant locations and chemicals manufactured at each site. A cross-referenced list of 13 000 products gives producer names and plant locations and, for some products, plant capacities. Additional indexes of plant location by geographic region and company names, including abbreviations, common and former name, are also included. Subscribers to SRI's *Chemical Economics Handbook (CEH)* will shortly be able to access the data in *CEH* online through Dialog. *CEH* covers producer and capacity data by location, future demand, trade statistics, manufacturing processes, and price and other information. ICIS-LOR provides information on the operating status of ethylene crackers including capacities and dates of scheduled shutdowns.

CHAPTER THIRTEEN

Health and safety

S. PANTRY

The Control of Substances Hazardous to Health (COSHH) Regulations came into force in the UK on October 1, 1989, and imposed upon the management of organizations many new responsibilities towards their employees and the general public in respect of chemicals used in the workplace. Sources of information on health and safety therefore became especially important to chemists. The Toxic Substance Control Act (TSCA) in the USA has a similar importance.

Public libraries in the UK and abroad not only hold a significant amount of information themselves, but also are the key link to other sources of information. In addition to these library and information services the major libraries such as British Library Science Reference Information Service in London and also the British Library Document Supply Centre at Boston Spa, which is a huge warehouse of information, can be used via the public library system or via an organization's own library and information service. The Health and Safety Executive Library and Information Service, which has its headquarters in Sheffield with branches at London (Baynards House) and Bootle, Merseyside, for public enquiry service, has an extensive collection of abstracting and indexing periodicals in health and safety in all disciplines. Various parts of the HSE specialize in information such as occupational hygiene and medicine, explosives, explosions, harmful dusts and gases, etc.

There is a wealth of information available — arguably too much — but the various systems, services and publications described below should help to cut through the mountain of data and enable the practitioner to get hold of authoritative information reasonably quickly. Mention of any product or service in this chapter does not imply approval of it by the HSE.

Online and compact disk read-only memory services (CD-ROMs)

Searching for and keeping up to date with occupational health and safety information can be a problem. Using some of the world's computerized services described below will help to overcome some of the problems.

Many of the printed versions of catalogues, indexes and abstracts to the literature are also available on computers throughout the world. It is not now necessary to spend many hours travelling to sources of information, checking manually through periodicals, etc., or worse, not being able to access the information at all.

Commercially available online services have proliferated over the past decade making information more easily accessible. It is estimated that there are well over 4000 databases available for retrieval in the world covering many subjects and totalling over 100 million references. The following databases will give useful information. The addresses of many of the organizations that produce these files, and of the principal online hosts, are given in the final section of this chapter. The online and CD-ROM databases listed below are additional to the general chemical databases described in Chapter 5.

HSELINE (Health and Safety Executive)*

This file covers all aspects of health and safety associated with chemicals and the chemical engineering industry. It is updated monthly and is particularly useful for up-to-date information from worldwide sources, UK legislation and incidents information. There is no printed version. It is also available on the OSHROM CD-ROM.

The United Kingdom Health and Safety Executive Library and Information Service (HSE/LIS) database, HSELINE, was launched in December 1981 and is hosted on the European Space Agency Information Retrieval Service (ESA-IRS), Orbit and Data-Star. The latest development is CD-ROM. About 12 500 new references are added to HSELINE each year, with about 55 per cent taken from some 250 periodicals which are scanned on a regular basis. Currently, HSELINE contains about 150 000 references. These include all Health and Safety Executive publications as well as articles appearing in a wide range of national and international periodicals. Also included are books, conference proceedings, standard specifications, and decided cases, law reports and legislation in the UK and elsewhere relevant to health and safety at work. An important element is publication of the latest research results carried out by the Research Division of HSE and from many other countries. The COSHH Regulations require employers to assess the risk created by

---

*Databases with an asterisk are also available on CD-ROM.

work involving a hazardous substance, under Section 6 of the Health and Safety at Work, etc. Act 1974. All the information on handling chemicals in the workplace produced by HSE is indexed in HSELINE. HSE enjoys reciprocal agreements with similar organizations abroad and receives, on a regular basis, information from worldwide authoritative organizations which is indexed into HSELINE. HSELINE contains references to the work carried out by HSE's Working Group on Toxic Substances (WATCH). Single copies of these summaries can be obtained from HSE's enquiry points. The summaries contain details of the assessment of the chemical or substance.

HSE/LIS produces the *Publications in Series* twice per year in January and July. This lists all the priced and free publications. In addition, there are an increasing number of subject catalogues, e.g. COSHH, agriculture, chemicals, construction, and textiles, which list HSE publications which are industry related.

Copies of all HSE translations can be bought from HSE/LIS Translations Unit, Harpur Hill, Buxton, Derbyshire SK17 9LN, UK: this unit also produces a quarterly bulletin listing all new translations. All these translations are listed in HSELINE.

The addresses of the Health and Safety Executive's public enquiry points are as follows:

Health and Safety Executive, Library and Information Services,
Broad Lane, Sheffield S3 7HQ;
Telephone 0742 752539; Telex 54556; Fax 0742 755792.

Health and Safety Executive, Library and Information Services,
Baynards House, 1 Chepstow Place, Westbourne Grove,
London W2 4TF;
Telephone 071-221 0870; Telex 25683; Fax 071-727 2254

Health and Safety Executive, Library and Information Services,
St Hugh's House, Stanley Precinct, Trinity Road, Bootle,
Merseyside L20 3QY;
Telephone 051-951 4381; Telex 628235; Fax 051-922 5394

The list of HSE Area Offices is available from any of the public enquiry points.

BSI STANDARDLINE*

All British Standard specifications are listed in this database which is continually updated. It is especially useful to the chemist because it lists amongst the 12 500 standards about 500 or 600 which are particularly health and safety dominated. Printed versions of the database are the *BSI*

Catalogue, BSI News (monthly) and *BSI Sales Bulletin*. Also available on CD-ROM is PERINORM, which contains the DIN (German) and AFNOR (French) standards as well as BSI standards. It is available from the British Standards Institution at its Milton Keynes address.

CHEMICAL HAZARDS IN INDUSTRY **(Royal Society of Chemistry)**

This is updated monthly and covers all aspects of chemical hazards in industry. It is particularly useful for: chemical and hazard enquiries, including accident prevention, hazardous waste management, UK legislation. The printed version is *Chemical Hazards in Industry*.

CHEMICAL HAZARDS RESPONSE INFORMATION SYSTEM **(CHRIS) (US Department of Transportation, Coast Guard)***

The coverage is of chemicals for use in spill situations; there are 1000 references. It is particularly useful for details on nomenclature, molecular formulae and biological and fire hazard potential.

CISDOC **(Centre International d'Information de Sécurité et d'Hygiène du Travail, the International Labour Office, Health and Safety Centre, Geneva, Switzerland)**

This database embraces occupational hygiene, medicine, physiology, industrial toxicology, accident prevention and safety engineering. It contains 20 000 refe ences and is updated seven times per year. It is particularly useful for checking worldwide occupational health and safety references, especially legislation. The printed version is *CIS Abstracts*. It is also available on the OSHROM CD-ROM.

Environmental chemicals data and information network (ECDIN) (Commission of the European Communities)

This is a database on chemical substances of environmental importance, and is particularly useful for chemical structures, producers' plants, and health and safety including toxicological surveillance.

HAZARDOUS SUBSTANCES DATABANK

This covers 41 000 chemicals which are known to be potentially toxic, and is updated quarterly. It is particularly useful for identification, chemical and physical data, flammability, stability, explosive potential and emergency guidelines. There is no printed version.

HAZDATA FOR COSHH (National Chemical Emergency Centre)

This database covers 8000 product names covering more than 1100 chemicals, including those substances in the HSE Approved List and Guidance Note EH40. It is available from the National Chemical Emergency Centre.

LABORATORY HAZARDS BULLETIN (Royal Society of Chemistry)

This covers worldwide sources on hazards to be found in chemical and biochemical laboratories including handling of chemicals. It is updated monthly and contains 3500 references. Particularly useful for information on chemicals in laboratories and associated hazards, it also has a printed version.

OCCUPATIONAL SAFETY AND HEALTH (NIOSHTICS) (US National Institute for Occupational Safety and Health)

This is a bibliographic database; articles taken from several sources dating back to 1973 are entered into the database as well as important articles from the early nineteenth century onwards. The file is particularly useful for all aspects of the occupational safety and health field. It is also available on the OSHROM CD-ROM.

OHS MSDS (MATERIALS SAFETY DATA SHEETS)* ON DISC

This is a CD-ROM covering over 10 000 of the most frequently used chemicals. It is available from Occupational Health Services Inc., 450 Seventh Ave., Suite 2407, New York, NY 10123, USA.

RTECS (US National Institute for Occupational Safety and Health)

This file covers the toxic effects of chemical substances, contains 40 000 substances, and is updated quarterly. It is particularly useful for threshold limit values, recommended standards in air and aquatic toxicity levels, and toxicity effects. The printed version is the *Registry of Toxic Effects of Chemical Substances*, also available in microfiche, updated quarterly.

SAFETY SCIENCE ABSTRACTS (Cambridge Scientific Abstracts)

This is a bibliographic file covering safety science literature, and it is particularly useful for chemical hazards, pesticides, radiation, pollution and waste disposal systems. The printed version is *Safety Science Abstracts Journal*, published ten times per year.

236 Health and safety

SIGMA ALDRICH MATERIALS SAFETY DATA SHEETS

This database covers over 24 000 Sigma Aldrich data sheets containing chemical, physical and structural data and literature information on over 50 000 products. It is particularly useful for checking details of Sigma Aldrich products. The printed version is the *Sigma Aldrich Materials Safety Data Sheets*, available from Aldrich Chemical Co. Ltd, The Old Brickyard, New Road, Gillingham, Dorset SP8 4JL; telephone 0747 62211.

TSCA INITIAL INVENTORY (US Office of Toxic Substances)

This is a non-bibliographic dictionary listing chemical substances in commercial use in the USA. It is not a list of toxic chemicals. It contains 56 000 references, and is particularly useful for Chemical Abstracts Registry Numbers, preferred names, synonyms and molecular formulae. It is derived from the printed *Toxic Substances Control Act Chemical Substances Initial Inventory*.

Chemical and toxicological information from encyclopaedias, dictionaries, etc.

In addition to general reference books listed in Chapter 4, there are many other publications which are essential aids in trying to define a hazardous substance. Some titles are as follows.

Hazards in the Chemical Laboratory (4th edn), edited by L. Bretherick, is published by The Royal Society of Chemistry (1988). This book is easy to read and includes some good advice. The same author's *Handbook of Reactive Chemicals Hazards* (Butterworths, 1986) is an essential for those seeking information on laboratory planning, fire protection, reactive chemical hazards, toxicology, radiation, medical services and first-aid requirements.

Patty's Industrial Hygiene and Toxicology, by G.D. Clayton and F.E. Clayton, is published by John Wiley in a number of volumes. Volumes 2A, 2B and 2C are especially useful on toxicology data.

Croner Publications Ltd, Surrey, have a number of publications: *Dangerous Substances*, a loose-leaf reference book covering the legal requirements of handling dangerous substances, which has a bimonthly amendment service; *Dangerous Chemicals Emergency Guides*; and *Substances Hazardous to Health*.

Managing Safety in the Chemical Laboratory, by J.P. Dux and R.F. Stalzer, is published by Van Nostrand Reinhold.

Casarett and Doull's *Toxicology: The Basic Sciences of Poisons*, by John Doull and others, is published by Macmillan Publishing Company (1986) and is regarded as a good basic reference book.

Health and Safety Executive publications

The Health and Safety Executive Library and Information Services issue twice per year in January and July a full list of documents. It is called *Publications in Series* and is available free of charge from any of the HSE's enquiry points. There is also a subject catalogue entitled *Information Relevant to COSHH* and one entitled *Chemicals*, again free of charge from HSE's enquiry points.

Health and Safety Executive Guidance Note *Environmental Hygiene EH 40 Occupational Exposure Limits* is published annually by HMSO. Note that this is updated twice per annum by *Toxic Substances Bulletin* available on subscription from HSE Sales Point, Bootle. It should be noted that HSE has broken away from the US threshold limit values.

US National Institute of Occupational Safety and Health Publications

A number of publications are produced by the US National Institute of Occupational Safety and Health. The following will be of use to chemists.

NIOSH/OSHA Pocket Guide to Chemical Hazards, edited by F.W. Mackison and others, is a source of summary information for those who need a quick reference. It does not attempt to present exhaustively all pertinent data, but covers approximately 400 chemical hazards.

NIOSH and the US Occupational Safety and Health Administration, *Occupational Health Guidelines for Chemical Hazards* (1981).

NIOSH *Registry of Toxic Effects of Chemical Substances* (RTECS), published by the US Government Printing Office, is now in a number of volumes; it is difficult to use but does also refer to articles describing the lowest toxic dose or concentration of a substance, and gives over 200000 listings of chemical substances including synonyms. RTECS is updated by microfiche and is also available as a computerized database and on CD-ROM.

Handbook of Laboratory Health and Safety Measures, by S.B. Pal, is published by MTP Press, Lancaster, UK (1985), and covers electrical safety, hospitals, health care, patients, radiation protection and the clinical cytogenetics laboratory.

238 Health and safety

Health and Safety in the Chemical Laboratory — Where Do We Go from Here?, Special Publication no. 51 of The Royal Society of Chemistry (1984), provides an overview of health and safety developments in the chemical laboratory and workplace.

Dangerous Properties of Industrial Materials, by N.I. Sax and others, seventh edition, published by Van Nostrand Reinhold (1988), has concise data on thousands of common industrial and laboratory materials. The publishers are seeking to make this very important publication available in a more easily updated form. There is now a periodical complementary to the book, entitled *Dangerous Properties Materials Report*, which updates the information. Each issue has a number of articles plus a 'hazards material' list.

Rapid Guide to Hazardous Chemicals in the Workplace by N. Irving, now edited by N.I. Sax and R.J. Lewis and published by Van Nostrand Reinhold (1986), covers critical hazard information on chemicals, gives acute immediate effects and chronic long-term effects, as well as flammable and explosive properties, incompatibilities and instabilities.

Industrial Toxicology Safety and Health Applications in the Workplace, by P.L. Williams and J.L. Burson, is published by the American Conference of Governmental Industrial Hygienists (1989). See also *Elsevier Monographs on Toxic Agents* and Albert's *Selective Toxicity* in Chapter 10.

Audiovisual Resources in Occupational Health and Safety: Films, Videos and Tape–slides available from Distributors in the United Kingdom, 2nd edn, 1989 (ISBN 0 7176 0334 2) is available from HSE at £6.50.

Hazardous Materials: Sources of Information on their Transportation (N. Lees, BLSRIS, 1990) contains 259 entries from official sources and journals. Brief overviews of the international situation are given, as well as information on packaging and legislation in specific European countries.

POHSO Handbook is published by the Co-ordinating Committee for Professional Occupational Health and Safety Organisations, c/o BOHS, 1 St Andrew's Place, Regents Park, London NW1 4LB.

NAMAS Directory is produced by the National Measurement Accreditation Service; telephone 071-943 7112.

The Health and Safety Directory is published annually by Kluwer Publishing, 1 Harlequin Avenue, Great West Road, Brentford TW8 9EW.

Health and Safety at Work magazine publishes *Directories of Training Organisations and Consultants* annually.

Legal information

Working with substances hazardous to health will require a knowledge of the appropriate legislation. The Health and Safety Executive Library and Information Service (HSE/LIS) can be contacted for a list of all UK legislation in force, both before and since the Health and Safety at Work, etc. Act 1974, including the COSHH regulations. There are a number of publications which annotate the legislation. All these items are indexed in HSELINE. The following publications will also be of use.

Redgrave's Health and Safety in Factories, by I. Fife and E.A. Machin, is published by Butterworths, and a new edition was published in February 1990. *Health and Safety at Work*, published by Croner Publications Ltd, is a loose-leaf book updated every two months which gives a comprehensive summary of all the acts and legal requirements. In addition the book gives a thorough guide to how the health and safety law is administered in the UK.

Law on Health and Safety at Work: Law and Practice, edited by P. Allsop and published by Sweet and Maxwell, is continuously updated.

Croner issues two publications, *Dangerous Substances* and *Substances Hazardous to Health*, which will keep the health and safety practitioner updated with any changes in legislative requirements.

Translations

As searches are carried out for chemicals and hazardous substances information, some of the references may well be in foreign languages but may have been translated into English. The Health and Safety Executive Library and Information Service is a prolific source of translations, producing about 700 per year. Copies of these translations are all deposited with the British Library Document Supply Centre (BLDSC) at Boston Spa, Wetherby, West Yorkshire. Remember that many organizations need to have translations made of articles in periodicals, reports and even chapters of books. The translations unit of the HSE was mentioned earlier.

Journals in the health and safety field

There are a number of periodicals available in the health and safety field which will keep the practitioner up to date. Some of these are indexed on a regular basis by the HSE Library and Information Service into HSELINE. A list of journals indexed is available free of charge from HSE's public enquiry points. Amongst the chemical journals which the information seeker should be aware of are *Chemical Hazards in Industry* and *Laboratory Hazards Bulletin*, both published by the Royal Society of Chemistry, and containing a substantial amount of information of

240 *Health and safety*

Table 13.1 Software packages for health and safety information

Name	Scope	Supplier
ACCSYS	The software consists of four major sections: (1) accident recording incorporating accident report maintenance and accident diary maintenance; (2) reporting and analysis; (3) archive and restore facility; (4) utilities for altering the parameter table files. This package is also available as a permanently installed graphics facility for those who wish to produce high quality graphs and charts.	Financial Data Services Ltd, Seymer Road, Romford, Essex RM1 4LA; telephone 0708 752048.
BNFL chemicals database system	The package enables the user to input suppliers' health and safety details and to add data from other sources. It covers chemical properties, CPL labelling, fire and first aid, and is promoted as useful for COSHH assessments.	Mr J. Watson, British Nuclear Fuels plc, Corporate Management Services, Allday House, Risley, Warrington, Cheshire WA3 6AS; telephone 0925 832377
CAMAXYS — Camhealth COSHH Management System	This is a computerized management system. Camhealth is promoted as being able to help employers cope with the introduction and operation of COSHH. The stated object of Camhealth is 'to assist managers to comply with the COSHH regulations speedily, with a minimum of administrative effort'. By using the system, managers in occupational health and occupational hygiene and safety specialists who advise them would find the following activities — carrying out assessments, maintaining statutory records, seven modules containing known hazards database, work activities database, record of current assessments, a record of employees and their training, exposure monitoring records, and a record of all control measures and their maintenance and health surveillance record.	CAMAXYS Ltd, Baggator, Station Road, Harston, Cambridge CB2 5NY; telephone 0223 872420; fax 0223 872596.

Table 13.1 — *continued*

Name	Scope	Supplier
Cascade	This is an accident analysis system for workplace diseases and accidents; it can record and analyse accidents by location as well as type, and is suitable for organizations employing more than 500 people.	Eddie Trotter, Sedgwick Risk Control Services, Sedgwick House, The Sedgwick Centre, London E1 8DX; telephone 071-337 3247.
COSHH Manager System	This package is designed to store a substances inventory and data, and then provide the means to manage their use and keep essential records.	Applied Environmental Technology Ltd, Station Approach, Trent Valley Road, Lichfield, Staffs, WS13 6HE; telephone 0543 416550, fax 0543 415459.
Environmental Health Computer System (EHCS)	A comprehensive range of environmental health activities is covered by EHCS.	LAMSAC, Vincent House, Vincent Square, London SW1P 2NB; telephone 071-828 2333.
Flow Gemini	This program covers environmental information and occupational health information.	General Research Corporation, Woodcock Hill, Harefield Road, Rickmansworth, Herts. WD3 1PQ; telephone 0923 774666.
Hazdata for COSHH	The initial release of Hazdata contains information on over 8000 product names covering more than 1100 chemicals, including those substances in the HSE Approved List and EH40 publications. The database can be searched on product name/synonyms, CAS, EINECS, ARTECS and SI (UN) Numbers.	National Chemical Emergency Centre, B7.22 Harwell Laboratory, United Kingdom Atomic Energy Authority, Oxfordshire OX11 0RA; telephone 0235 432919; telex 83135; fax 0235 832591.
COSHH Companion	This is an interactive computer-based training and student-management programme which runs on any IBM or true compatible PC with hard disk, colour monitor and MicroSoft mouse.	Eurotech, Oakfield Road, East Wittering, Chichester, West Sussex PO20 8RP; telephone 0243 672891; fax 0243 672031.
COSHH Database		GKN Occupational Health, Washford House, Claybrook Drive, Redditch, Worcestershire; telephone 0527 517747.

Table 13.1 — *continued*

Name	Scope	Supplier
COSHH Datafax	'Records assessment details, automatic reminders for assessment updating, materials safety data sheets, maintains exposure records, ventilation plant examination, records of individuals, etc.'	Applied Environmental Technology Ltd, Station Approach, Trent Valley Road, Lichfield, Staffs, WS13 6HE; telephone 0543 416550, fax 0543 415459.
COSHH-DM	COSHH-DM can now be supplied with a database of substances as detailed in the Health and Safety Guidance Note EH40/90. The database contains data on over 550 substances and over 250 associated synonyms.	British Nuclear Fuels plc, Corporate Management Services, Allday House, Risley, Warrington, Cheshire WA3 6AS; telephone 0925 832377.
COSHH Administrator	This system maintains records of areas, processes, substances, personnel, training and equipment. 'Records are kept in a database and the programs allow for entry of information on clear and easy to use input screens, and then reporting on the information to show that the regulations are being complied with.'	CogniSoft Ltd, Carrington Business Park, Manchester Road, Carrington, Manchester M31 4DD; telephone 061-776 4239.
COSHH Manager	This program is designed to find information and automate cross-referencing; it uses artificial intelligence to aid searching. It requires minimal training, and provides risk assessment and cost effective.	Systematic Upgrade, 58–60 Edward Road, New Barnet, Herts. EN4 8AZ; telephone 081-449 9699; fax 081-441 2297.
COSHH Management Systems		OHS (Occupational Hygiene Consultants), 15–17 Campus Road, Bradford, West Yorkshire BD7 1HR; telephone 0274 735848.

Table 13.1 — continued

Name	Scope	Supplier
OH Programs	A number of programs in BASIC is available for statistics and hygiene.	H & H Consultants Ltd, 28 High Ash Drive, Leeds LS17 8RA; telephone 0532 687189.
Safechem II	Data are given for 1000 of the most common chemicals. Material safety data sheets can be generated from the stored information.	Sergeant Safety Supplies, P.O. Box 78, Kegworth, Derby DE7 2DN; telephone 0509 672656.
Safety Information System	This system is designed 'to assist with the implementation of COSHH and includes an extensive range of reporting facilities, e.g. hazard data sheets, health register, summaries of substances used.'	Royal Society for the Prevention of Accidents, Cannon House, The Priory Queensway, Birmingham B4 6BS; telephone 021-233 2461.
Staffware	'The system will help with COSHH record keeping. It requires little computer knowledge, and is easy to complete and manipulate. It uses free-text fields, and is designed for smaller business and local authorities. It is inexpensive.'	Staffordshire County Council, 15 Tipping Street, Stafford; 0785 223121; fax 0785 215153.
Linked to COSHH	This program is designed to provide information on chemicals and processes. Its 'Anteta Base' system uses a Psion Organiser II computer to provide employees with information on materials and processes at their place of work.	Advance New Technology; contact John Brooking, telephone 0342 824909.
Micro COSHH	This is an occupational hygiene data management system, which 'allows users to access and create chemical hazard data sheets, run a diary function, and record and analyse assessments.'	Bowring Protection Consultants, The Bowring Building, Tower Place, London EC3P 3B3; telephone 071-283 3100.

Table 13.1 — *continued*

Name	Scope	Supplier
Microsafe	This program stores a wide range of data on incidents (including personal injury and damage records) and analyses the information in a variety of ways. It displays the results as tables or in graphical form; it is compatible with many commercially-available software packages.	Bowring Protection Consultants (see Micro COSHH).
O'Heal Occupational Health Management System	This is a series of files covering site details, material safety data sheets, exposures, etc.	O'Heal (Occupational Hygiene Engineering Associates Ltd), 23 Highcroft Estate, Horndean, Hants, PO8 0BT; telephone 0705 594573.

interest to the chemist. Other journals such as the *Journal of Hazardous Materials, CA Selects (Chemical Hazards and Health and Safety), The Health and Safety Commission Newsletter, Occupational Safety and Health Safety Management, Safety and Health Practitioner, Health and Safety Information Bulletin, Occupational Health* and *Journal of Health and Safety* will also be useful.

Software for occupational health and safety

Since the advent of COSHH, many software houses have produced packages which run on industry-standard IBM-compatible office PCs and assist management in complying with the regulations.

Table 13.1 gives only a sample of the software packages available; others are described by J.F.J. Kibblewhite in *Microcomputer Applications in Safety Management*, ISBN 0951400002, published by the author (1988).

The HSE does not give approval to products and services. The data in Table 13.1 are for information only.

Organizations

In addition to the Health and Safety Executive, the following UK organizations may well be of help to the practitioner seeking health and safety information.

British Agrochemicals Association (Secretary: Miss Anne H. Buckenham, 4 Lincoln Court, Lincoln Road, Peterborough PE1 2RP) is a trade association. It promotes the responsible and safe manufacture and use of agrochemicals with due regard for the interest of the community and the environment. It publishes *Annual Report and Handbook*, including industry statistics.

British Industrial Biological Research Association (Woodmansterne Road, Carshalton, Surrey SM5 4DS; telephone 081-643 4411) is an independent non-profit-making research organization supported by its membership. BIBRA was founded in 1960 to promote research and other scientific work in toxicology, and has a large programme of research related to the fundamental problems of toxicology, much of it funded by the UK Government. With this strong research base, BIBRA offers industry advisory, safety evaluation and investigative research services. Member companies, who pay an annual subscription, also have access to an information and consultancy service. Occupying a position between government and industry, BIBRA plays an important role in maintaining communication between these two parties on all toxicological matters, particularly through informal closed meetings of the BIBRA Group Forum system. BIBRA acts as a centre for training in toxicology, organizing courses ranging from introductory to postgraduate specialist education for industrial and government personnel. BIBRA also contributes to a number of external university-based courses. Its publications include: *Food and Chemical Technology; Toxicology* in vitro; *Pesticides Screening for Safety* (booklet); *BIBRA Bulletin; Newsletter; Toxicity Profiles; Annual Report; Brochure*; and scientific papers.

British Occupational Hygiene Society (1 St Andrew's Place, Regents Park, London NW1 4LB; telephone 071-486 4860). This society brings together members from many professions working in the field of occupational hygiene. BOHS publications include the journal *Annals of Occupational Hygiene* as well as other guides.

British Standards Institution (Head office: 2 Park Street, London W1A 2BS; telephone 071-629 9000. Sales: Linford Wood, Milton Keynes MK14 6LE; telephone 0908 320033; and 3 Park Street, Manchester M2 2AT; telephone 061-832 3731/4) is an independent, non-profit-making body operating under a Royal Charter. BSI is the nationally recognized organization for preparing standards which are used in all industries and technologies. BSI represents the views of UK industry in discussions on standards with international bodies such as the International Organization for Standardization (ISO) and the International Electrotechnical Commission (IEC), and also CEN/CENELEC, the Joint Standards Institution which works towards harmonization of standards requirements. Technical Help to Exporters (THE) is a unique export advisory service which exists to give guidance to UK companies seeking to sell their products abroad. Publications: British Standard Specifications covering all subjects; *BSI Catalogue* (annual); *BSI Sales Bulletin* (six issues per year); *BSI News* (monthly); *BSI Buyers' Guide*; sectional lists.

Chemical Industries Association Ltd (Kings Buildings, Smith Square, London SW1P 3JJ; telephone 071-834 3399) is an association of chemical manufacturers and processors. The CIA promotes co-operation within the chemical industry and negotiates with government and other bodies on problems in the economic, social, commercial, labour and technical areas affecting members. Its publications fall into various categories: general publications; health, safety and environment; industrial relations; and the distribution industry. The health and safety publications highlight the major issues, e.g. guidelines for chemicals with major hazards, safety audits and storage of highly flammable liquids. CIA produces a publication price list.

Institute of Occupational Hygiene (University of Birmingham, P.O. Box 363, University Road West, Edgbaston, Birmingham B15 2TT; telephone 021-414 6030) is partly self-financing, partly university funded. Its principal functions and activities are: (1) teaching — mainly postgraduate doctors, nurses, safety managers and some undergraduates; (2) research — medical, hygiene and toxicological; (3) consultancy for industry — industrial surveys, help and advice on hygiene, medical and toxicological problems. Its publications include: *Annual Report; Pocket Consultant in Occupational Health*, by J.M. Harrington and F.S. Gill; *Monitoring for Health Hazards at Work*, by F.S. Gill and I. Ashton; various research papers.

Institute of Occupational Medicine (8 Roxburgh Place, Edinburgh EH8 9SU; telephone 031-667 5131, 8 lines) is a multi-disciplinary research centre with extensive experience and expertise in epidemiological studies of disease relationships in many industries; it provides a medical consultancy and advisory service on work-related problems, occupational hygiene and environmental analysis, and carries out ergonomic evaluations and basic research in pathology, physics and chemistry. Publications: booklets listing the publications arising from this research are available from the librarian.

Institution of Occupational Safety and Health (222 Uppingham Road, Leicester LE5 0QG; telephone 0533 768424) is an independent body, formerly the Institution of Industrial Safety Officers. Its principal functions and activities are to improve and raise the standards, status, and technical and general knowledge of persons engaged in the provision of a professional health and safety service at their places of work, to facilitate the exchange of information and ideas among members and others, and by successfully pursuing the above objectives to assist in collective efforts to improve the quality of the workplace environment. It works through branches which hold regular meetings. Its publication is *Safety Practitioner* (monthly).

London Hazards Centre Ltd (Polytechnic of the South Bank, 103 Borough Road, London SE1 0AA; telephone 071-261 9558) is a voluntary organization, funded until 1985–86 by the Greater London Council. It is concerned with the provision of advice and information of health and safety hazards to workplace and community groups, in London and elsewhere. Publications: *The Daily Hazard* (bi-monthly newsletter); *Hazards Information Packs* on particular topics.

National Chemical Emergency Centre (B7.22 Harwell Laboratory, UKAEA, Oxfordshire OX11 0RA; telephone 0235 432919; fax 0235 832591.

National Radiological Protection Board (Chilton, Didcot, Oxfordshire OX11 0RQ; telephone 0235 831600) was created by the Radiological Protection Act 1970, to establish a national point of authoritative reference in radiological protection for the UK. The Act came into effect on 1 October 1970. By direction of the Health Ministers, the Board assumed responsibility on 1 April 1971 for the Radiological Protection Service (hitherto provided jointly by the Medical Research Council and the Health Departments) and for the activities in the field of radiological protection hitherto carried on by the central Health and Safety Branch of the United Kingdom Atomic Energy Authority. The Board's statutory functions are: (a) by means of research and otherwise, to advance the acquisition of knowledge about the protection of mankind from radiation hazards; and (b) to provide information and advice to persons (including government departments) with responsibilities in the UK in relation to the protection from radiation hazards of either the community as a whole or particular sections of it. Publications: three-yearly reports; report series; miscellaneous publications. A list of publications is available on request.

Robens Institute of Industrial and Environmental Health and Safety (University of Surrey, Guildford, Surrey GU2 5XH; telephone 0483 509203 ext. 439). The Institute's principal functions and activities are to conduct research into areas of health and safety in which there are valid needs, to provide information, advisory and investigative services for organizations with health and safety problems, to be a training centre for those with particular involvement with health and safety, and to operate a membership scheme (RISC) incorporating all Robens Institute services plus a separate occupational health service. Publications: *Prospectus, Annual Report*, scientific papers, separate brochures on each unit and service, RISC prospectus and regular newsletter.

Royal College of Nursing of the United Kingdom (20 Cavendish Square, London W1M 0AB; telephone 071-409 3333) includes a trade union, a professional organization, the Institute of Advanced Nursing Education and a charity, incorporated by Royal Charter.

Royal Society of Chemistry (Burlington House, Piccadilly, London W1V 0BN; telephone 071-437 8656; and Thomas Graham House, Science Park, Milton Road, Cambridge CB4 4WF; telephone 0223 420066 ; fax 0223 423623) is a learned society and the professional qualifying body for chemists in the UK. In addition to many other activities the RSC provides advice and information on chemical aspects of health, safety and environmental matters by means of

Health and safety 249

symposia, publications, and computer-based information services. Its activities are organized by various specialist subject groups and by a Health, Safety and Environment Committee. Submissions are made to government and other bodies on proposals for new legislation. The society operates a Link Scheme whereby about 30 members are linked with individual Members of Parliament. This scheme allows for a two-way flow of views and information on health, safety and environmental matters. The Society has undertaken a number of studies on behalf of the Commission of the European Communities. These include reviews of the toxicity of a number of commonly used solvents (with recommendations for occupational exposure limits), and a long-term study of the morbidity and mortality of chemists. Publications: Two abstract publications on health and safety — *Chemical Hazards in Industry (CHI)* and *Laboratory Hazards Bulletin (LHB)*; both are available online through Orbit, Data-Star, STN and ESA-IRS. Other publications include the proceedings of symposia and meetings, the *Guide to Safe Practices in Chemical Laboratories* and the widely used book *Hazards in the Chemical Laboratory*. There are discounted publications prices for members. The Society's library at Burlington House holds a wide range of books and journals on chemistry and related subjects.

The addresses of the major online host services are as follows.

Data-Star, Plaza Suite, 114 Jermyn Street, London SW1Y 6HJ; telephone 071-930 5503; telex 94012671; fax 071-930 2581.

Dialog Information Services, P.O. Box 188, Oxford OX1 5AX; telephone 0865 730275; telex 837704 INFORM G; fax 0865 736354.

ESA-IRS Dialtech, 25 Southampton Buildings, London WC2A 1AW; telephone 071-323 7951; fax 071-323 7954.

Orbit, Achilles House, Western Avenue, London W3 0UA; telephone 081-992 3456; telex 8814614; fax 081-993 7335.

STN, Royal Society of Chemistry, Thomas Graham House, Science Park, Milton Road, CB4 4WF; telephone 0223 420066; fax 0223 423623.

CHAPTER FOURTEEN

The pharmaceutical industry

M. ARCHER, A. MULLEN AND W.R. PICKERING

Introduction

The pharmaceutical industry is essentially research-based despite the upsurge in generic (i.e. mainly out-of-patent) manufacture in recent years. In order to retrieve costs and generate profit, industrial research is increasingly market-orientated and related to ability to pay. Accordingly, considerable effort is placed into market evaluation and development. Despite the undoubted benefits of modern medicines, pharmacologically-active substances are potentially hazardous and the testing and use of drugs is subject to control in most countries. In fact, regulatory requirements designed to ensure safe and effective medicines largely determine the information that is generated about drugs and their usage. However, even the most rigorous testing can fail to reveal inadequacies and unwanted characteristics, and thorough post-marketing surveillance continues to add to the available knowledge. The modus operandi of the industry and associated information requirements are discussed at length in the companion volume *Information Sources in Pharmaceuticals* (W.R. Pickering, ed., Bowker-Saur, 1990) to which the reader is referred. Chemical information in its various guises is crucial to the main processes of discovery, development and production and influences the outcomes of these processes. Early on, an enterprise is concerned with identifying potential products and with safeguarding its intellectual property through the patents system (see Chapter 11). As work progresses, product licensing and launch become the principal targets until, ultimately, product improvements and extensions take over.

The role of the chemist in pharmaceutical research is central. The challenge is illustrated by a quoted success rate of one successful drug for every 10 000 test compounds synthesized and an estimated 12 years (DiMasi, 1990) to develop an ethical drug to market launch.

When a new chemical entity has been identified as having potentially valuable pharmacological properties it may be selected for further devel-

252 The pharmaceutical industry

opment. Development comprises the interrelated avenues of work involved in progressing from research through to production (Figure 14.1). In process development, the research synthesis of the active agent is scaled-up to give a safe, commercially viable process. Analytical methods are required for the product, its impurities and decomposition products during both the process and the pharmaceutical development phases. Pharmaceutical development leads to a dosage form acceptable to the patient, the physician, the manufacturer and the regulatory authorities. If successful, the project will culminate in the production process, which will be subject to regulatory compliance, quality control, packaging, engineering, safety and environmental control.

Figure 14.1
The work associated with pharmaceutical development and production

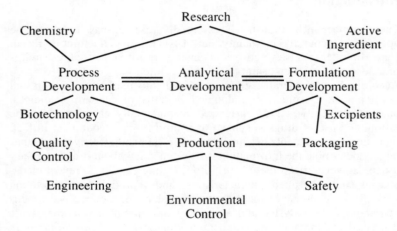

Leading pharmaceutical companies invest an average of 14.4 per cent of their turnover in R & D (SCRIP, 1989) and this proportion is growing. Not surprisingly, most concentrate their effort in defined therapeutic areas, as it is too costly, as well as ineffective, to attempt to be present in every field. The focus of R & D investment in the USA in 1988 (SCRIP, 1990*b*) is shown in Figure 14.2. R & D investment is also spread over the different phases involved in development (Figure 14.3). Synthesis and/or extraction of biologically active substances accounted for 10.3 per cent of the total outlay in 1988 (SCRIP, 1990*b*) whereas process development and quality assurance, both with a marked chemical content, account for a further 9.1 per cent. Thus, the chemistry-related contribution to pharmaceutical R & D is substantial. On average, some 50 New Chemical Entities (NCEs) are launched as ethical drugs each year

The pharmaceutical industry 253

(Figure 14.4). These products stem mainly from advanced industrial nations and, as in other fields of technology, Japanese companies are emerging strongly.

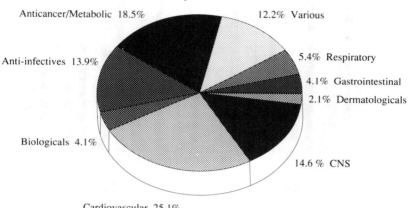

Figure 14.2 Distribution of R&D investment (USA)
Source: *PMA Annual Survey Report 1988–90*
Total Outlay = $5.05 billion

Anticancer/Metabolic 18.5%
12.2% Various
Anti-infectives 13.9%
5.4% Respiratory
4.1% Gastrointestinal
2.1% Dermatologicals
Biologicals 4.1%
14.6 % CNS
Cardiovascular 25.1%

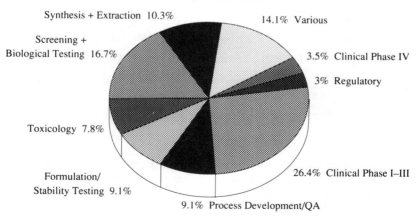

Figure 14.3 R&D investment according to development phase in USA
Source: *PMA Annual Survey Report 1988–90*
Total Outlay = $5.05 billion

Synthesis + Extraction 10.3%
14.1% Various
Screening + Biological Testing 16.7%
3.5% Clinical Phase IV
3% Regulatory
Toxicology 7.8%
Formulation/Stability Testing 9.1%
26.4% Clinical Phase I–III
9.1% Process Development/QA

Figure 14.4 Sources of NCEs 1982–90.
Source: *Scrip Review Issues 1984–90*

Year:	1982	1983	1984	1985	1986	1987	1988	1989	1990
NCEs:	46	49	43	57	47	61	53	35	43

■ Europe ☐ USA ▨ Japan ▨ Various

The key role of the chemist in several aspects of this large and complex industry is reflected in the evolution of sophisticated and comprehensive information systems designed with the medicinal chemist very much in mind.

Discovery of active agents

Nature of research
A new project can derive from several considerations:

  Increased incidence of a particular disease, e.g. AIDS

  New pharmacological rationale regarding biological mechanisms involved in a particular disease

  Changing demographic patterns in certain countries

  Need for a new or better therapy for a given disease

Basic pharmaceutical research is generally conducted by an interdisciplinary team using a wide range of computational techniques to study the structures of pharmaceutical molecules and their receptors or their biological target molecules if these are known and are available. These computational techniques comprise molecular/macromolecular modelling (including high-quality computer graphics and visualization techniques), conformational analysis, molecular orbital theory, structure–activity relationships and so on. Proprietary and commercial software are employed in drug design as well as experimental techniques (NMR, X-ray analysis) for structure elucidation (see also Chapter 6).

Because innovative research is more likely to lead to clinical advantages and thus to better ethical drugs, most organizations have invested in computerized information systems to husband their own proprietary data (structures of test compounds, chemical reaction files, biological test results, reports) as well as to access published results from external information sources.

Ideas and concepts in pharmaceutical research

The principal established routes to NCE are as follows:

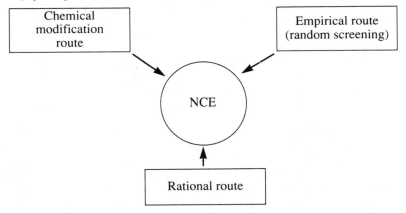

However, creativity and intuition — important components of the drug development process — are still highly prized attributes in the medicinal chemist.

Molecular modelling and QSAR studies
Molecular modelling (Marrer and Brueggmann, 1989) and quantitative structure–activity relationship (QSAR) studies (Hansch, 1989; Dunn, 1990) are devoted to investigating the pharmacological activity of a substance and its relationship to physical and chemical properties. Consequently, knowledge of these molecular properties is indispensable to rational drug design which, in recent times, has largely replaced large-scale random screening of test compounds.

A relatively new and promising tool for molecular design and discovery is three-dimensional (3D) searching of databases of molecular structures, thereby retrieving molecules containing a particular pharmacophore, i.e. a 3D arrangement of atoms which is considered to be responsible for a molecule evoking a particular biological response (Krieger, 1990).

In line with this development, a leading chemical software company — Molecular Design Ltd (MDL) of the USA — has upgraded its MACCS II software (for the graphical storage and retrieval of chemical structures) to MACCS 3D.

Crystallography and sequence files
Crystallography files. The Cambridge Structural Database contains published crystallographic data on some 70 000 smaller organic compounds. The file is generally used on in-house computer systems as it is currently not available via any major online host. This database, along with other related files (Mullen, 1990), is an important source of high-quality experimental 3D structural data, stemming mainly from X-ray and neutron diffraction studies.

The Brookhaven Protein Data Bank makes available for 3D studies the atomic coordinates of over 1000 proteins and nucleic acids originating from single-crystal X-ray or neutron diffraction studies.

There are computer programs available, such as Concord (Ruskino *et al.*, 1989), which use various techniques for fragments and energy minimization procedures to generate 3D models and coordinates of smaller organic molecules from their two-dimensional representation. The CA Registry File contains the Concord coordinates of over four million substances for use in molecular modelling studies. These coordinates can be downloaded using, for example, the STN Express software and exported to suitable molecular modelling programs such as Sybyl, Nitro and Alchemy II from Tripos Associates (see Chapter 6).

One of the roles for chemists in pharmaceutical research is to develop pharmacophore hypotheses for 3D-database searches and to thereby contribute to the development of 3D-structure–activity relationships and improve the pharmacological properties of lead compounds or, at least, to develop guidelines for the selection of the most promising candidates amongst compounds as yet unsynthesized.

Sequence files. Over recent decades, major advances have been made in developing procedures for determining the sequence of the building blocks — amino acids and nucleotides — in proteins and nucleic acids. It soon became apparent that these data could only be effectively manipulated by using computers. Reviews of the subject have recently appeared (Doolittle, 1990; Barker *et al.*, 1990).

In Europe and the USA, the EMBL Data Library and GENBANK project were established in 1980 and 1982, respectively, with the objective of collating all known biological sequences and making data available to scientists in a suitable computerized form.

In-house sequence files. In a review of this nature, only an overview can be presented of the available in-house files dealing with sequences relating to nucleic acids and proteins (see Table 14.1). The files listed are representative of the information sources now available to R & D scientists working in the molecular biology area. Retrieval from these specialist files is generally conducted by the scientists themselves.

Table 14.1. Information sources for R&D scientists

File	Version	Number of sequences	Number of residues	Supplier
Nucleic acids				
EMBL	24.0	37 784	47 354 438	EMBL
GENBANK	65.0	39 533	49 179 285	IntelliGenetics
HIVN	89.1	42	241 988	G. Myers
LOSNUC	62.0	16 650	16 238 363	NRBF
NBRF NUCLEIC	36.0	3 355	8 128 496	NBRF
PROMEGA		22	65 840	Promega Corp.
VECBASE	4.0	140	579 676	F. Pfeifer
Proteins				
HIPV	89.1	205	64 847	G. Myers
MIPSX	17.0	32 800	9 217 368	MIPS
NBRF PROTEIN	26.0	25 814	7 348 950	JIPID/NBRF/MIPS
NRL 3D	2.0	875	138 664	NRL via MIPS
SWISS PROT	15.0	16 941	5 486 339	A. Bairoch
YEASTPROT	1.0	571	258 004	MIPS

Other initiatives such as DDBJ (DNA Data Bank of Japan), MINE (Microorganism Information Network Europe), BRENDA (enzyme database with reaction data, isolation procedures, structural properties amongst other data) should also be noted (Bucher and Zink, 1989). Besides mainframe versions of these files, a number of CD-ROM equivalents have recently been introduced (Mullen, 1990).

Finally, reference should be made to the ambitious Human Genome mapping project — the Genome Data Base (GDB) which utilizes the Sybase relational database system based on DUN hardware. Access is currently available online via the Johns Hopkins University School of Medicine (C&EN, 1990).

External online sequence files
Chemical Abstracts Registry File. Towards the end of 1990, it became possible to retrieve almost all the protein sequences — ca. 150 000 — from the CA Registry File using single-letter or three-letter codes. A similarity searching feature has also been introduced. The file covers patents and other publications from 1957 onwards (CAS, 1990).

Geneseq File (Derwent). Derwent Publications Ltd has launched a database containing data on protein and nucleic acid sequences originating from the patent literature. It is planned to include data going back to 1981, considerably expanding the present patent file which coyers the period from 1986. IntelliGenetics software is used for retrieval, there being two main routes to access the data:

  online access to the IntelliGenetics Timesharing Service (updated bimonthly) in the USA;

  in-house interrogation dependent upon purchasing the Geneseq file together with IntelliGenetics software or making the file compatible with available in-house retrieval software.

All nucleotide sequences longer than nine bases, all protein sequences longer than three amino acids, and probes of any lengths, are recorded (Derwent, 1990).

Chemical and other important databases for pharmaceutical R & D

A report on the primary, secondary and tertiary sources of literature of paramount interest to the medicinal chemist recently appeared (Mullen, 1990). At the beginning of 1992, there were some 12.5 million discrete chemical compounds known, over ten million being present in the CA Registry File. For present purposes, emphasis is placed on specific online files of particular interest to pharmaceutical R & D.

Chemical Abstracts **files**
About half of the 80 sections in *Chemical Abstracts* are of relevance to the medicinal chemist: notably, those sections relating to pharmacology, enzymes, immunochemistry, alkaloids, carbohydrates, amino acids and pharmaceuticals. Biochemical genetics has experienced a 90 per cent growth since 1984 during which time immunochemistry has expanded by some 36 per cent (FIZ-Chemie, 1990). The overall number of abstracts in *CA* has remained, nonetheless, fairly constant.

There are a number of PC-based software products available which facilitate graphic chemical structure retrieval from the CA Registry File, for example: ChemLink; ChemTalk Plus; Genesys; Infolog; Molkick; PC Plot; STN Express (*STNews*, 1990c).

MDL's ChemTalk Plus is of particular interest to the pharmaceutical chemist who is generally familiar with proprietary databases running inhouse under MDL's MACCS system. The product enables the chemist to use the same searching tools for internal and external chemical structure files (Warr and Wilkins, 1990). No doubt, there will now be increased usage of ChemTalk Plus by end-users to access the CA Registry and BEILSTEIN files on STN.

DARC's Chemlink PC-software is available for structure retrieval in the *CA* files via the Questel host.

BEILSTEIN ONLINE file

Besides the established Messenger software used by the STN host for structure retrieval, the S4 software developed by Beilstein/Softron is implemented by the Dialog host. Two more hosts — Orbit and Data–Star — will also be launching BEILSTEIN ONLINE (*STNews*, 1990a,b).

The comparative performance of available software for substructure retrieval — MACCS, DARC, HTSS, Messenger and S4 — was recently studied (Hicks and Jochum, 1990), S4 being apparently the most effective in terms of search time. The capabilities of PC-based chemical substructure search software for personal computers — ChemBase, Chemfile II, Chemsmart, HTSS and PSIDOM — were also evaluated recently (Heller and Meyer, 1990).

Although the BEILSTEIN ONLINE file is about one-third the size of the CA Registry (*ca.* 4.5 million substance records of which 2.8 million possess CA Registry Numbers), it reaches much further back in time, covering the literature from the period 1830–1979 in a thorough manner. It is the major factual database in the chemical area and is indispensable to the pharmaceutical R & D scientist in view of its excellent presentation of organic chemistry. Over 200 fields of information are foreseen for each substance record. By 1994, it is anticipated that the BEILSTEIN ONLINE file will contain over six million discrete chemical compounds.

Patent files

Although pharmaceutical patents represent only about 5 per cent of the annual coverage of Derwent's World Patent Index Latest (WPIL) file (Table 14.2) they are an extremely important source of information. An overview of patent files of relevance to the pharmaceutical chemist was recently presented (Eisenschitz, 1990; Mullen, 1990). Selected highlights are discussed below; see also Chapter 11.

Table 14.2. New pharmaceutical patents in Derwent's WPIL file* 1991

Area	Number	Percentage of total
Chemistry	161639	42·9%
Pharmaceuticals	17301	4·8%

*Total WPIL patent coverage   376609

Derwent Publications Ltd
The Farmdoc section of Derwent's World Patent Index Latest (WPIL) file is one of the major sources of information on pharmaceutical patents, due to the in-depth coding to the content (conceptual and structural) of pharmaceutical patents. Derwent Publications Ltd have been using Markush DARC software for generic structures in parallel with a fragmentation code. It is planned to drop fragmentation coding once Markush DARC gains acceptance by users of the database.

On account of the present lack of currency of the WPIL file — due to manpower constraints resulting from the present parallel coding of chemical structures — Derwent recently introduced a hard-copy product, *Patents Preview*, initially reporting in abstract form on the following areas: cancer, chemotherapy, and endocrine system; cardiovascular system; central nervous system; immune system. This service is similar to the Current Patents Fast Alert service (see below) introduced by Current Patents Ltd, bulletins appearing 1–2 weeks after the receipt of the original patent application.

DIOGENES
The online DIOGENES file, produced by FOI Services, provides data on the expiry of US patents relating to ethical drugs on the US market, together with other pertinent patent information.

DRUG PATENTS INTERNATIONAL
The DRUG PATENTS INTERNATIONAL online file, available via the Orbit host, is produced by Imsworld Publications. It provides a patent profile analysis on some 500 drugs which are either marketed or undergoing development. Subscribers to the printed equivalent can additionally access data related to the expiry period. This file is a useful tool for identifying future potential generic competition from patent-free products. The database will no doubt grow in importance parallel to its enhanced coverage.

MARPAT— Markush Patent File from Chemical Abstracts Service
MARPAT (*STNews*, 1990b) is a CAS Online patent file which covers patents containing Markush structures published on or after January 1988. Markush structures are frequently to be found in pharmaceutical patent applications. Over 25 000 chemical structures, extracted from some 12 000 patent documents, will be added annually to the MARPAT file. Generic searching is an innovative feature of this patent file, which has the usual *CA* patent country coverage, with the exception of the former USSR.

CURRENT PATENTS FAST ALERT/CURRENT PATENTS EVALUATION
The new online file from Current Patents Ltd — CURRENT PATENTS FAST ALERT — focuses on pharmaceutical patents, the following areas being currently covered: anticancer agents, metabolic disease therapy, hormones; anti-inflammatory agents, anti-allergic agents, respiratory aspects; antimicrobial agents; biotechnology, immunology; cardiovascular agents; CNS agents. A ChemBase version of this service entitled Fast-Chem has been made available for in-house use since 1991.

An important asset of CURRENT PATENTS FAST ALERT, also available via the Data-Star host, is the currency of coverage. Another related hardcopy product as well as online file — CURRENT PATENTS EVALUATION — assesses the potential impact of significant new pharmaceutical patents. The upgrading of information in this way will make this file attractive, particularly as its coverage expands.

MPHARM
MPHARM (formerly known as PHARMSEARCH) is an online file, produced by INPI, which covers pharmaceutical patents from the European Patent Office (EPO), France, Japan and the USA. Access to chemical structures is via Questel's Markush DARC software. Updating ensues rapidly — 6 weeks for European patents, 10 weeks for US and Japanese patents. The specific peptide code developed by INPI and Questel is included. European and US pharmaceutical patents are recorded from 1986. It is planned to extend coverage back to 1980.

There is considerable interest in submitting patent and other online files to statistical analyses to obtain an insight into R & D trends in the pharmaceutical area as well as assessing the strengths and weaknesses of competitor companies and products (Möller *et al.*, 1983; Blunck, 1984; Mullen *et al.*, 1988; Blunck *et al.*, 1989).

Reaction retrieval files
The various systems and databases available for reaction retrieval on external hosts and in-house systems of relevance to the medicinal chemist have recently been reviewed along with computer assisted syn-

thesis planning and metabolic planning software (Mullen, 1990). During 1992, the computerized version of *Cheminform* — CHEMINFORM RX — was made available as a reaction file with some 60000 records, both ORAC and REACCS formats being provided from the outset. CHEMINFORM RX is expected to have an annual growth of *ca.* 60000 reactions. This file, based on a well-established printed product with an experienced editorial team, will no doubt become a significant source for information on chemical transformations in future.

Chemical Abstracts Service recently announced that their CASREACT file now contains over one million single-step reactions extracted from the post-1985 literature. These reaction files are particularly relevant for optimizing the preparation of NCEs.

Biotechnology information

Biotechnology has made a major impact during the 1980s on pharmaceutical R & D as it presents a new route to the production of ethical drugs. Some products have already been launched on the market e.g. human insulin, interferon, somatropin and TPA. Reviews of information sources in this area can be found in the relevant literature (Bruce *et al.*, 1989; Crafts-Lightly, 1990).

The intense R & D activity in this innovative field is pictured in Figure 14.5, which presents an overview of products based on recombinant DNA/RNA technology which are currently undergoing development. A number of interesting products will emerge on to the market

Figure 14.5 Pharmaceutical development products based on recombinant technology
Source: *Pharmaprojects*

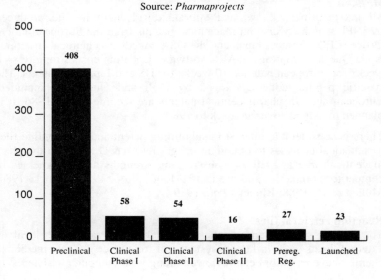

over the next few years as candidate drugs, at present in the preclinical phase (*ca.* 70 per cent), advance. Consequently, it is advisable to have a good overview of the online sources covering biotechnology. A selection — by no means exhaustive — of some of the major online files dealing with basic research in this field are BIOTECHNOLOGY ABSTRACTS, CAS ONLINE, CURRENT BIOTECHNOLOGY ABSTRACTS, CURRENT PATENTS FAST ALERT, SUPERTECH and WORLD PATENTS INDEX.

The burgeoning impact of recombinant DNA/RNA technology is illustrated by Figure 14.6, derived from Derwent's *Biotechnology Abstracts* in which it can be seen that the traditionally strong antibiotics sector has been overtaken.

Figure 14.6 Biotechnology patents in the pharmaceutical area. 1982–90
Source: *Biotechnology Abstracts*

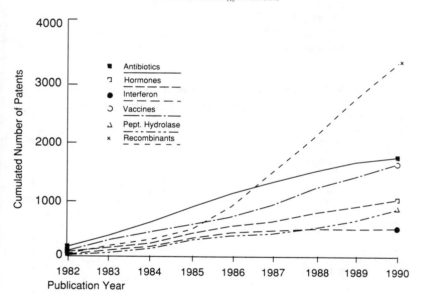

Competitor information sources

In view of the long development period for pharmaceutical products, it is advisable to monitor the competition. There is a whole range of specific information sources available which give an insight into the activities of competitors.

External files
Besides printed sources, such as *Drug Data Report, New Drug Commentary* and broker-type studies from Fleming, Woodmac and PJB

264 The pharmaceutical industry

publications, there are several relevant online databases available which yield information on pharmaceutical R & D: DE HAEN DRUG DATA; DRUG LICENSE OPPORTUNITIES; DATABASE PHARMA MARKETING SERVICE; PHARMAPROJECTS; PHARMCAST (soon to be available online outside Japan); PTS-PROMPT; SCRIP FULLTEXT; TEXTLINE. An online search involving a combination of these files will generally provide an excellent overview of the current competitive situation. A matrix showing the online files which can be used for competitor intelligence and the type of data retrievable is shown in Table 14.3. The symbol (+) indicates that this source may yield useful information on a particular topic.

Files for in-house use

Within pharmaceutical R & D, MDL's MACCS software has become a *de facto* standard for graphically storing and retrieving chemical structures of proprietary test compounds. These chemical files are generally linked to their associated biological data, present in a relational database system such as Oracle, under the MACCS II platform. Consequently, the

Table 14.3 Application of online files for competitor intelligence.

Online files	1	2	3	4	5	6	7	8	9	10	11	12	13	14	15	16
BIOBUSINESS			(+)		(+)		(+)	(+)		(+)			(+)			(+)
BIOCOMMERCE ABSTRACTS											(+)	(+)		(+)	+	+
BIOTECHNOLOGY ABSTRACTS			+	+	+							(+)				
CHEMICAL ABSTRACTS	+	+	+	+	+											
CHEMICAL BUSINESS NEWSBASE							+						(+)			(+)
CLAIMS			+	+	+	+										
COMLINE							(+)			(+)			(+)			(+)
CURRENT BIOTECHNOLOGY ABSTRACTS							(+)	(+)	(+)	(+)		(+)	(+)			
DATABASE PHARMA MARKETING SERVICE							(+)	(+)		(+)	(+)	(+)	(+)			(+)
DIOGENES			+	(+)	(+)		(+)		+							
DISCLOSURE											+			(+)	+	
DRUG INFORMATION FULLTEXT			+						(+)							
DUN AND BRADSTREET											+			+	(+)	+
IMS MARKETLETTER	(+)						(+)		(+)	(+)		(+)	(+)			(+)
INVES TEXT											(+)	(+)	(+)	(+)		+
MOODY'S CORPORATE NEWS											+	(+)	+		(+)	+
PHARMAPROJECTS			+	(+)			+	+	+	+		+				
PTS–ANNUAL REPORTS											+	(+)	(+)	(+)	(+)	+
PTS–PROMPT	(+)						(+)	(+)	(+)	+	(+)	(+)	(+)	(+)	(+)	(+)
SCRIP	(+)						(+)	(+)	(+)	(+)	(+)	(+)	(+)	(+)	(+)	(+)
WORLD PATENTS INDEX (DERWENT)			+	+	+	+			+			(+)		(+)		

Key:
1 = chemical structure
2 = substance name
3 = patent no
4 = priority data
5 = inventors
6 = patent citations
7 = development products
8 = status of development
9 = therapeutic indication
10 = licence data
11 = company organization
12 = main R & D areas
13 = R & D investment
14 = number of employees
15 = management
16 = financial data
+ = applicable
(+) = useful for information on particular topics.

chemist is accustomed to using this graphic chemical structure retrieval software for everyday work.

Development process

Transition from research phase

Much of the information about NCEs is generated in-house and the files initiated in research become a prime source which is expanded as time progresses. Development projects are frequently lengthy and a well-organized information base is vital to success. Since relevant work may take place in several locations, good communications are also essential. However, there remains a requirement to gather information from outside in response to the specific needs of development scientists and technicians.

Over the last few years, database files have become increasingly available in MACCS format for use within the R&D environment. A brief overview of the currently available files on competitor products for PC and mainframe use is presented in Figure 14.7. As mentioned earlier, some of these files (e.g. CMC, MDDR) are now available in 3D format, thereby encouraging their use in QSAR type studies.

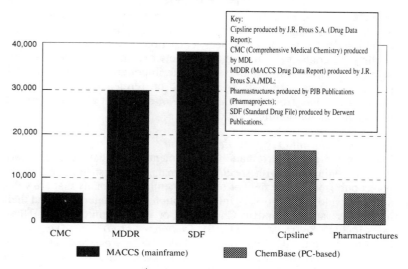

Figure 14.7 Commercially available mainframe and PC-based files on pharmaceutical development products.

Key:
Cipsline produced by J.R. Prous S.A. (Drug Data Report);
CMC (Comprehensive Medical Chemistry) produced by MDL
MDDR (MACCS Drug Data Report) produced by J.R. Prous S.A./MDL;
Pharmastructures produced by PJB Publications (Pharmaprojects);
SDF (Standard Drug File) produced by Derwent Publications.

*Multiple entries possible

Process development

Process development enables the laboratory synthesis of a pharmacologically active substance to be upgraded for large-scale manufacture of the active ingredient. The method of manufacture can be based on either chemical synthesis or biotechnology and information needs are affected accordingly.

Candidate drugs are passed to process scientists to devise processes which are economical, produce high yield, are safe, use starting materials that are readily available, preferably use equipment already installed, are patentable and not already patented, and produce material of the required quality consistently. High yield is desirable not only because the amount and cost of starting materials can be thus reduced but also because the work involved in extraction and purification is minimized together with the quantity of waste that has to be removed.

The process must be safe and, if hazardous materials or reactions have to be used, additional expense is incurred in providing suitable equipment and housing. If starting materials are not readily available, processes for producing them from precursors that are will have to be devised. Investment in new plant is undesirable if market penetration is uncertain at the outset. Finally, if any stages in the proposed production process are under patent, then licensing agreements will probably have to be negotiated.

The process development scientist, therefore, needs initially information on alternative methods of preparation, whether chemical or biotechnological, or, if information is unavailable on the specific material, on compounds of similar structure. In this respect, the chemical entity may be either the new drug substance or a suitable precursor. It is important to know, at the same time, whether the starting materials have been registered under hazardous substances legislation. The potential hazards of starting materials, solvents and byproducts become greater as the scale of preparation increases and are of prime importance at the production stage. Reactions information, chemical or fermentation, is likewise essential not only from the point of view of safety but also to determine the conditions for optimum yield.

Literature relevant to process development

Many of the above information requirements are satisfied by standard texts in conjunction with specialist journals. Amongst the hundreds of books published in the field of organic chemistry are many that are valuable in development work. A full description of the major titles and their usefulness can be found in Chapter 10 or in Pickering (1990). These titles include *Beilstein* (Chapter 9), *Dictionary of Organic Compounds* (J. Buckingham, ed., Chapman & Hall, 1982), *Rodd's Chemistry of Carbon Compounds* (S. Coffey, ed., Elsevier), the *Compendium of Organic Synthetic Methods* (I.T. and S. Harrison, Wiley), Pizey's

Synthetic Reagents (J.S. Pizey, Ellis Horwood, 1983) and Fieser and Fieser's *Reagents for Organic Synthesis* (L.F. and M. Fieser, Wiley, 1967–), as well as the invaluable Aldrich Chemical Company's *Catalogue*.

Hazardous chemical reactions are well documented in L. Bretherick's *Handbook of Reactive Chemical Hazards* (Butterworths) (a fourth edition and computer-readable version of which are available), the *Manual of Hazardous Chemical Reactions NFPA 491M* published by the National Fire Protection Association, and the Association of The British Pharmaceutical Industry's recently revised *Guidelines on Chemical Reaction Hazard Analysis*. There are many texts covering sources of chemicals, often produced by national chemical manufacturers' associations such as, in the UK, the Chemical Industries Association and, in France, the Union des Industries Chimiques. Starting materials or intermediates may be checked to see whether they are registered under the hazardous substances regulations in the European Inventory of Existing Chemical Substances (EINECS) in Europe and in the USA under the Toxic Substances Control Act (see also Chapter 13).

A comprehensive coverage of information sources in biotechnology is available from the Biotechnology Information Service of the British Library.

The journals referred to in process development are numerous, including most of the general organic chemistry and biotechnology titles. *Chemical Abstracts* covers all the major areas of interest in pharmaceutical development, including much in the biotechnology field and it is best used as a database or a subset e.g. *CA Selects*. Other secondary sources include *Current Contents, Methods in Organic Synthesis, Journal of Synthetic Methods* and *Current Chemical Reactions*. Chemical supplier information can be found in *Manufacturing Chemist* (UK), *European Chemical News, Chemical and Engineering News* and *Chemical Marketing Reporter* (USA). In the biotechnology field there are *Biotechnology Abstracts* (Derwent Publications) and the Royal Society of Chemistry's *Current Biotechnology Abstracts*. Further information can be found in Pickering (1990) or via the index to this book.

Online services
Chemical Abstracts files (Chapter 5) are of great utility. Other sources worth investigating are CHEMQUEST and the BEILSTEIN database now available on several hosts. However, it is difficult to differentiate in the searching process between reactions which will be commercially viable and those which are merely of research or academic interest.

Another frequent information requirement is the patent situation of preparative routes. This is well covered by WORLD PATENTS INDEX produced by Derwent Publications Ltd and available online from several vendors. Structure searching has been possible using the Derwent

Chemical Coding, preferably simplified by use of the graphical input computer program Topfrag; however, the trend now is for Markush searching in this and the CAS MARPAT databases.

Databases specifically relating to chemical reaction information include the CHEMICAL REACTIONS DATA SERVICE from Derwent Publications Ltd, available online through the Orbit system, and Chemical Abstracts Service's database called CASREACT, available through the STN online system. Other reaction retrieval systems (see Chapter 6) are available for use on externally supplied databases and for compilation of in-house reaction records. The best known are REACCS (REaction ACCess System) from Molecular Design Ltd, San Leandro, California, SYNLIB (SYNthesis LIBrary) from Distributed Chemical Graphics Inc., Philadelphia, Pennsylvania, and ORAC (Organic Reactions Accessed by Computer) from ORAC Ltd, Leeds, UK. The databases available with these systems, prepared by academic chemists abstracting the primary literature, and using Theilheimer's *Synthetic Methods of Organic Chemistry*, now contain tens of thousands of reactions (Mullen, 1990). A comparison of the three systems has been undertaken by Borkent *et al*. (1988).

Computer programs and expert systems are being developed to assist in the design of chemical syntheses and to predict products and yields from reactions. LHASA (Logic and Heuristics Applied to Synthetic Analysis) is a program developed at Harvard and Leeds Universities which will, starting from the product required, perform a retrosynthetic analysis and suggest synthetic pathways, intermediates and starting materials. The program is highly interactive, with the user using graphical input and output and suggesting pathways by use of its database of types of reactions, called reaction transforms. CAMEO (Computer Assisted Mechanistic Evaluation of Organic reactions), on the other hand, is used to predict the outcome of chemical reactions. It does not use a database, but performs a detailed mechanistic evaluation of reactions to estimate what they will yield.

BIOTECHNOLOGY ABSTRACTS and CURRENT BIOTECHNOLOGY ABSTRACTS are both available online, the former on Dialog and Orbit, the latter on Data-Star. The Life Sciences Collection which includes biotechnology and industrial microbiology information from Cambridge Scientific Abstracts is searchable through Dialog and is available on compact disk for in-house use. Databases giving information on the properties, growth, preservation and storage and use of organisms listed in various culture collection catalogues, including the MICIS database from the Laboratory of the Government Chemist, are available from the Information Centre for European Culture Collections (ICECC) using their Microbial Information Network Europe (MINE) system.

In the area of chemical source databases, there is a need for a comprehensive database listing the suppliers of chemicals in bulk. The databases available are dominated by the suppliers of small quantities of materials for laboratory use rather than production. For instance, CHEMQUEST from MDL available online, on microfiche or on tape for loading on to computers for in-house searching is more suited to laboratory synthesis. The STN service offers an online version of CHEM SOURCES–USA, with the associated database CSCORP giving the names and addresses of suppliers. Information on manufacturers of chemicals may also be found in hazardous chemicals databases such as ECDIN (Environmental Chemicals Data and Information Network) on DIMDI, and HAZARDLINE on BRS.

As mentioned earlier in this chapter and in Chapter 13, bulk hazardous chemicals are controlled under European and US legislation based on inventories of existing chemicals. Both inventories are available as databases, the European EINECS directory on compact disk, the US TSCA directory online via Dialog.

Analytical development
Analytical procedures must be developed for the active ingredient, for starting materials in its synthesis and for other ingredients in the formulations. Specifications for all significant ingredients are required to include confirmation of their identity, quality and significant impurities together with criteria for acceptance of batches. There must be quality control at each stage in the manufacturing process and evidence has to be provided that the finished product itself is of the correct chemical and physical structure and purity. Other tests which may be required by some authorities concern solubility, crystal form, polymorphism, particle size, solvation, oil/water coefficients and pK_a/pH values where applicable. Methods of analysing the active ingredient, its impurities and degradation products in the final formulation are also used in stability and shelf-life tests carried out at various temperatures, humidities and in various light conditions. Test methods are necessary to indicate any reactions between the ingredients or between the ingredients and the packaging and also to monitor any sorption of the formulation ingredients to the container or leaching of any pack component into the medicine. Uniformity of dosage and release rates of controlled-release formulations are other criteria of importance. The methods used in these analyses include high-performance (pressure) liquid chromatography (HPLC), thin-layer chromatography (TLC), gas–liquid chromatography (GLC), infrared (IR) and ultraviolet (UV) spectroscopy, and thermal analysis.

Literature relevant to analytical development
The analytical area is well covered in several books, including Beckett and Stenlake's *Practical Pharmaceutical Chemistry* (A.H. Beckett and J.B. Stenlake, Athlone Press, 1988), *Clarke's Isolation and Identification*

of Drugs (A.C. Moffat, ed., Pharmaceutical Press, 1986) and the annual publication Analytical Profiles on Drug Substances (K. Fleney, Academic Press) as a supplement to the official compendial standards. Further details of these can be found in Pickering (1990) as can details of the major journals in this area and discussion of the secondary sources, such as Chemical Abstracts, Analytical Abstracts (see also Chapter 3) and Ringdoc. Chemical Abstracts, Analytical Abstracts and International Pharmaceutical Abstracts are the main databases used in analytical development, all of which have sections dedicated to the subject.

Formulation development

Drug substances are rarely given to patients in their pure form, but are usually administered as a formulated medicine. The formulation of a medicine is designed to make available to the patient the correct quantity of drug at the correct rate in a stable, safe and acceptable form. For the manufacturer, the formulation must be capable of being readily produced in bulk at the right price and remaining stable during storage under various or specific conditions. For the prescriber, the formulation to hand must be relevant to the case — oral formulations are useless if the patient is unconscious. Ideally, therefore, the formulation must be stable, safe, reproducible, amenable to large-scale manufacture, free from contamination (microbiological or foreign body) and, above all, in satisfying these requirements, the bioavailability of the active ingredient must not be compromised. These various criteria shape the information needs of the formulation scientist. The pre-formulation stage includes collecting information on the fundamental properties of the drug substance which might affect decisions on its formulation. Formulations consist of the active therapeutic agent and various non-therapeutic agents otherwise known as 'excipients'. Amongst these are listed antioxidants, binders, coating agents, colouring agents, diluents, dispersing agents, emulsifiers, fillers, flavours, lubricants, perfumes, preservatives, propellants, solubilizers, stabilizers, surfactants, sweeteners, tablet disintegrants, thickeners and wetting agents. Collecting basic information on the properties of intended excipients is also part of pre-formulation.

Guidance on the choice of excipients can be obtained by comparison with previous in-house formulations or those of competitor products or by considering the properties of new materials. If new excipients are to be used they will be considered in the same way as new drug substances by the regulatory authorities and will require the same lengthy and expensive testing unless they happen to be already permitted in the food industry. It is, therefore, advantageous for formulation scientists to be able to access information on chemicals used in the food industry.

Once candidate excipients have been selected they must be shown not to react with the active agent or with the other excipients. The parameters which might affect the bioavailability of the active ingredient are particle size, crystal form, solubility, the excipients present, tablet shape and hardness, coatings and salt form of active agent. Much work is undertaken on using the dosage form to control or sustain the release of drugs or to carry drugs to where they are needed before release. Other tasks which are the responsibility of the Formulation Development Department are the conduct of scale-up and pre-production studies and, from the experience thus gained, the issuing of manufacturing instructions to the production departments and contributing to trouble-shooting.

Literature relevant to formulation development
Of the books available in this area, the most relevant is the *Handbook of Pharmaceutical Excipients* published in 1986 jointly by the Pharmaceutical Society of Great Britain and the American Pharmaceutical Association with 134 monographs describing commonly used pharmaceutical excipients. Other listings of excipients are available, particularly from countries requiring their disclosure by law. Texts of interest are Florence's *Materials Used in Pharmaceutical Formulation* (A.T. Florence, ed., Blackwell Scientific Publications for the Society of Chemical Industry, 1984), *Pharmaceutics: The Science of Dosage Form Design* (M.E. Aulton, ed., Churchill Livingstone, 1988), and Ansel's *Introduction to Pharmaceutical Dosage Forms* (H.C. Ansel, Lea and Febiger, 1985) and *Pharmaceutical Dosage Forms (Dispense Systems* — H.A. Lieberman, M.M. Reiger and G.S. Banker, eds, 1988; *Tablets* — H.A. Lieberman and L. Lachman, eds, 1982; *Parenteral Medications* — K.E. Avis, L. Lachman and H.A. Lieberman, eds, 1986), a series of practical books by Marcel Dekker. Another relevant series of books is published under the general title of *Drugs and the Pharmaceutical Sciences* by Marcel Dekker. Particularly noteworthy are *Vol. 6, Sustained and Controlled Release Drug Delivery Systems* (J.R. Robinson, ed., 1978) and *Vol. 14, Novel Drug Delivery Systems: Fundamentals, Developmental Concepts, Biomedical Assessments* (Y.W. Chien, ed., 1982). The *Encyclopedia of Pharmaceutical Technology* (J. Swarbrick and J.C. Boylan, eds, Marcel Dekker, 1988–) is a running series encompassing everything from the properties of the drug to the delivery system and production and providing 'the most up to date and easily accessible reference work available on pharmaceutical technology'. Volumes 1–3 cover 'Absorption of Drugs' to 'Dermal Diffusion and Delivery Principles' and the latest, Volume 4, covers 'Design of Drugs: Basic Concepts and Applications' to 'Drying and Dryers'.

Trissel's *Handbook of Injectable Drugs* (L.A. Trissel, American Society of Hospital Pharmacists, 1988) summarizes the primary research on the stability and compatibility of injectables. In general, colouring agents used in the food industry are permitted in oral drugs and information on these can be obtained from The British Food Manufacturing Industries Research Association who provide lists of permitted natural and synthetic agents. In the area of general food additives that may be allowable in drugs, the publications of the joint Food and Agriculture Organization of the United Nations and the World Health Organization Expert Committee on Food Additives is the authoritative source. Their *Food Additives Data System* (FAO Food and Nutrition Paper 30) gives names, synonyms, functional class and evaluation status, and lists of 'Generally Regarded as Safe' (GRAS) substances for food use in the USA can be found in the *Code of Federal Regulations (CFR)*, title 21 (section 182).

For journals that have been found particularly valuable in pharmaceutical development see Pickering (1990).

Chemical Abstracts has the widest coverage of any comparable abstracting service or database, the pharmaceuticals section (63) and the pharmaceutical analysis section (64) being of particular interest to formulation scientists.

International Pharmaceutical Abstracts produced by the American Society of Hospital Pharmacists covers the pharmacy and drug-related international literature pertinent to all phases of drug development and use. *Index Medicus* or its online equivalent MEDLINE indexes the medical literature, and the detailed indexing of this database makes it possible to obtain very specific results from searches. Likewise, the *Excerpta Medica* database (EMBASE) may be interrogated. The FARMDOC section of Derwent's WORLD PATENT INDEX dates from 1963. Also available are online and CD-ROM full-text versions of Trissel's HANDBOOK OF INJECTABLE DRUGS, the AMERICAN HOSPITAL FORMULARY SERVICE — DRUG INFORMATION and MARTINDALE. The RINGDOC database from Derwent Publications Ltd addresses a substantial part of the pharmaceutical literature and has the major benefit of informative abstracts.

Production

Nature of information needs

The information needs of the production area are extremely varied and because of this they will be discussed by areas of work and types of information rather than consideration of the nature of the materials involved. The new pharmaceutical is now a marketable product and economic manufacturing processes will have been devised to enable its production. The internal information generated earlier is of major importance at this stage. Documentation describing how the active ingredient and its formulation must be manufactured are basic not only because the process has been registered with the regulatory authorities and must be closely adhered to but also because, by this time, the procedure is being passed on to non-technical staff for whom operating procedures must be closely defined. The manufacturing procedure has been agreed and a product licence, awarded on evidence of safety, efficacy and quality, has been granted. However, manufacture can be carried out only at production sites which have Manufacturing Licences. Once the product goes into production, the scale of work is increased such that safety, environmental control and pollution control considerations become paramount. The environmental impact must be controlled both within the factory to preserve the health of the workforce and in the immediate vicinity of the factory to preserve the health and goodwill of the local inhabitants.

Downstream processing in production areas to retrieve expensive reagents and the use of advanced separation technology, whether filtration, membrane separation or distillation, to recover extra product from waste streams can make a significant difference to the economics of production. Furthermore, the quality of the product and the requirements for further processing can be affected together with the amount and quality of the waste, all of which are factors which influence disposal decisions.

Information sources relevant to production
Medicines Act regulations, statutes and guidelines may be monitored either directly through HMSO publications or by using a secondary publication such as *Butterworth's Law of Food and Drugs*. All members of the Association of British Pharmaceutical Industry (ABPI) receive a 'Weekly Service Sheet' which discusses and reviews new legislation in the pharmaceutical sector. US legislation on drugs can be followed through the *Federal Register* or a secondary publication such as *Food Drug Cosmetic Law Reports* from Commerce Clearing House Inc.

All the regulations actually in force in the USA are published each year in the *Code of Federal Regulations (CFR)*, Food and Drug Administration legislation, title 21, with background documentation being available under the Freedom of Information Act.

In order to obtain a manufacturer's licence, the manufacturer has to show compliance with Good Manufacturing Practice (GMP). Many countries issue guides to what they consider is good manufacturing practice; international guidelines are prepared by the CEC, the World Health Organization and the Pharmaceutical Inspection Convention, European Free Trade Association. A useful series of books is the *Rules Governing Medicinal Products in the European Community* published by the Commission of the European Communities (CEC), Volume IV of which is the *European Guide to Good Manufacturing Practice*. Frank Wells of the ABPI has edited *Medicines Good Practice Guidelines* which is a compilation of ABPI policy guidelines on various aspects of good practice from GMP to disclosure of active ingredients. For the engineering professional, the standard reference works are Perry's *Chemical Engineers Handbook* (Chapter 7) and Kempe's *Engineers Year-book* published by Morgan-Grampian. For detailed list of information sources in the engineering sector, the book by L.J. Anthony (1986) in this series

274 The pharmaceutical industry

is recommended and, for the chemical engineering discipline specifically, *Chemical Engineering Data Sources* (D.A. Jankowski and T.B. Selover Jr, eds, American Institute of Chemical Engineers, 1986) provides relevant data. Access to relevant national and international standards is of particular importance to engineers and so membership of national standards institutes is useful. Online databases affording access to equipment suppliers information are DEQUIP and DETEQ on STN International. There are several bibliographic databases covering the various engineering disciplines: COMPENDEX and EI ENGINEERING MEETINGS provided by Engineering Information Inc., CAS ONLINE, CHEMICAL ENGINEERING AND BIOTECHNOLOGY ABSTRACTS produced jointly by the Royal Society of Chemistry and DECHEMA, and other discipline-specific databases including ISMEC (Information Service in Mechanical Engineering) and INSPEC (Information Services for the Physics and Engineering Communities).

A frequent information requirement of chemical engineers is physical property data on chemicals (see Chapter 7) and a valuable reference source is published by the Science Reference and Information Service: *A Guide to Sources of Physical Property Data held by the Science Reference and Information Service* (J.M. Sweeney, 1987). DIPPR (Physical Property Data on Pure Chemicals), a database on the STN system, contains physical property data on 766 commercially-important chemicals.

There are an increasing number of safety information sources for production areas (see Chapter 13). UK legislation and guidance is covered by the updated *Barbour Index* microfiche. The databases collected by the Health and Safety Executive in the UK (HSELINE), the International Labour Organization (CISDOC) and the National Institute for Occupational Health and Safety (NIOSHTICS) are all searchable online and all three sources are available together on one compact disk (OSH-ROM) from SilverPlatter Inc.

Systems which afford safety data on chemicals in the form of datasheets are accessible both online and on compact disk systems as the Environmental Chemicals Data Information Network (ECDIN), the OHS system and TOXNET. The major periodical in this area is the *American Industrial Hygiene Association Journal* which usually publishes very practical papers about the factory environment. Medical contingencies are covered by the *British Journal of Industrial Medicine* and more general hazard information appears in the *Dangerous Properties of Industrial Materials Report* (the updating journal to Sax's book of similar name) (see Chapter 13).

Pollution and environmental control are matters of prime concern on any large production site. Garner's *Control of Pollution Encyclopaedia* (J.F. Garner, ed., Butterworths, 1977–) is designed as a guide to the rules and regulations prevailing and consists of two loose-leaf volumes and a supplementary updating service. The *Environmental Data Service* (ENDS) reviews environmental and pollution issues. There are in addition several bibliographic databases including POLLUTION ABSTRACTS produced by Data Courier, APTIC (Air Pollution Technical Information Centre) and ENVIROLINE from the Environment Information Centre, all of which are available on Dialog and may prove useful.

General

As well as providing a focus for accessing information sources in specific subject areas, an information centre on a pharmaceutical development and production site will hold the standard reference texts needed time to time by all personnel. Perhaps the most important requirement in the development sector is an up-to-date set of pharmacopoeias (see Chapter 4 for details of the major ones), e.g. the *British Pharmacopoeia (BP)* the *British Pharmacopoeia (Veterinary)*, the *United States Pharmacopoeia (USP)* and the *European Pharmacopoeia (EP)*, with updates in the journals *Pharmacopeial Forum* and *Pharmeuropa* for the *USP* and *EP*, respectively.

The WHO has compiled a list of current pharmacopoeias from all over the world, over 30 countries having their own publications in addition to the several regional and international versions. This index can be obtained from the WHO in Geneva.

Other essential works include *Remington's Pharmaceutical Sciences* (A.R. Gennaro, ed., Mack Publishing Co., 1990), *Martindale, the Extra Pharmacopoeia* (Chapter 4), the *Merck Index* (Chapter 4) and the *Handbook of Chemistry and Physics* (Chapter 7). For a comprehensive list of publications of use in this area the *Keyguide to Information Sources in Pharmacy* (B. Strickland-Hodge, ed., Mansell, 1989) contains a large international bibliography.

An information centre serving the wide subject interests involved in pharmaceutical manufacture and production benefits from a library of general encyclopaedic texts. The *Encyclopedia of Chemical Technology*, usually referred to by the names of its original editors, *Kirk-Othmer* (see Chapter 4), ranges over most areas of industrial chemistry and technology and nearly all the contributions are very thorough with extensive bibliographies. It is recommended as a primary source for enquiries seeking general background information. KIRK-OTHMER too is available online through vendors such as Dialog. A compact disk version can be purchased and, for libraries with the appropriate hardware, is an attractive proposition since searching is rendered far easier and more versatile.

Conclusions

The chemist in pharmaceutical R & D has a high information requirement which can generally be covered by a host of sources, some available via in-house mainframes, others via external computers or even via conventional hard-copy sources. It is, and will continue to be, the task of information management in the pharmaceutical industry to place these resources at chemists' disposal in a well-packaged manner to help support the innovative process in drug design and, ultimately, the introduction of new ethical drugs.

The trend towards providing easy-to-use computer systems will doubtless continue in order that the medicinal chemist can access some of the information needed directly and not via an intermediary. The task will then evolve more and more to encompass the provision of suitable files, training end-users and providing systems tailored to the organization's needs.

Nonetheless, there are many who prefer to rely upon information specialists and central services will persist albeit in a reduced and modified form. Moreover, some searching is complex and it remains more economic to focus the skills required in specialized units. It is difficult to envisage the demise of the information centre in either the research or the development section of the industry.

References

Barker, W.C., George, D.G. and Hunt, L.T. (1990), Protein sequence database. *Methods in Enzymology*, **183**, 31–49.

Blunck, M., Möller, E. and Mullen, A. (1985), Statistical and graphical evaluation of patent analyses. 2. Vortragstagung der GDCh Fachgruppe Chemie-Information.

Blunck, M., Busse, W.D., Meister, G., Möller, E., Mullen, A. and van Rooijen, L.A.A. (1989), Online patent searching — rapid answers to critical questions. *Naturwissenschaften*, **76**(3), 96–98.

Borkent, J.H., Dukes, F. and Noordik, J.H. (1988), Chemical reaction searching compared in REACCS, SYNLIB and ORAC. *Journal of Chemical Information and Computer Science*, **28**, 148–150.

Bruce, N.G., Arnette, S.L. and Didner, M.D. (1989), Finding Biotech information: databases and vendors. *Biotechnology*, **7**, 455–458.

Bucher, R. and Zinke, H. (1989), Chemische Datenbanken. *Labor*, **2000**, 60–72.

Bush, T., Blunck, M., Mullen, A. and Möller, E. (1990), Comparison of scope and application of currently available drug information files. In *Proceedings of the Montreux 1990 Conference*, pp. 159–176.

CAS (1990), Search protein and peptide sequences online. CAS 1794, September.

C&EN (1990), Genome Database goes online. *Chemical and Engineering News*, 33. November 5.

Crafts-Lightly, A., (1990), Biologicals and biotechnology information. In *Information Sources in Pharmaceuticals*, ed. W.R. Pickering, pp. 52–96. London: Bowker-Saur.

Derwent (1990), *Patent Sequence Databases*. London: Derwent Publications.

DiMasi, I. (1990), New Chemical Entity (NCE) now costs $231 mill. PHARMACEUTICAL AND HEALTHCARE INDUSTRIES NEWS DATABASE, accession no. S00236912, 27 April.

Doolittle, R.F. (1990), Searching through sequence databases. *Methods in Enzymology*, **183**, 99–100.

Dunn, W.I. (1989), Quantitative structure–activity relationships (QSAR). *Chem. Intell. Laboratory Systems*, **6**, 181–190.

Eisenschitz, T. (1990), Intellectual property. In *Information Sources in Pharmaceuticals*, ed. W.R. Pickering, pp. 144–170. London: Bowker-Saur.

FIZ-Chemie (1990), CAS ONLINE-Statistik. *FIZ-Chemie Aktuell*, (4), 6.

Hansch, C. (1989), Comparative structure–activity relationships. *Progress in Clinical and Biological Research*, **291**, 23–30.

Heller, S.R. and Meyer, D.E. (1990), Chemical substructure search software for personal computers. *Chemistry International*, **12**(3), 89–94.

Hicks, M.G. and Jochum, C. (1990), Substructure search systems. 1. Performance comparison of the MACCS, DARC, HTSS, CAS-Registry MVSSS, and S4 Substructure Search Systems. *Journal of Chemical Information and Computer Science*, **30**, 191–199.

Jankowski, D.A. and Selover, T.B, eds, (1986) *Chemical Engineering Data Sources*, AIChE Symposium Series 247, Volume 82. American Institute of Chemical Engineers.

Kreiger, I. (1990) New dimensions in chemical software. *Chemical and Engineering News*, September 24, 26–27.

Marrer, S. and Bruggemann, I. (1989), Molecular modelling in drug research. *PZ Wissenschaft*, **2**, 145–152.

Mills, J.E., Maryanoff, C.A., Sorgi, K.L., Scott, L. and Stanzione, R. (1988), REACCS in the chemical development environment. 1. *Journal of Chemical Information and Computer Science*, **28**, 153–155.

Möller, E., Mullen, A. and Blunck, M. (1983), Support for strategic planning and innovation in the pharmaceutical/chemical industries via information and documentation activities. 1. Vortragstagung der GDCh Fachgruppe Chemie-Information.

Mullen, A. (1990), Chemical and physicochemical information. In *Information Sources in Pharmaceuticals*, ed. W.R. Pickering, pp. 11–52. London: Bowker-Saur.

Mullen, A., Möller, E. and Blunck, M. (1984), Applications of PC 350 (DEC) for online searching and evaluation and upgrading of results from patent and literature files. In *Proceedings of the 8th International Online Information Meeting.*

Mullen, A., Blunck, M., Busch, T. and Möller, E. (1988), From technological current awareness to strategic information. In *3e. Colloque sur l'Information en Chimie [CNICI].*

Mullen, A., Blunck, M., Busch, T. and Möller, E. (1990), Drug information files for inhouse use. *Derwent Literature Subscriber Meetings 1990*, pp. 45–59.

Pickering, W.R., ed., (1990), *Information Sources in Pharmaceuticals.* London: Bowker-Saur.

Ruskino, A., Sheridan, R.P., Nilakantan, R., Harhari, K.S., Baumann, N. and Venkataraghavan, R. (1989), Using Concord to construct a large database of three dimensional coordinates from connection tables. *Journal of Chemical Information and Computer Science*, **29**, 251–255.

SCRIP (1989), Glaxo heading for number one? *SCRIP Review Issue 1989*, 8–9.

SCRIP (1990a), Glaxo positions for future growth. PHARMACEUTICAL AND HEALTHCARE INDUSTRIES NEWS DATABASE, accession no. S00259904, 5 December.

SCRIP (1990b), Pharmaceutical Manufacturers Association (PMA) firms: allocation of research and development (R & D) budget. PHARMACEUTICAL AND HEALTHCARE INDUSTRIES NEWS DATABASE, accession no. S00259408, 29 November.

STN International (1990a), BEILSTEIN (factual searching) — database description. *CAS* 1802-1190, November.

STN International (1990b), BEILSTEIN (structure searching) — database description. *CAS* 1863-1190, November.

STNews (1990a), More than 1 million reactions! *STNews*, **6**(6), 10–11.

STNews (1990b), MARPAT — An innovative way to search chemical patents. *STNews*, **6**(5), 1–2.

STNews (1990c), STN Express. *STNews*, **6**(3), 7.

STNews (1990d), Protein sequence searching. *STNews*, **6**(5) 1, 4–6.

Warr, W.A. and Wilkins, M.P. (1990), Graphics front ends for chemical searching and a look at ChemTalk Plus. *Online*, 50–54.

CHAPTER FIFTEEN

Agrochemical and food industries

H. KIDD AND J.F.B. ROWLAND

This chapter describes the specialized sources of chemical information that are used in agriculture and the agrochemical industry. Books, journals, abstracting and indexing services, review journals, grey literature and computer-based databases are listed; these are additional to the more general chemical sources of these kinds listed in Chapters 2 to 5. The addresses of organizations involved in the provision of information services in this specialized area are listed as an appendix to the chapter.

Primary Sources

In addition to the general primary journals of chemistry, the agricultural chemist will make use of the specialized journals; not all of the content of the journals in Table 15.1 is chemical, but all of them will contain some relevant information. As this is a less 'academic' area of chemistry than most, some of those listed are trade, rather than scholarly, journals.

Table 15.1. Agricultural chemistry journals

Agricultural and Biological Chemistry
Annals of Applied Biology
Aquatic Toxicology
Archives of Environmental Contamination and Toxicology
Archives of Environmental Health
Bulletin of Environmental Contamination and Toxicology
Chemical and Engineering News
Chemistry and Ecology
Chemistry and Industry
Chemistry in Britain
Crop Protection
Ecologist
Exotoxicology and Environmental Safety
Environmental Health

Environmental Pollution, Series A and B
EPPO Bulletin
Fundamental and Applied Toxicology
International Journal of Environmental and Analytical Chemistry
International Journal of Environmental Studies, Section A
International Pest Control
Journal of Agricultural and Food Chemistry
Journal of Applied Ecology
Journal of Ecology
Journal of Environmental Science and Health, Part B, Pesticides
Journal of Pesticides Reform
Journal of the Science of Food and Agriculture
Journal of Toxicology and Environmental Health
Pest Control
Pest Control Technology
Pesticides Biochemistry and Physiology
Pesticide Science
Toxicology and Environmental Chemistry
Tropical Pest Management
Weed Science

Books

Reference books
In this subject area, there is also a large number of reference books which contain information that does not change rapidly and therefore remains of value in the time-scale of book production. The first important category of books is those which provide factual data on pesticides, including their chemical, trade and trivial names, physical, toxicological and environmental data, and appropriate analytical methods for their detection and estimation.

Agricultural Chemicals (4 vols, W.T. Thomson, Thomson Publications, 1988–90)

Agrochemicals Handbook (D. Hartley and H. Kidd, eds, 2nd edn, RSC, 1987)

Farm Chemicals Handbook 1990 (O.H. Willoughby, Meister Publishing)

Handbook of Environmental Data on Organic Chemicals (K. Verschueren, 2nd edn, Van Nostrand Reinhold, 1983)

IPCS (International Programme on Chemical Safety) Environmental Health Criteria Series (available from the World Health Organization)

Pesticide Fact Sheets (WHO, FAO Series; available through the AGROW Bookshop)

Pesticide Index (H. Kidd and D. Hartley, eds, RSC, 1988)

Pesticide Manual (C.R. Worthing, ed., 9th edn, The British Crop Protection Council, 1991)

Pesticide Users' Health and Safety Handbook (A. Watterson, Gower Publishing, 1988)

Directories

A second category of significant books in the agrochemical area is the directories of organizations and persons who have responsibilities or expertise in the areas of crop protection and the control of chemicals in the environment.

Crop Protection Directory 1992–93 (Elaine Warrell Associates, biennial)

Directory of European Agricultural Organizations (Kogan Page, 1984)

Directory of Research Workers in Agricultural and Allied Sciences (R. Vernon, ed., CABI, 1989)

Green Pages (J. Button, Macdonald–Optima, 1990)

Green Research Directory

International Crop Protection Directory (AGROW, annual)
Part I, Europe: Part II, The Americas: Part III, Rest of the World

Pest Management: A Directory of Information Sources (Vol. 1, C.J. Hamilton, CABI, 1990–)

World Directory of Pesticide Control Organisations (RSC, 1989)

Legal information

The next group of books give legal information; they list the agrochemicals that are officially approved for use in different countries with details of the precise uses for which the chemicals are approved.

European Directory of Agrochemical Products (3rd edn, RSC, 1988) — lists 25 000 registered pesticides in 25 European countries

Pesticides 1990 (Ministry of Agriculture, Fisheries and Food) — complete listing of pesticides registered in the UK for use in crop protection and public health

Regional Agro-Pesticide Index — Asia and the Pacific (1988–89 edn; International Co-operation Centre of Agricultural Research for Development, Bangkok, Thailand)

282 *Agrochemical and food industries*

UK Pesticides Guide, 1990 (G.W. Ivens, ed., British Crop Protection Council– CABI) — principal approved UK pesticides with usage information

Pesticide use handbooks
The next group of publications in book form provides advice on the correct use of pesticides.

FAO/WHO Guidelines

GIFAP Guidelines

Pest and Disease Control Handbook (N. Scopes and L. Stables, eds, 3rd edn, British Crop Protection Council, 1989)

Pesticides (6th edn, Council of Europe, 1984)

Weed Control Handbook (8th edn, Blackwell Scientific Publications, 1989)

General books
There are a great many books, aimed both at the specialist reader and at the general public, on the topical subject of environmental pollution. Any list of these is bound to be selective, but Table 15.2 gives some of the more recent ones together with a few of the classics.

Table 15.2. Environmental pollution texts

A Better Mousetrap: Improving Pest Management for Agriculture (World Resources Institute, 1985)
A Growing Problem: Pesticides and the Third World Poor (D. Bull, Oxfam, 1982)
Drinking Water Health Advisory: Pesticides (US Environmental Protection Agency, Office of Drinking Water Advisories, 1989)
Farm Conservation Guide (Schering, 1988)
Farming and Wildlife (K. Mellanby, Collins, 1985)
Pesticide Poisoning (Ministry of Agriculture, Fisheries and Food, 1983)
Pesticides and Human Welfare, (D.L. Gunn and J.G.R. Stevens, Oxford University Press, 1976)
Pesticides and Nature Conservation: The British Experience 1950–1975 (J. Sheail, Oxford University Press, 1985)
Pesticides: Minimizing the Risks (N. Ragsdale and R. Kuhr, American Chemical Society Symposium Series, no. 336, 1986)
Pesticides and Pollution (K. Mellanby, Collins, 1967)
Risk Assessment of Chemicals in the Environment (M.L. Richardson, ed., RSC, 1988)

Silent Spring (R. Carson, Fawcett, 1962)
Silent Spring Revisited, (G.J. Marco, R.M. Hollingworth and W. Durham, eds, 1987)
Since Silent Spring (F. Graham Jr, ed., Pan/Ballantine, 1970)
The Pesticide Conspiracy (R. van den Bosch, Doubleday, 1978)
The Pesticide Handbook (P. Hurst, A. Hay and N. Dudley, Journeyman, 1991)
This Poisoned Earth (N. Dudley, Piatkus, 1987)
Toxic Hazard Assessment of Chemicals (M.L. Richardson, ed., RSC, 1986)

Reviews

Review journals, which summarize and bring together the findings of many primary studies, and give a structure to a subject area, are an important source of information in any subject area. Relatively few exist in this subject area, but the following should be noted. *Outlook on Agriculture* (CABI); *Pesticide Outlook* (RSC); *Shell Agriculture* (Shell, Sittingbourne).

Grey literature

There is also a substantial amount of grey literature and trade newsletters that must be noted as a source of information in agricultural chemistry.

Ag Chem (British Agrochemicals Association)

BCPC Bulletin (British Crop Protection Council)

CABI News (CABI)

Circuit Cultures

Farm Chemicals

GIFAP Bulletin

MAFF Bulletin (Ministry of Agriculture, Fisheries and Food, UK)

ODNRI News (Newsletter of the Overseas Development and Natural Resources Institute)

Pesticides in Perspective, a series of seven leaflets (British Agrochemicals Association)

Pesticide News (Pesticides Trust)

Phytoma

284 *Agrochemical and food industries*

In the agrochemicals area, legal information is of considerable significance since virtually all countries have legal restrictions on agrochemicals manufacture and use, for reasons of health and safety of workers, protecting the environment or ensuring the safety of food. Furthermore, business, as opposed to technical, information is of greater significance in this field than in many other areas of chemistry. The following sources provide access to business and legal information: *Agrow* (PJB Publications); *Chemical Business Bulletin (Agrochemicals)* (RSC); *Pesticide Related Law* (British Crop Protection Council).

Abstracts and indexes

As in other areas of chemistry, there are specialized printed abstracts journals covering agrochemical areas. CABI (CAB International) (formerly the Commonwealth Agricultural Bureaux) has for many years been a major source of abstracting services for all areas of agriculture. Originally, a number of separate bureaux, each one responsible for a separate abstracts journal, were sited around the UK, usually attached to a relevant academic or research establishment, and the overall organization was a semi-official entity responsible to a number of Commonwealth governments. It is now an international, inter-governmental organization registered with the United Nations and has a number of non-commonwealth countries, such as Hungary, as members. With the advent of computerization, the organization became more centralized and the overall abstracts database is now marketed online; the separate printed abstracts journals, however, still exist and are derived from the database. The organization has become a non-profit company.

Of particular relevance to this chapter are:

Biocontrol News and Information

Weed Abstracts

Environmental Abstracts

Pollution Abstracts

Chemical Abstracts Service provides the following relevant services in its *CA Selects* series:

Environmental Pollution

Fungicides

Herbicides

Insecticides

Novel Pesticides and Herbicides

Online databases

Many of the abstracts journals, as in most fields nowadays, also exist as online databases, and there are also databases which do not exist in printed form. There are in fact a large number of databases concerned with chemistry, agriculture and the environment, which are available from the major online hosts via telecommunication networks. Charging is complex, but is usually on a pay-as-you-use basis (see Chapter 5). Chapter 5 also provides comprehensive information on computer-based information retrieval in chemistry, so here are simply listed those databases which are specific to the areas of agriculture and the environment. Bibliographic databases covering these areas include the following:

AGRICOLA	United States Department of Agriculture
AGRIS	Food and Agriculture Organization
AQUALINE	Water Resource Centre
CAB ABSTRACTS	CABI
CAS ONLINE	CAS
ENVIRON	
ENVIRONLINE	Bowker-Saur
HSELINE	Health and Safety Executive
PASCAL: AGROLINE	Centre National de la Recherche Scientifique
PESTDOC	Derwent Publications Ltd
POLLUTION ABSTRACTS	Cambridge Scientific Abstracts
TOXLINE	National Library of Medicine

In addition to these databases which provide access to scientific and technical information, in the agrochemical field there is increasing interest in databases which provide abstracts and indexing of the many sources of business information in this area. Among these are: AGRIDATA NETWORK; AGROW ONLINE; CHEMICAL BUSINESS NEWSBASE; PREDICASTS.

Databanks provide numeric or factual information rather than simply bibliographic access to printed publications, and thus are in many ways more valuable in less-academic fields such as agrochemicals. Major ones available online include: ECDIN (Commission of the European Communities, available through Datacentralen); AGROCHEMICALS HANDBOOK and EUROPEAN DIRECTORY OF AGROCHEMICAL PRODUCTS (see above for details of the printed products) (RSC, available through Data-Star and Dialog); NPIRS (West Lafayette, Indiana, USA).

CD-ROMs and floppy disks

In addition to services available on the international online networks, many databases today are also available for searching on the user's own premises from compact disk–read only memory (CD-ROM), though these databases, unlike online ones, are not of course continuously updated. However, update disks are issued at intervals, and the services are more economical than online ones because the licence fee allows for unlimited usage of the disk. Some small databases are issued on floppy disks; these have the further advantage that no compact disk drive is required and therefore they can be searched on a standard office PC without extra hardware investment.

CD-ROM databases

    CAB ABSTRACTS (CABI) through SilverPlatter

    PEST-BANK (US Pesticide Information) through SilverPlatter

    PESTICIDES DISC (RSC) through Pergamon Compact Solution

Floppy disk databases

    ACE (Hoechst)

    COMPUT-A-CROP

    HESTA

    MICROHERB

    OXFORD AGRICULTURAL CONSULTANTS

    PESTICIDE RELATED LAW (British Crop Protection Council)

    UK PESTICIDES PC (RSC)

Appendix: useful addresses

The addresses of a number of the organizations mentioned in this chapter, and of other organizations which can act as sources of information about chemicals, agriculture and the environment, are given below.

British Agrochemicals Association (BAA), 4 Lincoln Court, Lincoln Road, Peterborough PE1 2RP (telephone 0733 49225)
British Crop Protection Council, 49 Downing Street, Farnham, Surrey GU9 7PH (telephone 0252 733072)
British Standards Institution, 2 Park Street, London W1A 2BS (telephone 071-629 3000)
CAB International, Wallingford, Oxford OX10 8DE (telephone 0491 32111)
Commission of the European Communities, Directorate-General VI, 200 Rue de la Loi, B-1049 Brussels, Belgium
Council of Europe, Palais de Europe, B.P. 431 R6, F-67006 Strasbourg Cedex, France (telephone 88 61 49 61)
Department of the Environment, 2 Marsham Street, London SW1 (telephone 071-212 3434)
EPPO, 1 Rue le Norte, F-75016 Paris, France (telephone (1) 45 20 77 94)
Food and Agriculture Organization, Via delle Terme di Caracalla, I-00100 Roma, Italy (telephone (6) 57971)
Friends of the Earth, 26–28 Underwood Street, London N1 7JQ (telephone 071-490 1555)
GIFAP, Avenue Albert Lancaster 79A, B-1180 Brussels, Belgium (telephone (2) 375 68 60)
Greenpeace, 30–31 Islington Green, London N1 8XE
Health and Safety Executive, Enquiry Point and Library, Broad Lane, Sheffield S3 7HQ (telephone 0742 768141)
International Labour Organization, 4 Route des Morillons, CH-1211 Geneva 22, Switzerland (telephone (22) 996 7116)
International Referral System for Sources of Environmental Information (INFOTERRA), United Nations Environment Programme, P.O. Box 30552, Nairobi, Kenya (telephone (2) 333930)
Ministry of Agriculture, Fisheries and Food, Pesticides Safety Division, Room 325, Ergon House, c/o Nobel House, 17 Smith Square, London SW1P 3JR (telephone 071-238 6305)
Pesticides Trust, 20 Compton Terrace, London N1 2UN (telephone 071-354 3860)
Royal Society of Chemistry (RSC), Thomas Graham House, Science Park, Milton Road, Cambridge CB4 4WF (telephone 0223 420066), and library enquiries to Burlington House, Piccadilly, London W1V 0BN (telephone 071-437 8656)

SilverPlatter Inc., 37 Walnut Street, Wellesley, MA 02181, USA (telephone (617) 239 0306)

Society for Chemical Industry, 14 Belgrave Square, London SW1X 8PS

United Nations Environment Programme, ALCOA Building, Room A-3630, 866 United Nations Plaza, New York, NY 10163, USA (telephone (212) 754 8139)

World Health Organization, Avenue Appia 20, CH-1211 Geneva 27, Switzerland (telephone (22) 912111)

For other database vendors and online hosts, see Appendix 2 of Chapter 5.

Acknowledgement

The lists of publications, databases and organizations in this chapter are largely identical with those in the chapter 'Information sources on chemistry, agriculture and the environment' by H. Kidd, in *Chemistry, Agriculture and the Environment*, edited by M.L. Richardson and published by the RSC (1990). We gratefully acknowledge the permission of the editor and publisher to reproduce this information in this chapter.

CHAPTER SIXTEEN

National and international governmental information sources

J.A. DESCHAMPS

Many official sources of information of value to the chemist are available. Most are easily accessible. A significant number of less obvious sources can also be used to good effect, especially when it is important to explore all possible avenues. This chapter considers the sources available in the UK and other countries, as well as those emanating from international bodies.

The UK

Methods of publication

Although UK government documents are officially divided into two groups, parliamentary and non-parliamentary, and many of them are published by Her Majesty's Stationery Office (HMSO), in practice the neat division into these groups is increasingly being eroded. Also, although HMSO continues to publish most government information, changes in the administrative structure of the public sector, especially the continuing creation of agencies, and the transfer of bodies from the public to the private sector are leading to more diverse sources of publications and difficulties in tracking down items produced in the past by bodies that no longer exist.

There is an increasing number of publications (mostly unpriced) issued by various parts of a Ministry or Department without reference to any central control point or to the organization's library. This is indeed 'greyer than grey' literature. This trend is exacerbated by the increasing availability of desk-top publishing facilities. In addition to 'published' information there is much unpublished information of value. Examples are photographs, film and video, as well as specimens and samples.

To return to the main groups, parliamentary and non-parliamentary: parliamentary publications are those needed for, or directly the result of, the work of both Houses of Parliament. These documents are usually

290 Governmental information sources

dealt with in guides to government publications. The chemist is, however, likely to find non-parliamentary publications the more interesting. They include output from various government departments and agencies that have interests in a wide range of topics, including agriculture, environment, health, scientific and medical research, trade, industry and social matters.

The output from HMSO is enormous and extremely varied. As well as publications produced as part of the legislative process, it also has first option on a Department's publications. This means that it publishes items that it considers will sell well, plus a selection of less popular (but often more important) documents. The end result is an incomplete but reasonable mix. Most Departments issue only some of their publications through HMSO and many produce their own lists of documents. Good examples of these are the lists produced by the Departments of Environment and Transport Library and by the Health and Safety Executive (see Chapter 13). Because of differences in publishing policy, because not all Departments issue a list, and because compilers of lists are not always completely informed by their colleagues, no source of information of this kind can be guaranteed to be complete. The enquirer may, therefore, find it useful to get to know the responsibilities of the various government departments and to write to their libraries for details of publications as well as consulting HMSO lists. The *Civil Service Yearbook* provides a useful introduction. The commercial publisher Chadwyck-Healey produces a catalogue of non-HMSO government publications.

The work of Parliament
Parliamentary proceedings are recorded in the *Journals of the House of Commons* and the *Journals of the House of Lords*. These set out what is done, rather than what is said. Less detailed records are given in the *Votes and Proceedings of the House of Commons* and the *Minutes of Proceedings of the House of Lords*. Debates are recorded in *Parliamentary Debates (Hansard)*. There are separate issues for the Commons and for the Lords. Written answers to questions from Members of Parliament are included in the Commons *Hansard*, which can be entertaining reading. An online database, POLIS, covers the proceedings of Parliament.

House of Lords and House of Commons *Papers* are usually responses from government departments to requests from parliament for information. Other *Papers* are produced because reports or accounts are required to be laid before parliament under the provisions of specific Acts. Reports of Select Committees and minutes of proceedings of Standing Committees appointed to examine Bills are also published in this series. Some titles will be of interest to the chemist, for example *Reports of the*

Select Committee on Science and Technology and *Reports of the Select Committee on the Environment*.

Command Papers are so called because, in theory, they are presented to Parliament by command of the sovereign; they introduce business which did not originate there. Examples are White Papers, State Papers, Annual Reports, Reports of Royal Commissions, Reports of Departmental Committees, Reports of Tribunals and Commissions of Enquiry, and Statistical Reports. Recent titles of interest to the chemist are the White Paper on the Environment (*This Common Inheritance*) and the follow-up to it (*This Common Inheritance: The First Year Report*). The reports of the Royal Commission on Environmental Pollution: *Managing Waste: The Duty of Care* and *Best Practicable Environmental Option* are also of direct relevance to the chemist.

Legislation

Both Bills, which are drafts of proposed Acts of Parliament, and Acts come into the category of parliamentary publications. Bills may be Public Bills which have general application or Private Bills which confer special powers on companies, corporations and, occasionally, private persons. Acts of Parliament are Bills which have passed both Houses and received the Royal Assent.

Statutory Instruments are, strictly speaking, non-parliamentary publications as they are 'delegated legislation' — that is, legislation not directly enacted by Parliament but consisting of regulations made by Ministers under the authority of Acts of Parliament or Orders in Council. They are of two kinds, General and Local.

A good deal of this material is of direct relevance to the chemist. For example, the Environmental Protection Act 1990 consolidates earlier legislation made under the Control of Pollution Act 1974 and points the way ahead in such areas as waste management and pollution control.

An examination of the Statutory Instruments listed in the *Tables of Government Orders and Index to Government Orders* will reveal much of interest. *Statutes in Force,* also published by HMSO, contains the text of largely Public General Acts. It is a continuing series and comprises a large number of volumes arranged by subject, e.g. medicines, poisons and drugs; environment; and water. The commercial publisher, Butterworths, produces separate indexes to the Statutes and to Statutory Instruments.

Statistics

The Government Statistical Service comprises the statistics divisions of all major departments, plus the Business Statistics Office and the Office of Population Censuses and Surveys. The Central Statistical Office co-ordinates the system. The chemist who needs to know about industrial production and sales will find much of interest in the *Business Monitor* series, including the annual *Census of Production*. Detailed information on trade in named chemical substances is included in the *Special Chemicals Return*. Publications such as the *Digest of UK Energy Statistics*, which contains tables and charts of UK energy production and consumption of individual fuels, oil and gas reserves, fuel prices and foreign trade in fuels, and the *Digest of Environmental Protection and Water Statistics*, which includes information on air and water quality, waste disposal and levels of radioactivity amongst other data, are of value.

Statistical information guides include *Government Statistics — A Brief Guide to Sources* published annually and available free from the Central Statistical Office (this lists major publications and contact points for further information), and the more detailed but less frequent *Guide to Official Statistics*, which is published every few years (6th edn,1990).

Research and development

Government participation in research goes back to the seventeenth century with the establishment of the Royal Observatory at Greenwich in 1675. Other important developments include the setting up of the Meteorological Office in 1854, the Department of the Government Chemist in 1842, and the National Physical Laboratory in 1900. Subsequently, progress was rapid, marked particularly by the establishment in 1916 of the Department of Scientific and Industrial Research (DSIR) whose existence until 1964 symbolized a very positive attitude on the part of government to research and development. The period up to 1964 also saw the setting up of the Medical Research Council in 1931 and the United Kingdom Atomic Energy Authority in 1954 (now the Atomic Energy Authority).

During the mid-1960s, a major reorganization took place leading to a much wider dispersal of government responsibilities for research and development. Under the Science and Technology Act 1965, central responsibility for basic civil science passed to the Secretary of State for Education and Science, advised by the Advisory Board for the Research Councils. Responsibility for technology rests mainly with the Secretary of State for Trade and Industry. Other government departments are responsible for research and development related to their executive responsibilities; individual departments have appointed Chief Scientists to co-ordinate their research need. Interdisciplinary research and development is co-ordinated through the Advisory Council for Science and Technology (ACOST).

In 1972 the government announced its decision in the White Paper *Framework for Government Research and Development* to extend the 'customer–contractor' principle to all its applied research and development. This meant that the 'customers', government departments with policy responsibility for the subject area, determined the requirements

Governmental information sources 293

and funding available for research they wish to be done by the 'contractor', who might be from the public or the private sector. The research objective should be attained at reasonable cost.

The Department of Education and Science is responsible for research carried out through five councils: the Science and Engineering Research Council, the Economic and Social Research Council, the Medical Research Council, the Agricultural and Food Research Council and the Natural Environment Research Council. Much useful information can be found in the reports issued by these Councils.

The Department of Trade and Industry is responsible for a number of industrial research establishments. Some of these have recently acquired agency status and hence more control over their own budgets. Examples of agencies are the Laboratory of the Government Chemist, which offers analytical and advisory services, including advice on chemical nomenclature, and the Warren Spring Laboratory, which is particularly concerned with materials recycling and recovery and with the development of clean technology. The Warren Spring Laboratory operates a public enquiry point for information on industry and the environment.

The Building Research Establishment is an agency of the Department of Environment and the Transport and Road Research Laboratory is a Department of Transport agency. Both organizations produce many useful reports which can be purchased.

The Atomic Energy Authority (AEA, formerly UKAEA) publishes its own documents and lists of documents which are available to the public. It operates the Waste Management Information Bureau which is the national centre for advice on this topic. It also runs the National Chemical Emergency Centre which offers immediate-response advice on how to deal with chemical spills and other accidents. AEA, as a whole, has diversified increasingly over the years and has expertise in areas such as ceramics, cryogenics, environmental sciences, and alternative energy sources.

The Meteorological Office, an agency of the Ministry of Defence, publishes journals, charts and meteorological data. The recently-established Hadley Centre for Climate Change comes within its jurisdiction.

The Ministry of Agriculture, Fisheries and Food (MAFF) carries out research into the causes, pathology and control of animal and plant diseases and into harmful insects, mites, fungi, birds and mammals. The Food Sciences Division carries out work on the interaction of food additives and components, and on heavy-metal contamination. It has also done work on the migration into food of plasticizers in clingfilm. MAFF issues many publications such as *Advisory Leaflets, Booklets* and from the Fisheries Laboratory, *Technical Reports* and *Technical Bulletins*. MAFF's Fisheries Laboratory is concerned, amongst other topics, with studies of the pollution of the North Sea.

294 Governmental information sources

The Health and Safety Executive is concerned, amongst its many responsibilities, with work on the safe use of agricultural pesticides. It is also the Joint Competent Authority (together with the Department of the Environment) for work on the notification and assessment of new chemical substances. It reviews the toxicity of existing chemicals and issues an ongoing series, *Toxicity Reviews*. Further details of its work and publications are given in Chapter 13.

As a final example, the Department of Health carries out work on environmental health and contributes its expertise in the preparation of *Environmental Health Criteria* (see p. 296 for further information). It is also responsible for Poisons Units which hold information on chemicals, formulations and products in order to advise medical practitioners in cases of emergency.

Indexes to British Government publications
Daily List of Government Publications is published by HMSO. It lists parliamentary publications first, then non-parliamentary publications, followed by those issued by the European Community and other bodies for which HMSO is an agent. Statutory Instruments, arranged by number, end the list.

The monthly catalogue, *Government Publications*, is arranged similarly to the *Daily Lists*. It does not include Statutory Instruments. It also has an alphabetical index which is cumulated annually. The *Annual Catalogue* covers the whole year, and, once again, excludes Statutory Instruments which are listed separately in the monthly *List of Statutory Instruments* and also annually. A CD-ROM listing of Statutory Instruments 1987–90 is also available from HMSO.

HMSO also produces *Sectional Lists* which are catalogues of current, largely non-parliamentary publications. They are free and are updated regularly. Titles include: *Agriculture and Food; Health and Safety Executive;* and *Environment.*

A monthly newssheet, *Advance Information on Government Publications*, lists important new titles in preparation. There is also a weekly *List of Non-Parliamentary Publications sent for Press*.

A bibliographical database, HMSO DATABASE ONLINE, covers the period from 1976 to date. It is available on the British Library's online database service Blaiseline and on Dialog.

Further information
Further information on UK government publications is given in the following sources.

Johansson, E. (1978), *Current British Government Publishing*.
London: Association of Assistant Librarians (SE Division).

Richard, S. (1984), *Directory of British Official Publications: A Guide to Sources* (2nd edn). London: Mansell.

Howard, B. (1989), Introduction and Overview of British Official Publishing, in *Who Publishes Official Information on Health, Safety and Social sciences? Proceedings of a One-day seminar held by ISG SCOOP, 20 June 1988*, ed. V.J. Nurcombe. London: Library Association Information Services Group.

Howard, B. (1991), What is Government Information? The Variety and Complexity of Government Publishing. *Government Libraries Journal*.

Nurcombe, V.J., ed, (1990), *British Official Publications Online: A Review of Sources, Services and Developments*. London: Library Association Information Services Group.

Goodier, J. (1991), *The Green and the Grey: Government Environmental Publications*. London: Refer, Library Association Information Services Group.

Butcher, D. (1991), *Official Publications in Britain*, 2nd edn, London : Library Association Publishing.

Non-UK Government publications and expertise

The services covered in this section are sponsored by the United Nations (particularly the United Nations Environment Programme, UNEP), OECD (the Organization for Economic Co-operation and Development) and the Commission of the European Communities. Some account will also be given of sources of information on US and Japanese government publications. These services provide access to much grey literature, as well as producing a number of well-established series.

United Nations Environment Programme (UNEP)

UNEP came into being after the Stockholm Conference of 1972, which represented the first worldwide attempt to get to grips with environmental problems. It was given a catalytic role, since other agencies, such as the Food and Agriculture Organization, existed already.

The International Register of Potentially Toxic Chemicals (IRPTC)
This UNEP service has a central databank at Geneva which contains information on a large number of chemical substances in areas such as toxicity and ecotoxicity, waste management, biodegradation and methods of analysis. The databank also has information on regulatory aspects of chemicals control. It works through a network of National Correspondents. It publishes a *Bulletin* about twice a year. IRPTC also

publishes monographs such as *Treatment and Disposal Methods for Waste Chemicals* (1986). An important title is its *Legal File* (1987) which is currently being updated. The 1987 edition contains regulatory information from 12 countries, including the UK and USA, and six international organizations on around 600 substances. Each record has full details of related documentation. The *Legal File* can also be accessed online on ECDIN (see p. 296).

IRPTC cooperates with the USSR State Committee for Science and Technology and its successor body in the production of a series of monographs reviewing the Russian literature on toxicity and hazards of chemicals. Over 100 titles have been produced so far; many deal with pesticides. A number of 'one-off' monographs have also been produced in cooperation with the former USSR. Examples are *Long Term Effects of Chemicals on the Organism* and *Principles of Pesticide Toxicology*.

International Programme on Chemical Safety (IPCS)
IPCS is a joint programme of UNEP, the International Labour Office and the World Health Organization. It evaluates the effects of chemicals on human health and on the environment, develops guidelines on exposure limits and methodology for toxicity testing and related studies, and develops information needed for dealing with chemical accidents. It publishes an extensive series of monographs, *Environmental Health Criteria* together with a companion series of *Health Guides*. IPCS has recently started to issue, in cooperation with the European Communities, a series of *Chemical Safety Cards*. These are designed for use at workplace level.

In addition to the above, IPCS produces, in cooperation with IRPTC, the *Computerised Listing of Chemicals being Tested for Toxic Effects* (CCTTE). This is updated periodically, the most recent update being that of July 1990. CCTTE lists, by chemical name, work currently being done, and gives the name and organization of the research worker(s). It also lists reviews that have recently been completed. Data are exchanged with OECD's EXICHEM database.

Industry and Environment Office (IEO)
The IEO, which is located in Paris, was set up in 1973 to bring government and industry together to cooperate in reducing the adverse effects of industries on the environment. It has a central computerized databank, the *Industry and Environment File*, which contains, for example, information on pollution and abatement control technologies and their costs. IEO also has links to key organizations in specific industries such as pulp and paper, and petroleum. It produces a range of publications on individual industries.The IEO does not provide an online service, but its staff will answer enquiries using the range of resources available to them.

Governmental information sources 297

A quarterly newsletter, *Industry and Environment*, is available on subscription. Each issue focuses on a particular area, such as low-waste and non-waste technology, and recycling and recovery in the metals industry. IEO issues guidelines for emergency awareness and preparedness; it is a partner in UNEP's APELL (Awareness and Preparedness at Local Level) Programme in which it co-operates with OECD, ILO, EC and others.

INFOTERRA (International Environmental Information System)

INFOTERRA is different from IRPTC and IEO because it covers the whole environmental subject area and operates on the referral principle; it refers enquirers to sources of information or expertise. However, factual information is often supplied, much as grey literature. INFOTERRA is a valuable source of information on chemicals and the environment. Many organizations included in its *International Directory* (see below) have expertise on chemical topics such as pollution and wastes. It is of particular value for multidisciplinary topics.

INFOTERRA has two components: the *International Directory* containing descriptions of around 7000 information sources worldwide from a wide range of organizations, including government bodies, research groups, academic institutions and private consultants, updated every two years; and the network of National Focal Points (NFPs), each member of which has links with and knows about information sources in its own country. Requests to INFOTERRA are usually made through NFPs. In some areas of the world, there is a supporting network of Regional Service Centres.

INFOTERRA also has a number of Special Sectoral (subject) Sources who provide substantive information, particularly to developing countries. Examples of these, in addition to IRPTC and IEO, are CAB International (see Chapter 15), the Waste Management Information Unit at Harwell (see p. 293), and the Environmental Law Information Service of the International Union for the Conservation of Nature and Natural Resources.

International Labour Office (ILO)

ILO is the United Nations body with responsibility for the protection of workers' health and safety. It has close working links with UNEP. Its publications include *Chemical Safety Data Sheets, Codes of Practice* (such as *Safety and Health in Agricultural Work), Manuals* and *Technical Guides*. The key publication, *Encyclopedia of Occupational Health and Safety,* is being updated at present. It contains much information on chemical topics. ILO's Information Service publishes a bulletin, *Safety and Health at Work*.

Organization for Economic Co-operation and Development (OECD)

EXICHEM

OECD has a special programme for the control of chemicals. As part of this activity, it coordinates a database, EXICHEM, which contains details of work being carried out on existing chemicals. The database is available as a printed version and on diskette. EXICHEM, in common with other records of current work, can be used to avoid duplication and to identify possibilities for co-operation.

Work on High Production Volume Chemicals
OECD is carrying out work on filling gaps in data on high production volume chemicals that are considered likely to have any adverse effects. The results lead to the preparation of *Screening Information Data Sheets (SIDS)*. The information obtained will be supplied to IRPTC. Lead countries have been appointed for specific substances. (Details of this work are given in *International Environment Reporter*, June 1990, pp. 263–70.)

European Communities

ENVIRONMENTAL CHEMICALS DATA AND INFORMATION NETWORK *(ECDIN)*

ECDIN was initially designed as a network-based system. However, for a number of reasons, it is at present the online equivalent of a chemical handbook. The main data categories included in the commercial version, which can be accessed via DIMDI, are: identification; physicochemical properties; production and use; legislation and rules (supplied by IRPTC); occupational health and safety; toxicity; concentration and fate in the environment; waste management (supplied by IRPTC); detection methods. Most categories are broken down into a number of separate files, such as effects on soil microorganisms and aquatic bioaccumulation. Basic information is included on around 65 000 chemicals (although the numbers included in individual files vary considerably). ECDIN also operates the online version of EINECS, the EUROPEAN INVENTORY OF EXISTING COMMERCIAL CHEMICAL SUBSTANCES.

COMMUNITY R & D INFORMATION SERVICE *(CORDIS)*

CORDIS is a new database and is available on ECHO, the European Communities Host. It contains information on Community Research and Technological Development (RTD) activities, particularly those included in the EC's R & D Framework Programme. Other information in areas such as energy, agriculture and environment is also included. The details of project reports listed in *European Abstracts (EABS)* form part of CORDIS.

System for Information on Grey Literature in Europe (SIGLE)
SIGLE is an EC-wide database on grey literature and contains records from 1981 onwards. Although it covers all subject areas, much scientific and technical information is included. The UK input is provided by the British Library and is the online equivalent of *British Reports, Translations and Theses*.

Further information
Further information about European Communities publications and databases can be found in the following sources:

> Thomson, I (1989), *The Documentation of the European Communities: A Guide*. London: Mansell.
>
> *1991 Directory of EEC Information sources*. Genval, Euroconfidential.
>
> *The European Communities Encyclopedia 1992* (1991), Europa Publications.
>
> *European Access* (The Current Awareness Bulletin to the Policies and Activities of the European Communities). This periodical is published six times a year by Chadwyck-Healey in association with the United Kingdom Offices of the European Commission.

USA
The US government is responsible for many publications of interest to the chemist. A comprehensive compilation of sources which includes information on technical reports, indexes and abstracts, databases, and specialized information centres is: R. Aluri and J.S. Robinson (1983), *A Guide to US Government Scientific and Technical Resources* (Littleton, CO, USA: Libraries Unlimited). The chief way by which details of US government publications are made known is through the US National Technical Information Service (NTIS) which publishes *Government Reports Announcements and Index (GRA&I)* twice monthly with an annual index. NTIS also publishes a series of weekly *Abstract Newsletters* on topics such as chemistry, environmental pollution and control, and agriculture and food. Another NTIS publication is an extensive series of *Published Searches*; individual titles are updated periodically. Other listings are also produced, for example, *Selected Water Resources Abstracts*. The US Environmental Protection Agency (EPA) issues a number of guides to its publications. Microinfo Ltd of Alton, Hampshire, is the UK sales agent for NTIS publications.

The following report series will also be useful:

> *Toxicological Profiles* (Agency for Toxic Substances and Disease Registry and EPA)

300 Governmental information sources

Pesticide Fact Sheets (EPA)

Ambient Water Quality Criteria (EPA)

Drinking Water Health Advisories (EPA)

More detailed information concerning NTIS publications and services is given in C.P. Auger (1990), *Information Sources in Grey Literature* (London, Bowker-Saur), a companion volume in this series.

Japan

NTIS in the USA has produced a *Directory of Japanese Technical Resources* which lists US sources giving access to Japanese technical information. The directory includes a list of technical reports that have been translated into English. The British Library's Science Reference and Information Service (SRIS) offers a special Japanese information service.

Other non-UK government publications

The London embassies of major nations maintain counsellors and attachés, some dealing specifically with science, who will supply information on publications including government ones from their respective countries. Their addresses can be found in the *London Diplomatic List* (HMSO), the London *Telephone Directory, Kelly's London Directory, Whitaker's Almanack*, etc.

Other useful sources are *HM Diplomatic Service Overseas Reference List*, issued by the Foreign and Commonwealth Office, and *The Diplomatic Service List* (HMSO). These give details of staff and their responsibilities in British embassies, etc., abroad who can be approached for help.

INFOTERRA (see above) and other international networks can be used to obtain overseas contact points.

Some general comments

Apart from the conventionally published literature mentioned in this chapter, many references have been made to the grey literature and to contact points for information. These non-conventional sources may be divided into three broad categories. They are not, however, mutually exclusive. Each has its strengths and weaknesses, particularly with respect to currency, comprehensiveness and ease of access.

Databanks

Examples are IRPTC, UNEP INDUSTRY AND ENVIRONMENT FILE and ECDIN. These systems are of most value in giving an overview of a named chemical. They have good coverage of data taken from key works produced by international agencies. However, often comprehensive information is available for only a comparatively small number of substances.

Referral services
IRPTC and INFOTERRA are such services. Whether or not they hold data, referral services give access to a wide range of sources through their national and international networks. Their main weakness is an inbuilt delay, but their strength is that they give access to publications and data that are not generally available, and to specialist advice.

Bibliographic records
CORDIS and GRA&I are examples of bibliographic records. Such records do not usually contain data. The original documentation will have to be obtained or the research worker contacted. Coverage will vary depending on the commitment of, and resources available to, contributors. However, such records are a valuable source of information on the grey literature.

Section D: Conclusion

CHAPTER SEVENTEEN

Practical use of chemical information sources

R.T. BOTTLE AND J.F.B. ROWLAND

Chapter 1 indicated the structure of this book, showing how the various chapters serve to answer certain problems. A number of cross-references have, of course, been necessary and this chapter will contain many more. It is thus intended to be a broad index to the book's use. There is, however, little point in parading a vast number of titles with annotations before the reader and saying much treasure awaits him or her in the literature. The key is practice; the more one uses the literature and its electronic surrogates, the more information one will gather from it. Information is too valuable a crop to be picked in an inefficient and wasteful manner; some sample search topics have therefore been included in the section on the detailed survey (Table 17.1). This is so that the novice can gain search experience. Other literature exercises will be found in earlier editions of this book.

The ensuing sections of this chapter are, in effect, the various purposes for which the literature is searched (apart from acquiring background information, as is described in Chapter 4). These divisions are, at best, somewhat arbitrary, and there is considerable overlap of one with another depending on the depth of knowledge required.

Quick reference

This is probably the purpose for which the literature is most often used. Many of the necessary books should be found in, or adjacent to, any well-equipped laboratory (although it will probably keep only those testing standards, etc. which are indispensable to its routine work). Undergraduate practical manuals will, of course, solve many of the simpler queries in the categories mentioned below.

Physical properties such as melting point, refractive index, solubility, and so on, are frequently required and can usually be found in the ubiquitous *Handbook of Chemistry and Physics*, the 'Rubber Bible'

(Chapter 7). With new editions now appearing annually, libraries often pass on the previously held edition to one of the laboratories as it contains still useful information. If, however, one requires more critically compiled data then reference should be made to the books of tables discussed in Chapter 7. Trade literature available on commercially obtainable chemicals is dealt with briefly in Chapter 12, or more fully by Bottle in the chemicals and pharmaceuticals chapter of *Finding and Using Product Information* (R.A. Wall, ed., Gower, 1986).

Physico-chemical techniques are now so widespread and varied that there are few encyclopaedic works in this field and each distinct method is usually described by one or more monographs. A large number, however, are described with ample literature references in B.W. Rossiter and J.F. Hamilton's *Physical Methods of Chemistry* (2nd edn, 1985–). These volumes stem from the *Techniques of Chemistry* series, which originated under A. Weissberger's general editorship. Spectroscopic data sources are discussed in Chapter 7, while spectroscopic and other techniques for structure determination are the subject of a section in Chapter 10. From 1959 to 1962 the *Journal of Chemical Education* carried a most useful basic survey on chemical instrumentation and since 1962 this has been replaced by 'Topics in Chemical Instrumentation', which describes the operation, principles, circuits, etc., of commercially available physico-chemical instruments (see the annual indexes). (The same journal is a useful source of advertisements and editorial matter which show what new instruments and apparatus are available in the USA.) Applications of the above methods on the plant are dealt with in Considine's *Process Instruments and Controls Handbook* (3rd edn, McGraw-Hill, 1985) and its companion volume, *Handbook of Applied Instrumentation* (D.M. Considine and S.D. Ross, eds, McGraw-Hill, 1964, reprinted Krieger, 1984).

Preparative details for inorganic and organic substances will usually be found in the books mentioned in Chapters 8 and 10, otherwise reference will have to be made to the original publication (located through the formula or chemical substance indexes of *Chemical Abstracts*, see Chapter 3, or via online searches, see Chapter 5). *Dictionary of Organic Compounds* (Chapter 10) gives the characteristics of some 150 000 organic compounds with literature references to their original preparation and is also available online.

Biochemistry is a large discipline in its own right, and the biochemical sections make up about one-third of the content of *Chemical Abstracts*; Chapter 8 of *Information Sources in the Medical Sciences* in this series (L.T. Morton and S. Godbolt, eds, 8th edn, Bowker-Saur, 1992) covers the biochemical area in detail. *Biochemistry* (C.K. Mathews and K.E. van Holde, Benjamin Cummings, 1990) is a recent general textbook of the subject which looks at it from a chemical viewpoint; *Introduction to Practical Biochemistry* (D.T. Plummer, McGraw-

Hill, 1987) covers practical aspects. A well-known popular introduction to the subject is *The Chemistry of Life* (S. Rose, Penguin, 3rd edn, 1991). In an academic library one should be able to find the series *Laboratory Techniques in Biochemistry and Molecular Biology*, currently edited by R.H. Burdon and P.H. van Knippenburg (Elsevier); Volume 20 appeared in 1990. Much broader in its scope than its title suggests — indeed, covering all areas of the biochemical sciences — is the massive series *Methods in Enzymology*, currently edited by J.N. Abelson and M.I. Simon (Academic Press). Volume 201 appeared in August 1991. Volume 183 is of great interest from an information-retrieval point of view, since it is devoted to computer analysis of protein and nucleic acid structures, especially in connection with the US government's Human Genome project, and therefore describes many advances and innovative information sources valuable in the molecular biology field.

Analysis and testing are frequent operations in laboratories and while for most purposes textbook methods are quite adequate (see Chapters 8 and 10), in certain circumstances the official or standard method must be used. The pharmacopoeias mentioned in Chapter 4 give methods to be used and specifications to be met by substances of pharmaceutical interest. In the UK the British Standards Institution (BSI) (Linford Wood, Milton Keynes MK14 6LE; telephone 0908 220022, general; 0908 221166, enquiries; 0908 320856, fax) issues specifications for materials, standard methods of testing, codes of practice, etc., as and when there is a generally recognized need for them. Standards are revised from time to time. The *BSI Standards Catalogue* divides the available standards specifications into subject categories, such as chemical engineering, chemicals, fats and oils, scientific apparatus, etc. BSI also publishes the periodical *BSI News*, and also *Bibliotech — the BSI Library Magazine*. There is also an online database for standards, STANDARDLINE, and a CD-ROM called PERINORM which contains British, French and German standards. BSI at the above address can supply information about all of these services.

The equivalent international body, the International Organization for Standardization (ISO) (1 rue de Varembe, Case postale, CH-1211 Geneva, Switzerland; telephone (41) (0) 22 34 12 40, fax (41) (0) 22 33 34 30; telex 41 22 05 iso ch), publishes its *ISO Catalogue* which lists ISO international standards, according to its own classification scheme. The *American National Standards Catalog* is available in this country from ILI of Ascot, Berkshire. An English translation of *DIN — The German Standards Catalogue* is available from Beuth Verlag of Köln.

Specifically chemical methods of analysis are listed in an authoritative bibliography originally sponsored by the Society of Public Analysts, *Official Standardised and Recommended Methods of Analysis*, edited by N.W. Hanson (2nd edn, Society of Analytical Chemistry and The Chemical Society, 1973); a new edition is in preparation by The Royal

Society of Chemistry as successor to the two societies that published the second edition, and was expected to appear in 1992. The equivalent US publication, which is revised every five years, is *Official Methods of Analysis* (Association of Official Analytical Chemists, Washington, DC). Both these volumes list methods of analysis that are legally required to be used for certain purposes.

Minor theoretical or practical points can be verified by looking in the appropriate monograph or in one of the chemical encyclopaedias or dictionaries discussed in Chapter 4.

Biographical details of a particular chemist are occasionally needed, or sometimes just an address is required. On such occasions the lists of members published by learned societies are useful. The *Royal Society of Chemistry Register of Members* (latest edition, 1987), the *Royal Society Yearbook* and the membership lists issued from time to time by the American Chemical Society are cases in point. International organizations sometimes sponsor directories, e.g. *International Union of Crystallography Membership Directory* (4th edn, 1971), while all academic staff in Commonwealth universities are included in the *Commonwealth Universities Yearbook* (Association of Commonwealth Universities), a four-volume annual publication. *College Chemistry Faculties* (ACS, 1989) is a triennial listing of 18 000 staff in 2000 North American colleges, whilst *Chemical Research Faculties: An International Directory* (ACS, 1988) lists 10 000 chemists and their research areas in 960 departments worldwide. If more than academic qualifications and addresses is required, then publication of the 'Who's Who' type give details of very eminent chemists. For example *Debrett's People of Today* (Debrett, 5th edn, 1993) is an adequate, recently updated source for the UK. Somewhat more detail of those included is given by *Who's Who in Science in Europe* (7th edn, 4 vols, Longmans, 1991). This has a separate volume for UK scientists. *American Men and Women of Science: Physical and Biological Sciences 1992–93* (8 vols, 18th edn, Bowker, 1992) is the most complete directory of this type available, arranged by broad subject headings. Chapter 4 deals with biographies of famous chemists of the past. (This is about the only chemical field in which the popular, door-to-door peddled encyclopaedias are up-to-date!)

Consultants can be located through the *Directory of Consulting Practices in Chemistry and Related Subjects* (RSC, 1986) or, for chemical engineering, in the Institution of Chemical Engineers' *List of Consultants 1987*.

The detailed survey

In any survey or search the first requirement is a clear definition of the subject, nature and scope of the search. If the search is being carried out for someone else, it is essential to understand clearly the purpose of the survey. One problem is to know how far back to go; in chemistry, the nineteenth-century literature cannot always be ignored. It is possible to search back to 1871 with the precursors of *British Abstracts*, or earlier with *Chemisches Zentralblatt*. The pitfalls of searching the chemical literature before 1875 have been described by Dyson (1961), but in most cases an adequate summary of nineteenth-century research can be obtained from *Gmelin* and *Beilstein* (see Chapters 8 and 9).

Chemical Abstracts started in 1907, but before 1930 it was less comprehensive than *Chemisches Zentralblatt* and searches in the 1907–1930 period should use both abstracts journals.

Organic chemical searches often centre on a particular compound or group of compounds. For these searches, *Beilstein* is the prime source of information up to 1930, or in some cases more recently (see Chapter 9). Similarly, *Gmelin* serves if the interest is in specific inorganic compounds. Recent compound-based information can be very readily obtained from *Chemical Abstracts* if one knows the CA Registry Numbers of the compounds in question, as CAS and many other databases now contain Registry Numbers as searchable data-elements. In general, printed *Chemical Abstracts* is the comprehensive source of fairly recent information; the Collective Indexes (quinquennial in recent years) bring large amounts of information together, but the period since the most recent Collective Index, which may be seven or eight years owing to the publication delays of the Collective Indexes, requires searches through the Volume Indexes which appear every six months. This can be very tedious and nowadays one would tend to use the online search service instead for this most recent period.

The various alternative versions of the CHEMICAL ABSTRACTS database, available through different hosts, are described in Chapter 5. If the interest is in specific compounds, CAS ONLINE, available from CAS itself through the STN network, would be the preferred version as it includes the facility for searching for compounds by means of their structures.

A useful feature of *Beilstein* in the printed version is the fact that the classification system brings together large numbers of related compounds within a few pages, and thus if one browses on either side of one's entry point one can get further useful information. *Beilstein's* classification scheme is complex and the occasional user may be better off using the indexes for compounds which have been known for a long time. Unlike *Chemical Abstracts*, however, *Beilstein* uses trivial names in its indexes. Nomenclature can be a problem in many fields. Within each Collective Index, *Chemical Abstracts* standardizes names of compounds, but they can differ between Collective Index periods. Substances other than pure, synthetic compounds — natural products, for example — may only have trivial names, and biochemists are in general not good at nomenclature and often use acronyms. Since the advent of CAS ONLINE, the problems for the non-specialist in nomenclature have lessened because one can now search by means of structure diagrams, the language of choice of the working organic chemist. For natural products whose structure has not yet been determined, however, the trivial names have to suffice for searching, and searches become similar to those in other subject areas.

The same can be said for searches for a process, property, measurement technique or theoretical concept — anything which is not related to specific compounds. Considerably more background knowledge is required, and the task becomes an information retrieval activity as in other subjects. The bench chemist, who would be the actual end-user of the information required, should normally seek help from an information scientist, especially if searches in these areas are infrequent. Even so, the information scientist will need to interview the end-user with care in order to ensure that (s)he has understood precisely what subject matter is required. The availability of online and CD-ROM databases for recent information has altered the search activity radically in recent years. The major changes that have resulted from the availability of these forms of databases are as follows. Firstly, the searcher now has a choice: either the search can use the structured indexing provided by the publisher of the database, or the authors' words in the titles and (in many cases) the abstracts can be used as search terms — a free-text search. There has tended to be some controversy in the information world about the relative effectiveness of these two approaches, but it must be helpful to have both available.

Secondly, the search can combine different data-elements; for example, one could ask for work which was in English, published between 1970 and 1979, about charge-transfer complexes, and *not* published in certain journals. (The last-named type of term can conveniently exclude the journals which the user routinely scans anyway.) Such a combined type of search is almost impossible to carry out using traditional printed abstracts and indexes publications.

Thirdly, many of the online hosts now make it possible to perform cross-database searches; thus, if one was interested in a biomedical topic, one could perform searches across CHEMICAL ABSTRACTS, BIOLOGICAL ABSTRACTS and MEDLINE, for example, to ensure that one had achieved complete coverage of the required field. This is helpful because many investigations have shown that a single database rarely provides 100 per cent coverage of a topic (see Martyn and Slater, 1964), even though chemistry is more fortunate than most disciplines in the comprehensiveness of *Chemical Abstracts*.

Furthermore, the host STN (Scientific and Technical Network), which is part-owned by the American Chemical Society, provides a further facility: in addition to the CAS ONLINE file included chemical structures, STN also carries the CHEMICAL JOURNALS ONLINE (CJO) files, which contain the full texts of leading chemical primary journals published by the ACS, the RSC, and other major publishers of chemistry. Hence the cross-file searching capability enables one to combine searches of the structures (in the CAS ONLINE file) with searches of the full text (in CJO), for papers that are in these major journals.

The search may reveal a recent bibliography or review on precisely the subject of interest; if this happens the searcher is fortunate, as much of the work has then already been done. It is unlikely to be fully up to date, owing to publication delays, but it can be used a source of references for use in a search of *Science Citation Index* (printed or online) which will lead forwards to more recent references, though it covers only the major core journals of science (see Chapter 3).

If the subject in question is one in which patents are important — pharmaceuticals, for example — then the more specialized patents sources (such as those of Derwent Publications Ltd) mentioned in Chapter 11 must be consulted for complete coverage. *Chemical Abstracts* covers patents quite well nowadays; however, the new MARPAT online service from CAS enables one to search for Markush structures, the kind of generalized structure diagrams that are prevalent in chemical patents.

There are various sources of advice on hardware, software and skill necessary for getting started with online searching, for example:

The book *Going Online 1989* (G. Turpie, Aslib)

Going Online with RSC Databases — A Beginners' Guide, a booklet available free from The Royal Society of Chemistry

Computer Database Advice Centre, at the Thames Valley University, Ealing

The UK Online Users' Group (UKOLUG)

The final product of the detailed search should contain the references located, all cited in one of the standard formats; several software packages for personal computers exist today which will convert bibliographic references between the various standard styles, and the searcher will therefore be able to collate references downloaded from online services on to the local PC and cite them in a standard format regardless of which database and host they came from. The bibliography produced should indicate what sources have been used in its compilation and should, of course, be dated. Some topics for practice searches are listed in Table 17.1.

312 *Practical use of chemical information sources*

Table 17.1. Suggested topics for practice searches

Detection and determination of acrolein, furfural and glutaraldehyde in (working place) air.

Methods of analysis of morphine, or codeine or heroin in blood and urine.

Synthesis of calix[6]arene derivatives

Fluorine-substituted sulphines

Synthetic perovskites

Substitution reactions of cyclophosphazenes

Toxic combustion products from burning polymers (Note: Limitation to specific polymers, e.g. polyurethanes, considerably shortens the search)

Recovery of copper from waste water (shorter search)

Recovery of heavy metals from waste water (longer search)

Environmental hazards of radon gas

Thermochromic organic compounds

Yeast-mediated reduction of (mono-)carbonyl compounds

Structure of sarafotoxins (snake venoms)

Applied chemistry

The pharmaceutical industry is probably the most information-conscious sector of the chemical industry. Indeed, the earliest computer-based selective abstracts service from CAS, *Chemical Biological Activities (CBAC)*, was aimed at this area. Thus the industry's information requirements are discussed in Chapter 14, and more fully in a companion volume in this series, *Information Sources in Pharmaceuticals* (W.R. Pickering, ed., 1990). Another heavy user of chemical information is the plastics and fibres industry where the best literature guide is also in this series, *Information Sources in Polymers and Plastics* (R.T. Adkins, ed., 1989). The production of specific fibres, plastics, resins and rubbers is well covered by the *Encyclopedia of Polymer Science and Engineering* (H.F. Mark *et al.*, eds, 2nd edn, 17 vols, Wiley, 1985–90). This is also a good source of analytical and testing methods and each article is well documented. A CD-ROM version is also available.

The most useful single-volume compilation for the production chemist is Perry's *Chemical Engineer's Handbook* (see Chapter 7). Two other useful quick reference works are *Reigel's Handbook of Industrial Chemistry* (J.A. Kent, ed., 8th edn, Van Nostrand Reinhold, 1982) and Faith, Keyes and Clark's *Industrial Chemicals* (F.A. Lowenheim and M. Moran, eds, 4th edn, Wiley, 1975), which reviews (US) processes and economics. *Kirk-Othmer's* and *Ullmann's* encyclopaedias (see Chapter 4) are invaluable reference works for the chemical engineer as well as the industrial chemist. There is, however, one work which is virtually a treatise in this field. This is *Chemical Engineering* (J.M. Coulson and J.F.R. Richardson, eds, Pergamon Press, 1977–83). Most of the six volumes have a third edition and now use SI units. Nearly 40 volumes have so far appeared in the *Encyclopedia of Chemical Processing and Design* (J.J. McKetta, ed., Dekker, 1976–).

Much of the primary publication on industrial chemistry is in the form of patents. The patent literature as a source of information is discussed at length in Chapter 11, but obviously Information Sources in Patents (C.P. Auger, ed., Bowker-Saur, 1992) is more detailed.

Much useful information is also to be found in such journals as *Industrial & Engineering Chemistry, Chemical Age* and so on. One should also peruse *Chemical and Engineering News, Chemical Week* and *European Chemical News*. This field is, of course, a commercial publisher's paradise. There are at least two or three 'technological glossies' for each sector of the chemical industry and one should endeavour to scan those appropriate to the particular sector with which one is connected. Many of these are 'controlled circulation' journals, i.e. they are sent free on request to anyone who might influence the purchase of things advertised in them. Suitable trade journals can be located through *British Rate and Data Services* or its US counterpart *Standard Rate and Data Services*. Some articles in this type of publication are not abstracted by *Chemical Abstracts* but may be located through *Engineering Index, Chemical Industry Notes* or, in the case of UK papers, through the *British Technology Index*.

A bibliographical guide, *Chemical and Process Engineering Unit Operations* (K. Bourton, Macdonald, 1967) covered the literature from 1950 to 1966. It is, in effect, updated by *Chemical Engineering Bibliography, 1967–1988* (Noyes, 1990). A well-documented biennial review of progress was *Fortschritte der Verfahrenstechnik* (Verlag Chemie, 1952–85). A number of consultants, especially the larger US ones, publish surveys of the scientific, technical and economic background to particular chemical processes for limited circulation among their clients. (Needless to say, such surveys are very expensive.)

Sources of chemical prices and also of suppliers, from whom appropriate trade literature may be obtained, are discussed in Chapter 12.

Government regulations for the production and use of chemicals (see Chapter 16) intimately concern the industrial chemist. Health and safety information sources and services are, however, discussed in Chapter 13.

Subjects peripheral to chemistry

The chemist's information needs are not confined to chemistry, even to chemistry defined as widely as by the coverage of *Chemical Abstracts*. With experience, perhaps assisted by this book, the chemist will come to know what are the major sources in the particular field of interest. Much of the greatest growth in chemistry in the post-war era has been at its interfaces with other disciplines, especially with physics and the biomedical sciences. The chemist has two problems when starting to work in such areas: one is knowledge of the information resources available and the other is the need for background information and terminology. In this section we therefore suggest good literature guides and also some specialist dictionaries and quick reference sources.

The Bowker-Saur series (formerly published by Butterworths) *Guides to Information Sources* is generally recognized as much better than the average literature guides, many of which are quite useless. It is of interest to note that this series was initiated by one of us with the second edition of this book, then titled *Use of Chemical Literature* (R.T. Bottle, ed., 2nd edn, Butterworths, 1969). The general guide also from Bowker-Saur, *Information Sources in Science and Technology* (C.C. Parker and R.V. Turley, eds, 2nd edn, 1986) is an easy-to-use first port of call.

The McGraw-Hill *Encyclopedia of Science and Technology* (see Chapter 4) is good for background information and some data for most topics. Chambers' *Scientific and Technical Dictionary* (P. Walker, ed., rev. edn, 1988) is similarly useful for general terminology problems.

The literature of physics is quite well documented, though not so well as that of chemistry. Fortunately the series guide, *Information Sources in Physics* (D. Shaw, ed., 2nd edn, 1985) is still quite up to date. Most of the data sources described in Chapter 7, especially the *Landolt-Börnstein* series, are just as useful for the physicist as for the chemist. An established single volume reference work is *Encyclopedia of Physics* (R.M. Besançon, ed., 3rd edn, Van Nostrand Reinhold, 1985). Over 3000 annotated items are listed in *Handbooks and Tables in Science and Technology* (R.H. Powell, ed., 2nd edn, Library Association, 1983). Though this is useful as a checklist, one is given little guidance on selecting the most appropriate data source other than the broad section headings.

Those concerned with metals at last have a good literature guide available, *Information Sources in Metallic Materials* (M.N. Patten, ed., 1990), which contains chapters on corrosion and metallurgy and perhaps surprisingly, one on ceramics. They will also find *Metallurgy and Materials Encyclopaedia* (C.R. Tottle, ed., Longmans, 1983) useful for quick reference and will find much of the data they need in *Roskill's Metals Databook* (6th edn, Roskill, 1985) or the Institute of Metals' *Metals Data Book* (C. Robb, 1988).

Engineering has a diverse literature which is not easy to organize. The only reasonably good guide is *Information Sources in Engineering* (L.J. Anthony, ed., 2nd edn, 1985), which includes a chapter on chemical engineering.

The chemist is probably more concerned with the chemical rather than the geological aspects of minerals and will therefore find Hey's *Index of Mineral Species and Varieties* (British Museum, 2nd edn, 1955, Appendix, 1963) still of interest. This index is arranged chemically and has an alphabetical list of accepted mineral names and synonyms. A related topic is *Industrial Minerals and Rocks* (S.J. Lelond, ed., 5th edn, SMM&E Inc., 1983). This is the most recent of several reference books in this area. The best guide to geological and related literature is *Information Sources in the Earth Sciences* (D.N. Wood, J.E. and A.P. Harvey, eds, 2nd edn, 1989). Background information can be obtained from *Encyclopedia of the Earth Sciences* (S.P. Parker, ed., 2nd edn, McGraw-Hill, 1988).

The chemist working in a medical field has more literature problems than most, that of terminology being one of the lesser ones. The best and most up-to-date literature guide is again in this series, *Information Sources in the Medical Sciences* (L.T. Morton and S. Godbolt, eds, 4th edn, 1992). This also covers the biochemical literature. It is deeper in coverage than Morton's original guide, *How to Use a Medical Library* (L.T. Morton and D.J. Wright, 7th edn, Bingley, 1990), which is aimed primarily at medical students. A rather more drug-oriented and biological approach is taken by *Medical Information: A Profile* (B. Strickland-Hodge and B. Allan, Mansell, 1986). The same pair of authors have also produced recent guides to the four most important abstracting services in this area. These are both published by Gower — *How to Use* Index Medicus *and* Excerpta Medica (1986) and *How to Use* Psychological Abstracts *and* Biological Abstracts (1987). Recent coverage of the life sciences literature is virtually completed by *Guide to Information Sources in Botany* (E.B. Davis, Libraries Unlimited, 1987).

Black's Medical Dictionary (C.C.W. Howard, ed., 3rd edn, 1990) provides an authoritative solution to the terminology problem. Much fuller information is given in *Cyclopedic Medical Dictionary* (C.L. Thomas, ed., 15th edn, Davis, 1985). Although long out of print, a useful function was provided by *Reversicon: A Medical Word Finder* (J.E. Schmidt, Thomas, 1958). This enabled one to generate an acceptable medical term from a layman's description.

Useful for background information, though really aimed at the layman, is the *Penguin Medical Encyclopaedia* (P. Wingate, ed., rev. edn, 1988). Penguin also publishes helpful dictionaries in the life sciences, e.g. *Penguin Dictionary of Botany* (1984) and *Penguin Dictionary of Food Additives* (E. Millstone and J. Abraham, 1987). Almost a textbook, but very useful for quick reference, is the *Cambridge Encyclopaedia of the Life Sciences* (A. Friday and D.S. Ingrams, eds, Cambridge University Press, 1985).

Agricultural libraries and information sources are discussed in detail, but with a US slant, in the Winter 1990 issue of *Library Trends* (K.W.

Russell and M.G. Pisa, eds, **38**(3), 327–638). The guide in this series, *Information Sources in Agriculture* (G.P. Lilley, ed., 1981) is now out of date (although a new edition will be available in 1993), but nevertheless, can supplement the agrochemical sources given in Chapter 15. As well as listing culture collections, *Introduction to Biotechnology Information* (M. Eusden, ed., BL, 1992) features patents and business information.

Outlined above are a selection of the major literature guides and reference works for some of the areas of knowledge where chemists may occasionally find themselves. The time-honoured expedient for getting information is, of course, to ask an expert in the relevant field. This section has been written not so much as a substitute for this procedure but for the many occasions when a good library is more accessible than the expert. For subjects not covered in this section the reader is referred to A.J. Walford's *Guide to Reference Materials* (Vol. 1, *Science & Technology*, 5th edn, Library Association, 1989), which contains about 6000 annotated entries classified by the UDC, or the 10 000 entry *Guide to Reference Books* (E.P. Sheehy, comp., 10th edn, American Library Association, 1986), formerly known as *Winchell's Guide*.

Communication to others

Much technical reading is undertaken so that a digest of the information obtained may be passed on to others, less well qualified technically than the reader, be they undergraduates or commercial management. For the former group, one must digest portions of suitable monographs and review articles (which are discussed in Chapter 4); for the latter, one must be familiar with the literature mentioned in the applied chemistry section of this chapter and especially the appropriate trade statistics, financial situation, and sources discussed in Chapter 12. (Daily reading of the *Financial Times* is most important in this context; one should also try to read regularly the magazines mentioned in the periodicals section of Chapter 12.)

For those concerned with industrial training schemes the video or film may well be preferred to the formal lecture for instruction. Suitable films may be located through the *RSC Index of Chemistry Films* (7th edn, 1984), which lists 1600 films and 400 film strips. New additions are noted in *Education in Chemistry*. Those particularly interested in this medium will find the British Universities Film and Video Council's quarterly *Videodisc Newsletter* (1983–) a useful source of current information. Although only four videos are listed in the RSC's 1992 catalogue of publications, many more are to be found in the ACS catalogue. The Open University is, of course, a major producer of teaching videos including many on chemistry.

It has often been said that there is little point in a scientist finding out anything, whether it is at the bench or in the library, if it is not possible to communicate his findings to other people. Several books have been written to help the inarticulate express ideas adequately. Even some well-known chemists have gone into print in this field. Technical writing is, however, little different from any other form of English composition. All too often it turns out to be less readable. As mentioned in Chapter 1, the readability of scientific literature deteriorated very badly in the first half of this century. Patents are the most unreadable form (Bottle, 1984). Readability (R) is conveniently measured by a modification of the Flesch formula:

$$R = 1.6m - S + 31.6$$

where m is the percentage of monosyllabic words and S is the average sentence length (Farr et al., 1951). Values of R above 50 indicate well above average readability for technical literature. We suggest that you compare the readability of the two forms of the advice given in Figure 17.1. One of the most effective users of the English language, Sir Winston Churchill, once said 'Old words are best and old words when short are best of all'. This ideal may not be easy to follow; as Sheridan remarked, 'Easy reading's curs't hard writing'!

Figure 17.1 Advice on readability

Scientists desirous of increasing the readability of their written communications should suppress their inherent predilection for polysyllabic terminology and minimise sentence length.	Scientists who want to make their papers easier to read should use short words in short sentences.
Reading Ease Score = 12	*Reading Ease Score = 76*

The finest book on English usage (and one of the cheapest) is Sir E. Gower's *The Complete Plain Words* (3rd edn, revised by Sir B. Fraser, HMSO, 1986), which is a reconstruction of Gower's famous *Plain Words* and *ABC of Plain Words*. If one, however, is stuck for the right word, Roget's *Thesaurus* (now available in a Penguin edition) will probably be more useful than a normal dictionary. In it words are arranged according to the ideas which they express, as well as alphabetically.

Many journals and publishers issue notes for the guidance of prospective authors. It saves time and trouble if one adheres strictly to these guidelines. Those from publishers often contain instructions for marking copy for printing and proofreading. The *Handbook for Chemical Society Authors* (1960) was useful in this respect and for its IUPAC nomenclature rules. It is unfortunately now out of print and somewhat dated. The ACS provides much more up-to-date guides in three formats. The book is *The ACS Style Guide: A Manual for Authors and Editors* (1986). For the more affluent there is an audio course, *Technical Writing and Communication* by H.E. Plotkin and C.M. Mablekos, or the five-cassette video *Technical Writing* by A. Eisenberg.

Those who have to present papers at a meeting will find much useful information on preparation of slides, speaking and many other aspects of the presentation of their material in *Oral Communication in Technical Professions and Businesses* (D.O. and R.D. Cassagrande, Wadsworth, 1986) or *Basics of Technical Communicating* (B.E. Cain, ACS, 1988). Sometimes, unfortunately, the actual delivery of a paper at a conference receives all too little attention from the author. The golden rules of oral communication are, however, neatly phrased in the old saw: Stand up, Speak up and Shut up!

The past two decades have seen an increase in the use of the poster session at conferences. A display of posters, each encapsulating the essentials of a paper and with its author standing by it, enables one to concentrate on those papers which are of particular interest. One can discuss points with the author in more detail than at a formal presentation — and one does not have to sit through numerous papers of little interest just to hear the relevant one. Useful advice on posters is contained in a survey of poster sessions by Eisenschitz *et al.* (1978/9).

The February 1990 issue of *Journal of Chemical Education* carried two papers giving excellent advice on how to write research proposals (Weissmann, 1990; Hirsch and Love, 1990) and one on applying for academic posts (Gould, 1990). Whilst these are specific to the US scene, the advice given could be read with profit by many UK chemists.

Research scientists, especially those working in academic institutions, have always made use of 'the invisible college' — informal communication with colleagues in other places by means of letters, visits, attendance at conferences and telephone calls, by which the community of a particular discipline coheres and passes on up-to-date information. This complements their retrieval of information by more formal means.

Recently, further new means of facilitating informal communication have been developed, using the international telecommunications networks. In most of the advanced nations, networks for the use of academics have come into existence, and increasingly these national networks are linked together throughout the developed world. Britain's is called JANET, the Joint Academic NETwork, and is administered by the Science and Engineering Research Council (SERC). Originally it linked SERC laboratories and the universities only, but now most of the polytechnics and other institutions of higher education in the UK have acquired JANET links as well. Furthermore, although most early users of the network were from scientific disciplines, usage by academics from all subject areas and by academic librarians is now widespread. JANET can be used for a variety of purposes. Some databases are available for search over JANET; these include some which can be accessed in other ways, such as the CIS services (see Chapter 5) which are made available to UK academics from the SERC's Daresbury Laboratory, and some which are exclusive to JANET. The latter include, for example, the services of NISS (National Information on Software and Services) based at the University of Bath, which provides a catalogue of software useful to academics as well as a bulletin board (see below). Another increasingly useful facility is the ability to search the University Library catalogues for other institutions; over 50 British academic libraries, at the time of writing, had already made their Online Public Access Catalogues (OPACs) accessible via JANET. Users in academic institutions can also make use of the gateways that exist between networks to gain access to the commercial online hosts (see Chapter 5); rather than each university or department having to maintain a separate account with the packet-switching network services of British Telecom or its competitors, users in academic departments can now make use of their own campus networks to reach JANET and then use the gateway facility to make onward connection to the public packet-switching services to reach the commercial online hosts.

However, it is in more informal communications that the academic networks, free of charge at the point of use, are most valuable. Communication between academics, within the same campus, across a country, and most especially internationally, can now be achieved by the use of electronic-mail facilities. Unlike telephone callers, the two participants do not both have to be at their desks simultaneously; and there is a permanent record of what was said. Unlike letters, e-mail messages pass instantaneously. By the use of mailing lists, one can send the same message to a number of recipients effortlessly. One can reply to a message readily, without needing to rekey the recipient's address. One can prepare messages beforehand on one's word-processor and then transmit them quickly over the network once a line into the recipient's mailbox has been established. And messages received by e-mail can, of course, be downloaded on to a floppy or hard disk by the recipient for further local computer processing. Thus one has a more flexible and versatile alternative to conventional mail, telex or even facsimile transmission, with the added advantage that almost all academics in the developed countries have, or soon will have, their own individual e-mail access.

Electronic mail is now used by academics for collaboration at a distance. Research workers and authors can undertake work at their own establishments and send drafts of papers, or even raw research data, to each other over the networks. If another university possesses superior computing facilities to one's own, one can do one's computation at a distance by using JANET to reach the remote machine. These ideas have led to various experiments in electronic publishing (see Chapter 3) where a learned journal is held on a computer and papers are submitted, and read, by academics online. This facilitates the appending of comments by others to papers; in a sense, every reader can become a referee of the paper. However, the use of electronic journals for truly scholarly publication has in general met with limited success; just like synopsis journals, they come up against the academic preference for 'proper publishing' as a measure of the quality of a research worker's contribution, and as a marker for academic priority.

JANET and its sister networks in other countries offer two further facilities which provide additional help in fostering informal communication. Bulletin board and computer conference facilities are well established on the commercial data networks, especially in the USA, and these facilities are now available to academics, and indeed are increasingly used by them, in the UK. Bulletin boards enable items of news to be posted for anyone to see who cares to call up the board; some also allow anyone to post items, but more typically there is a bulletin board manager who is responsible for vetting items which a user wishes to post on the board. The structure of bulletin boards varies, but they do not usually offer the sophisticated search software which is typical of online information-retrieval services. At the opposite extreme, they may only provide a tree structure like that of viewdata, without free-text search or classification. They are not suitable for the publication of complex or long documents but ideal for topical and up-to-date communications.

Computer conferences are more sophisticated; here, the organizer of the conference will announce that there is to be a computer conference on a particular topic, and users may then register for the conference. They can then read earlier messages and add their own; thus there is an ongoing dialogue between the participants. A 'conversation' ensues, but the participants do not have to be in the same place at the same time, and also have the advantage of being able at any time to refresh their memories of what has been said before. It may well be that these computer conferences, rather than electronic journals, will turn out to be the major advance in scholarly communication that results from the use of telecommunication networks. They will supplement actual conferences; often, a discussion that started a traditional conference will continue thereafter by means of computer conferencing methods.

Bulletin boards and computer conferences both have the disadvantage that one cannot ensure that a particular recipient will see the message that one is putting out; the person you want to reach may never again look at the board or participate in the conference. As people do, however, tend to look in their electronic mailboxes regularly, just as they tend to open their paper mail, e-mail has its place for communications that are for particular recipients while bulletin boards and computer conferences are best for messages aimed at everyone who is interested. As conference contributions tend to be signed and dated, you know who is looking; conversely, someone who looks at a bulletin board may leave no trace. Thus all three have their own particular roles in the communication process.

Keeping up to date

Most serious scientists have always kept their own card index of references relevant to their ongoing research work, and often a collection of reprints of the articles to which the card index refers. However, now that the scientist is likely to have a personal computer at the laboratory, very possibly another at home and perhaps a portable PC for use when travelling, the file of relevant references is likely to be an early candidate for computerization. A variety of programs suitable for this purpose is available; Ashton-Tate's dBase series is a leading brand in the DBMS (database management systems) field, but there are many others including the very straightforward Cardbox, which simulates the traditional card file. One useful labour-saving feature is the ability to download material directly from various electronic information sources into one's private database file. Thus output from a search of CAS ONLINE or other online information services (see Chapter 5) — or indeed even full text from the CHEMICAL JOURNALS ONLINE file (see Chapter 3) — can be directed on to a floppy or hard disk on one's own PC and then incorporated into one's private file, perhaps after a manual scan through the items to weed out irrelevant 'false drops'.

At one time, regular manual scanning of the incoming printed issues of the main abstracting and indexing journals (see Chapter 3) would have been recommended as a procedure for keeping up to date; however, these journals have become so voluminous that this activity has largely been replaced by computer searching in current-awareness mode. Graduate students, however, lacking the necessary funding to conduct online searches regularly, continue to have to search *Chemical Abstracts* and other abstracts journals by hand.

Items from one's incoming electronic mail, or from a bulletin board or computer conference, can also be copied on to a disk and added to a personal file. Careful design of the file format in the first place will pay dividends, since a DBMS offers a much greater variety of means of retrieval from the personal file than does a traditional card file.

An alternative to DBMS programs is offered by bibliographic control programs such as N.B.Ibid. This type of software is designed for the narrower purpose of controlling bibliographic citations of scholarly literature. The idea is that the scholar collects relevant references during the course of research and stores them in the program; when papers come to be written, the references relevant to the particular paper can be cited effortlessly, and the software has the capability of converting them into any one of the standard methods of citation, both in the text and in the reference list itself — and without transcription errors. This type of software seems to be more popular among scholars in the humanities than in the sciences at present, the latter tending to prefer the more versatile DBMS programs.

Whatever software one uses, it remains true that most of the publicly-available electronic information sources are themselves based on the traditional printed scholarly journals, even though these journals themselves may well by now be prepared and printed by computerized methods. There is usually a time-lag between the appearance of the primary journal article and its coverage in the secondary (abstracts and indexes) sources. Thus it remains necessary to scan regularly the incoming current issues of at least a few major primary journals. For any particular chemist, there will be three categories of journal that need to be looked at: the rapid-publication journals such as *Journal of the Chemical Society: Chemical Communications*; the main general journals of high prestige such as *Journal of the American Chemical Society*; and any very specialized journal targeted on one's own specific area of research interest.

General scientific journals such as *Nature* or *Science* give very quick publication to short papers of great topical interest; in general, though, the majority of papers in these journals tend to be in the life sciences and not much pure chemistry is published there. *Chemical Communications* has already been mentioned, and the Royal Society of Chemistry now publishes 'Communications' sections in both the *Perkin* (organic) and *Faraday* (physical) parts of *Journal of the Chemical Society*, as well as *Mendeleev Communications*, the recently launched English-language short-communications journal of chemical work from the former Soviet Union. *Tetrahedron Letters*, published by Pergamon, is probably the leading rapid-communication journal for organic chemistry. *Naturwissenschaften* and *Angewandte Chemie* also incorporate rapid-communication sections, though in the latter case there is a slight additional delay before the rapid papers appear in the English-language International Edition.

Other important sources of up-to-date chemical information are the chemical news journals. *Chemical and Engineering News*, published weekly by the American Chemical Society, is perhaps the most important, and its British equivalent *Chemistry and Industry* is published by the Society for Chemical Industry. The RSC's news journal, *Chemistry in Britain*, appears only monthly but contains many articles of interest for keeping up to date in areas outside one's narrow specialism; it is also probably the main medium for chemical job advertisements in the UK! There are also several newspapers such as *European Chemical News*, *Chemical Age* and *Chemical Week*, which concentrate on commercial

Practical use of chemical information sources 323

and industrial rather than strictly scientific news. The ACS has a database and abstracts journal, *Chemical Industry Notes*, which cover this area at the secondary level, and the RSC produces (in machine-readable form only) the comparable CHEMICAL BUSINESS NEWSBASE which has a much more European rather than North American orientation (see Chapter 12). General scientific magazines such as *New Scientist*, while not really valuable for work within one's own specialism, should be read regularly in order to retain a breadth of scientific vision; indeed, one of the main values of scanning current journals is the serendipity factor. One is much less likely to find new inspiration from outside one's current field in a computer-based information search than one is in a few minutes browsing through the current journals in the library.

References

Bottle, R.T. (1984), in *Representation and Exchange of Knowledge (Proceedings of IRFIS 5, 1983)*, ed. H.J. Dietschmann, pp. 175–184. Amsterdam: North Holland.
Dyson G.M. (1961) *Advances in Chemistry Series*, No. 30, pp. 83–91.
Eisenschitz, T.S., Knox, J., Oppenheim, C., Richards, K. and Wittels, P. (1978/1979), *Journal of Research Communication Studies*, **1**, 235–242.
Farr, J.M., Jenkins, J.J. and Patterson, D.G. (1951), *Journal of Applied Psychology*, **35**, 333–337.
Gould, E.S. (1990), *Journal of Chemical Education*, **67**, 123–126.
Hirsch, R.F. and Love, L.J.C. (1990), *Journal of Chemical Education*, **67**, 127–129.
Martyn, J. and Slater, E. (1964), *Journal of Documentation*, **20**, 212 (see also Martyn, J. (1967), *Journal of Documentation*, **23**, 45).
Weissmann, K.E., (1990), *Journal of Chemical Education*, **67**, 110–112.

Index-glossary of acronyms, databases, etc.

3DSEARCH	Structure handling system, 113
ABPI	Association of the British Pharmaceutical Industry, 273
ACCSYS	Accidents database, 240
ACMF	Augmented Connectivity/Molecular Formula (code), 107
ACS	American Chemical Society, 19, 24, 34, 61, 311, 318, 320
ADONIS	Document delivery service, 12
AEA	Atomic Energy Authority (formerly UKAEA), 292, 294
AFNOR	French standards, 233
AGRICOLA	(US) Agricultural database, 285
AGRIS	FAO Agricultural database, 285
AIChemE	American Institute of Chemical Engineers, 91
ALADDIN	Structure handling system, 113
APELL	Awareness and Preparedness at Local Level (UNEP programme), 297
API	American Petroleum Institute, see APIPAT
APIPAT	American Petroleum Institute Patents file, 73, 82, 97, 201
APTIC	Air Pollution Technical Information Centre, 275
AQUIRE	Aquatic Information Retrieval database (EPA), 92, 100
ASTM	American Society for Testing and Materials, 126, 131
BASF AG	German chemical company (Badische Anilin- und Soda-Fabrik), 90
BIBRA	British Industrial Biological Research Association, 245
BIOS	British Intelligence Objectives Subcommittee (reports on wartime German industry), 21
BIOSIS	Biosciences Information Service, 46
BITNET	US academic network, 86
BL	British Library, 12

326 Index-glossary of acronyms

BLAISE	British Library Automated Information Services, 57
BLDSC	British Library Document Supply Centre, 12, 28, 29, 61, 231
BLSRIS	See SRIS
BNB	*British National Bibliography*, 14
BOHS	British Occupational Hygiene Society, 246
BP	*British Pharmacopoeia*, 54, 275
BRS	Online host (Bibliographic Research Services), 97–100, 101
BSI	British Standards Institution, 233, 246, 287, 307
BUCOP	*British Union Catalogue of Periodicals*, 22
C13-NMR/IR	Spectral database, 90, 99
CA	*Chemical Abstracts*, 33, 72, 78, 97
CABI	CAB International (formerly Commonwealth Agricultural Bureaux International), 283, 284, 286
CAC&IC	*Current Abstracts of Chemistry & Index Chemicus*, 41
CAMAXYS	Computerised Management System for operating COSHH, 240
CAN/SND	Canadian Scientific Numeric Database Service, 87, 97–100, 101
CAOS	Computer Assisted Organic Synthesis, 261, 268
CAPF	*Chemical Age* Project File, 229
CAS	Chemical Abstracts Service, 33, 72, 78
CASREACT	CAS Chemical Reactions file, 84, 85, 98, 262, 268
CASSI	Chemical Abstracts Service Serials Index, 21, 35
CBAC	Chemical Biological Activities, 312
CBD	Directory Publishers, Beckenham, Kent, 221, 224
CBNB	Chemical Business NewsBase (file), 218, 323
CCDC	Cambridge Crystallographic Data Centre, 87, 98, 114, 116
CCR	*Current Chemical Reactions*, 42, 84
CCRIS	Chemical Carcinogenic Research Information System, 92, 100
CCTTE	*Computerized Listing of Chemicals being Tested for Toxic Effects*, 296
CD-ROM	Compact Disk-Read Only Memory, 34, 52, 95, 204, 232, 286, 310
CEC	Commission of the European Communities, 225, 226, 273, 287
CEFIC	European Council of Chemical Industry Federations, 228
CESARS	Chemical Evaluation Search and Retrieval System, 92
ChemDBS-3D	Structure handling system, 113, 116
CHEMLINK	Front-end system, 114
CHEMQUEST	(formerly) Fine Chemicals Directory file, 87, 98, 219, 267, 269

Index-glossary of acronyms 327

CHEMTRAN	Physical Properties Database, 91
CHETAH	Chemical Hazard Evaluation program, 126
CIA	(UK) Chemical Industries Association, 220, 228, 246
CIN	*Chemical Industry Notes*, 218, 313, 322
CIS	Chemical Information Systems, 79, 88, 97–100, 101, 319
CJxx	Complete journals file, 74
CJO	CHEMICAL JOURNALS ONLINE, 22, 311, 321
CNMR	NMR Database, 90, 99
COBRA	3D structural program, 112, 116
CODATA	Committee on Data for Science and Technology of ICSU, 127
CONCORD	3D structural program, 96, 112, 256
CORDIS	Community R&D Information Services (EC database), 298, 301
COSHH	Control of Substances Hazardous to Health (UK regulations), 161, 231, 232, 235, 237, 239–244
CPI	*Chemicals Patents Index* (Derwent), 82, 201, 205
CRC	(formerly) Chemical Rubber Company (publishers), 128, 129
CRDS	Chemical Reactions Documentation Service (Derwent), 84, 98
CRYSTIN	Inorganic crystals data file, 87
CRYSTMET	NRCC metals crystallographic data file, 87, 98
CS	The Chemical Society (now part of RSC), 38
CSCHEM	File from *Chem Sources USA*, 87, 98, 269
CSCORP	CSCHEM Suppliers file, 87
CT	*Chemical Titles*, 47
CT	Connection table, 106
CTCP	Clinical Toxicity of Commercial Products, 92, 100
CUP	Cambridge University Press
CUS	EC Customs Union and Statistics Directorate, 225
CWI	*Chemical Week International*, 215, 223
CZ	*Chemisches Zentralblatt*, 33, 309
DARC	Software for Questel, 78, 98, 114, 208, 259, 260
Data-Star	Swiss online host, 79, 97–100, 101
DDBJ	DNA Data Bank of Japan, 257
DDC	Dewey Decimal Classification, 10
DECHEMA	Deutsche Gesellschaft für Chemisches Apparatewesen, Chemische Technik und Biotechnologie, 40, 91
DES	(UK) Department of Education and Science, 293
DETHERM-SDC	DECHEMA Thermophysical Property Databank Calculation System, 91, 99
DETHERM-SDR	DECHEMA Thermophysical Property Databank Data Retrieval System, 91, 99
Dialog	US online host, 79, 97–100, 101

328 Index-glossary of acronyms

DIMDI	German online host, 97–100, 101
DIN	Deutsche Industrie-Norm (German standards), 233, 307
DIPPR	Design Institute for Physical Property Data of AIChemE (also databank), 91, 274
DOC	Dictionary of Organic Compounds, 89, 166, 306
DSIR	(UK) Department of Scientific and Industrial Research (1916–64), 292
EC	European Communities, 195, 225, 298
ECDIN	Environmental Chemicals Data & Information Network, 234, 269, 274, 286, 298
ECHO	European Communities Host, 298
ECN	European Chemical News (also file), 215, 223, 322
EINECS	European Inventory of Existing Chemical Substances, 63, 241, 267, 298
EI to EV	Supplements 1 to 5 (Ergänzungswerk) of Beilstein, 147–149
EMBASE	Excerpta Medica database, 272
ENDS	Environmental Data Service, 275
EP	European Pharmacopoeia, 275
EPA	(US) Environmental Protection Agency, 79, 80, 92, 132, 299
EPAT	European Patents file, 203
EPIDOS	(formerly) INPADOC (q.v.), 205
EPO	European Patent Office, 194, 195, 205, 261
ESA-IRS	European Space Agency Information Retrieval Service, 97–100, 101
ESCIMO	European Statistics Chemical Industry Monitor, 228
EURECAS	Database, 79
EXICHEM	OECD database on control of chemicals, 298
FAO	UN Food and Agriculture Organization, 272, 282, 285
FIAT	Field Information Agency, Technical (US reports on wartime German industry)
FPAT	French Patents file, 203
FT-IR	Fourier transform infrared (spectra), 130
GDB	Genome Database, 257
GEISCO	US online host, 229
GENBANK	Nucleic acid sequence file, 257
GFI	Gmelin Formula Index, 140
GIABS	Gastrointestinal Absorption Database, 93, 100
GRA&I	(US) Government Reports Announcements & Index, 295, 297
GRAS	Generally Recognised As Safe (for food use), 272
GREMAS	Fragment code system used by IDC, 82
HAIC	Hetero Atom in Context (former CA Index), 34

Index-glossary of acronyms

HIPV	Protein sequence file, 257
HIVN	Nucleic acid sequence file, 257
HMSO	Her Majesty's Stationery Office (UK government publishers), 289
HPLC	High-performance (pressure) liquid chromatography, 269
HSE	(UK) Health and Safety Executive, 231–233, 237, 239, 274, 294
HSE$_{LINE}$	Health and safety database, 232, 239, 274
HW	Hauptwerk (main series) of *Beilstein*, 147–149
IAEA	International Atomic Energy Agency, 145
ICRS	Index Chemicus Registry System, 42
ICSU	International Council of Scientific Unions, 127
ICT	*International Critical Tables*, 121
IDC	(German) International Documentation Centre, 82
IEO	Industry and Environment Office, 296
ILO	(UN) International Labour Office, 274, 287, 297
INFOTERRA	International Environmental Information System, 287, 297, 301
INIS	International Nuclear Information System, 145
Inkadata	German online host, 87, 97–100, 101
INN	International Nonproprietary Names, 225
INNM	International Nonproprietary Names and Modifications, 225
INPADOC	International Patents Documentation Centre (file)(now EPIDOS), 73, 97, 204
INSPEC	Information Service for the Physics and Engineering Communities, 46, 274
IPCS	International Programme on Chemical Safety, 280, 296
IRPTC	International Register of Potentially Toxic Chemicals, 295, 301
IRSS	Infrared Search System, 90, 99
ISBN	International Standard Book Number
ISHOW	Information System for Hazardous Substances in Water, 100
ISI	Institute for Scientific Information, Philadelphia, 41, 43, 61
ISO	International Organization for Standardization, 246, 307
ISSN	International Standard Serial Number
IUPAC	International Union of Pure and Applied Chemistry, 26, 63, 127, 129, 160
JANET	(UK) Joint academic network, 86, 318
JAPIO	Japanese Patents file, 203

Index-glossary of acronyms

JICST	Japan Information Center for Science and Technology, 73
JSM	File from *Journal of Synthetic Methods*
LASER	Crystallographic database, 112
LC	Library of Congress (Classification Scheme), 10
LHASA	Logic and Heuristics Applied to Synthetic Analysis (CAOS program), 268
LOSNUC	Nucleic acid sequence file, 257
MACCS	Molecular Access System, an MDL product, 78, 106, 110, 113, 116, 256, 259, 264, 265
MAFF	(UK) Ministry of Agriculture, Fisheries and Food, 283, 287, 293
MARPAT	CAS file of patents with Markush structures, 82, 209, 261, 311
MDDR	MACCS Drug Data Report, 265
MDL	Molecular Design Ltd, 84, 101, 113, 219, 256, 259, 264, 268
MINE	Microorganism Information Network Europe, 257, 268
MIPSX	Protein sequence file, 257
MITI	(Japanese) Ministry of International Trade and Industry, 221
MOLFILE	MDL software standard, 113
MOLKICK	Structure handling front-end, 114
MSDC	Mass Spectrometry Data Centre (part of RSC), 90, 132
MSDS	Materials Safety Data Sheets (Occupational Health Services), 102, 235
MSSS	Mass Spectral Search System, 90, 99
NAMAS	(UK) National Measurement and Accreditation Service, 238
NASA	(US) National Aeronautics and Space Administration, 12
NBRF	Nucleic acid and protein sequence file provider, 257
NBS	(formerly) National Bureau of Standards (now NIST), 126, 132
NCE	New Chemical Entity, 252–255, 262, 265
NIH	(US) National Institutes of Health, 79, 92, 132
NIOSHTICS	(US) National Institute of Safety and Health database, 235, 274
NISS	(UK) National Information on Software and Services, 319
NIST	(US) National Institute for Standards and Technology, 87, 91, 126, 132

Index-glossary of acronyms 331

NLM	(US) National Library of Medicine, 46, 74, 80, 97–100, 102
NMR	Nuclear magnetic resonance, 90, 187
NRCC	National Research Council of Canada, 87
NSDRS	*National Standard Reference Data Series*, 127, 130
NTIS	(US) National Technical Information Service, 299
OECD	Organization for Economic Co-operation and Development, 298
OHS MSDS	*See* MSDS
OPACs	Online Public Access Catalogues, 10, 319
ORAC	Organic Reactions Accessed by Computer, a software package for chemical reactions files (cf. REACCS) from ORAC Ltd, 102, 106, 116, 262, 268
Orbit	US online host, 79, 97–100, 102
OSAC	Organic Structures Accessed by Computer (from ORAC Ltd), 106, 110
OSHA	(US) Occupational Safety and Health Administration, 237
PCT	Patent Cooperation Treaty, 193, 194, 203
PESTDOC	Pesticides database, 285
PHYTOTOX	Toxicity of chemicals in plants file, 100
POHSO	(UK) Co-ordinating Committee for Professional Health and Safety Organizations, 238
POLYCAS	Database, 79
POLYMAT	German databank of plastics, etc., 94, 100
QCPE	Quantum Chemistry Program Exchange, 111, 116
QSAR	Quantitative Structure–Activity Relationships, 255, 265
R&D	Research and Development, 198, 252–255, 258, 262, 264, 292–294, 298
RAPRA	(formerly) Rubber and Plastics Research Association, 223
REACCS	Software package for chemical reactions, an MDL product, 84, 106, 116, 262, 268
RIC	Royal Institute of Chemistry (now part of RSC)
RSC	Royal Society of Chemistry, 13, 25, 38, 56, 216, 248, 287, 311, 323
RTECS	Registry of Toxic Effects of Chemical Substances, 93, 100, 235, 237
SAC	Society for Analytical Chemistry (now part of RSC)
SANDRA	Structure and Reference Analyser (for *Beilstein*), 160
SCI	Society of Chemical Industry, London, 26, 216, 288
SCI	*Science Citation Index,* an ISI product, 42–45, 56, 311
SDF	Standard Drug File, 265
SDI	Selective Dissemination of Information

Index-glossary of acronyms

SEMA	Stereochemically Extended Morgan Algorithm (code), 107
SI	Statutory Instrument (UK Government regulation), 291, 294
SIC	(UK) Standard Industrial Classification, 225
SIDS	Screening Information Data Sheets (from OECD), 298
SIGLE	System for Information on Grey Literature in Europe, 299
SITC	(UN) Standard Industrial Trade Classification, 223, 227
SMD	Standard Molecular Data, 113
SMILES	Simplified Molecular Input Line Entry System, a line notation, 105
SOCMA	(US) Synthetic Organic Chemical Manufacturers' Association (*Handbook*), 172
SORIS	Specialized Organics Information Service, 221
SPIR	Search Program for Infrared spectra file, 91, 99
SRI	Stanford Research Institute (consultants), 230
SRIS	Science Reference Information Service (part of BL), 12
STN	Scientific and Technical Network, 78, 97–100, 102, 114, 310
SYNLIB	Synthesis Library, software and file for chemical reactions, 85, 268
TECH DATA	Host system, 102
THE	Technical Help to Exporters (BSI), 246
THERMO	Database from NIST *Tables of Thermodynamic Properties*, 91
TLC	Thin-layer chromatography, 183, 269
Topfrag	Structure handling front-end, 208, 268
TOSCA	(US) Toxic Substances Control Act, 92, 221, 231, 236, 267
TSCA	*See* TOSCA
UDC	Universal Decimal Classification, 10, 316
UKAEA	*See* AEA
UKCIS	UK Chemical Information Service (part of RSC), 38
UNEP	United Nations Environment Programme, 288, 295
UPCAS	Database, 79
USAEC	US Atomic Energy Commission, 146
USAN	*US Adopted Names* (of pharmaceuticals), 64
USGPO	US Government Printing Office (Publisher)
USP	*US Pharmacopeia*, 64
VCH	Verlag Chemie, 41
WEFA	Online host, 229

WIPO	World Intellectual Property Organization, 196, 204
WLN	Wiswesser Line-Formula Notation, 42, 64, 105
WMSS	Wiley Mass Spectral Search System, 90, 99
WPI	*World Patent Information* or *World Patent Index*, 73, 82, 97, 201, 204, 206, 264, 267, 272
WPIL	World Patent Index Latest file, 206, 259

Index

Only the more important reference books and journals have been indexed by name and then only where there is descriptive matter pertaining to them. No entries have been made to pages where they are merely cited in lists or as examples.

As a separate Index-glossary of Acronyms, Databases, etc. has been provided, no entry has been made in this index. It follows therefore that if the reader cannot find the database, organization, etc., under its full form, he should look under its initials or acronym in the preceding section.

Abbreviations, journal, 35
Abstracts, 9, 32
Abstracts journals 31–42, 46, 284
  analytical, 33
  nuclear science, 146
  older, 33
  safety, 235
  technological, 46
Abstracts, writing, 32
Advertisements, *see* Journals, controlled circulation
Advances in....59
Advances in Chemistry Series, 55, 128
Addresses
  food and agriculture, 287
  lists of, 300, 308
  online services and hosts, 101, 249
  structure handling systems, 116
Agrochemicals, 93, 217, 245, 279–288, 315
Alchemy *see* Histories of chemistry
Alerting services, 41, 46, 149, 321

Alkaloids, 167, 170
American Men and Women of Science 5, 311
Analysis
  inorganic, 142, 307
  organic, 181–183, 269, 307
Analytical Abstracts, 39
Annual Reports of the Progress of Chemistry, 58
Applied chemistry, 311, 313
Associations, 221, 287
Atomic Energy, 292, 293
Atomindex, 146

BEILSTEIN ONLINE, 81, 125, 148, 259, 267
Beilsteins Handbuch der organischen Chemie, 81, 147–161, 309
Bennett's *Chemical Formulary,* 55
Bibliographic files, 71, 73, 97
Bibliographies, 60, 313, 321
Biochemistry, 85, 306

Index

Biographical details
  current, 308
  historical, 64
Biological Abstracts, 46
Biotechnology, 73, 262, 263, 267, 268
BLAISELINE, 57
Book reviews, 14, 55
Books, publication details of, 14, 56
  translated, 59
Boolean logic, 68, 69
British Abstracts, 33
British Library, 12 (*see also* BLDSC, SRIS)
British National Bibliography, 14
Browsing, 21
Bulletin signaletique, 41
Buyer's guides, 87, 214, 219

CA Selects, 14, 34, 35, 38
CAS ONLINE, 35, 37, 309
CA Search, 72
Catalogues
  library, 11 (*see also* OPACs)
  trade, *see* Trade literature
Chemical Abstracts, 32, 33–38, 139, 201, 267, 272, 309, 313
  collective indexes, 34, 37
  computer readable files, 72, 79, 258, 309
  early coverage, 309
  formula indexes, 34
  issues, arrangement in, 36
  journals covered, 21, 35
  languages covered, 27
  microfilm edition, 34
  patent coverage, 201
  Ring Index, 34
  searching, 37
  subject groupings, 36
  subject indexes, 34
Chemical engineering literature, 73, 274, 313, 314
Chemical Engineers Handbook (Perry), 123
Chemical Industry Notes, see CIN
Chemical literature
  doubling time, 4–5
  estimated value of, 4
  instruction in, 5
  other guides, 9
  use of, types of, 7, 8
Chemical manufacturing, 229, 230
Chemical Titles, 38, 46
Chemicals
  prices, 216, 223, 224
  suppliers, 87, 98, 219, 267, 269
Chemischer Informationsdienst, 41
Chemisches Zentralblatt, 33, 309
Chemistry and Industry, 216
Chemistry of the Carbon Compounds, Rodd's, 161
Chemists
  addresses of, 308, 311
  biographies of, 64, 308
  salaries of, 227–228
Chromatography, 183
Classifications, library, 10
Coding, chemical, 64, 105
Colour Index, 55
Communication by chemists, 4, 18, 316
  electronic, 318–321
Communication to others, 316–321
Company information, 207, 213
Comprehensive Organic Chemistry, 164
Computer readable files, *see* Databases and Online services
Conference proceedings, 12
Conferences, 60–62
Connection tables, 106
Consultants, 308
Controlled circulation journals, *see* Journals, controlled circulation
Corrosion, 93
Crystallographic data, 87, 98, 112, 129, 256
Current Abstracts of Chemistry and Index Chemicus, 41
Current awareness, 46, 47, 300, 321–323
Current Contents, 46, 144

Dangerous Properties of Industrial Materials (Sax), 238, 274
Data, physicochemical, *see* Properties, physical

Databases, 97–100, 285
  chemical substance, 77, 125, 268
  fulltext, 22, 74
  sequence, 85
Derwent Publications, 45, 82, 84, 205, 206, 258, 259, 267
Dictionaries
  chemical, 53, 166–168, 214
  other sciences, 314
Dictionary of Organic Compounds (Heilbron and Bunbury), 89, 166 (*see also* DOC)
Directories, 69, 213, 281
Dissertation Abstracts International, 29
Dissertations, *see* Theses
Drugs, *see* Pharmaceuticals
Dyestuffs, 55, 63

Education, 5, 134, 316
EEC documents, 298
Elastomers, *see* Polymers
Electrochemical data, 128, 133
E-mail, 319
Encyclopaedia Britannica, 50
Encyclopaedias, etc., scientific, 50, 89, 99, 165, 314, 315
Encyclopaedia of Chemical Processing and Design, 52
Encyclopaedia of Chemical Technology, *see* Kirk-Othmer's Encyclopaedia
Encyclopaedia of Polymer Science and Engineering, 51, 89
Encyclopaedia of Science & Technology, McGraw-Hill, 50
Energy Research Abstracts, 45
Engineering Index, 46
English, use of, 317
Environment, chemicals and the, 93, 274, 280, 296 (*see also* ECDIN, ENDS and EPA),
Eponyms, *see* Named reactions and laws
Factories Acts, *see* Health and Safety
Festschrifts, 62
Fibres, *see* Polymers
Files, 70, 267
  full text, 74, 92, 203

Films, scientific, 316
Financial information 213, 218, 219
Food science, 40, 272, 280, 285, 315
Formularies, 55, 271
Front ends, 113
Functioning classes (Beilstein), 150, 155

Genome sequence, 86, 257
Geology, 315
Gmelins Handbuch der anorganischen Chemie, 80, 140
Government publications, 289–302
  indexes, 291, 294
  non-British, 295–299
Government Reports Announcements and Index, (GRA&I), 299
Grey literature, 22, 283, 289, 297, 299, 300
Growth of science, 5
Guides
  to handbooks and tables, 123
  to libraries, 13
  to reports, 22
  to reviews, 59
  see also Literature guides

Handbook of Chemistry (Lange's), 124
Handbook of Chemistry and Physics, 124, 145, 305
Handbuch der anorganischen Chemie see Gmelins...
Handbuch der organischen Chemie, see Beilsteins...
Hazards, chemical, 144, 172, 173, 233, 238, 267
Health and safety, 92, 126, 173, 231–247, 269, 280, 296, 297
Heibron and Bunbury's *Dictionary of Organic Compounds*, see Dictionary...
Heterocyclic chemistry, 154, 155, 164, 170, 176, 184
Hill system, 37, 149
Histories of chemistry, 64
Hosts, database, 68, 72, 97–100
Houben-Weyls Methoden der organischen Chemie, 177

Index

Index Chemicus, see CAC&IC
Index Medicus, 46
Indexes
  author, 37, 44
  citation, 42, 56
  conference proceedings, 61
  cumulative, 34
  formula, 34, 37
  government publications, 291, 294
  interdisciplinary, 43
  patents, 34, 201
  reviews, 59, 60
  ring, 34
  spectra, 131
  translations, 28
Industrial chemicals, 51, 89, 171, 214, 221, 224–229
Industrial chemistry, 311, 312
Infrared spectra, *see* Spectra
Instrumentation, 306
International Critical Tables, 121
Isotopes, 145
IUPAC nomenclature rules, 63

Japanese, literature in, 27, 73
Journal of the American Chemical Society, 19
Journal Citation Reports, 24
Journals
  analytical, 27
  commercially published, 19, 26, 215, 279, 313
  controlled circulation, 313
  electronic, 24
  guides to, 21
  history of science, 64
  house, 218
  industrial, 215, 219, 239, 279, 313
  inorganic, 27, 143, 145
  learned society, 18, 26
  machine readable, 22, 74
  news, 322
  nuclear chemistry, 145
  organic, 26
  polymer science, 26, 28
  quick publication, 21
  Russian, 27
  searching nineteenth century, 33
  selecting, 24, 44
  synopsis, 20
  trade, 215–219
  translated, 27–28

Kaye and Laby's Tables of Physical and Chemical Constants, 119, 122
Keeping up-to-date, 321–323
Kinetics, reaction, data, 127
Kirk-Othmer's *Encyclopaedia of Chemical Technology*, 51, 89, 214, 275

Laboratory safety, (*see also* Health and Safety), 225, 236, 237
Landolt-Boernstein's 'Tables', 120, 121
  contents, 134–138
  New Series, 122
Lange's Handbook of Chemistry, 124
Language distribution, 18, 27
Legislation, 239, 272, 273, 274, 281, 284, 290, 291
  patents, 191–195
Libraries, 10, 13
  for chemists, 12, 13
  depository, 12
Library catalogues, 11 *see also* OPACs
Library of Congress classification, 10
Literature
  growth pattern, 5, 20
  guides, 273, 275, 299, 314–316
  scientific, instruction in use of, 5
  survey, 308–311 (*see also* Searching)
  see also Chemical literature

Market research, 206, 207, 261, 263, 264
Markush structures, 81, 98, 199, 208, 209, 260
Mass spectra, 79, 90, 132
Mass spectrometry, 186
Materials and Technology, 50
McGraw-Hill Encyclopaedia of Science and Technology, 50
Meetings, *see* Conferences
Medical literature, 46, 315
MEDLINE, 46
Melting points, 121, 124, 166, 182

Index 339

Merck Index, 53, 89
Metallurgy, 93, 314
Microforms, 11, 20
Minerals, 315
Miniprint, 20
Molecular modelling, 255
Monographs, 55, (*see also* Books)
Mutagenicity, 93

Named reactions and laws, 53, 173
Natural products, 54, 167, 170–172, 176
Networks, 318
New compounds, 48, 207
New reactions, 41
Nomenclature, 37, 63, 77, 79, 172, 173, 225, 309
Notations, 42, 64, 105
Nuclear chemistry, 144
Nuclear energy, *see* Atomic energy
Nuclear magnetic resonance, *see* NMR
Nuclear Science Abstracts, 146
Nucleic acid sequences, 85, 86, 96, 112, 256, 258, 262, 307

Official publications, *see* Government publications
Online services, 67–102, 309–311
Organic chemistry
 nomenclature, 63, 173
 treatises, 147, 163–165
Organic Reactions, 82, 175
Organic Syntheses, 82, 175, 178
Organometallic chemistry, 140, 142, 143, 144, 166, 167, 169, 177, 181

Patents, 81, 191–211, 259–261, 263, 311
 abstracts, 32
 concordance (CA), 34
 contents of, 196–199
 equivalent, 34, 195
 European, 195, 203
 files, 73
 literature, 201–206
 nature of, 191–196
 term, 191, 194
Periodicals, *see* Journals

Permuterm Index, 43
Perry's *Chemical Engineers Handbook*, 123
Personnel, *see* Biographies, current
Pharmaceuticals, 74, 167, 251–278, 312
Pharmacopoeias, 54, 275
Physical chemistry
 journals, 27, 126
 treatises, etc., 133
Physics, 46, 125, 314
Plant chemistry *see* Natural products
Plastics, *see* Polymers
Pollution, 235, 274, 282, 284
Polymer Reviews, 199
Polymers, 26, 85, 94, 165, 312
Preliminary communications, 21
Preparative methods, *see* Synthesis
Preprints, 17
Prices, *see* Chemicals, prices
Primary literature, 17–30, 322
Proceedings of conferences, etc., 62
*Progress in...*59
Properties, physical, 88–94, 119–132
Protein sequences, 85, 86, 96, 112, 256, 258, 307

Questel, 79, 97–100, 102
Quick publication journals, 21, 322 (*see also* Preliminary communications)
Quick reference sources, 49–55, 305–308, 312

Radiation chemistry, 145
Radiochemistry and isotopes, 145, 180
Radiological protection, 144, 248
Reaction mechanisms, 169, 175
Reactions, chemical, 42, 82, 98, 108, 175, 261, 262, 268
Readability, 9, 317
Records, personal, 311, 321
Referativnyi Zhurnal, Khimiya, 40
Refereeing, 17
Registry Number *(CA)*, 35, 79, 166, 219, 258, 309
Reports, 12, 22, 291, 299

Research
  academic, 29, 308
  government, 292–294
  industrial, 206 (see also R&D)
Research intelligence, 207, 208, 230, 264
Reviews and review serials, 57, 133, 142, 174–178, 283
  guides to, 60
Richter system, 149
Ringdoc, 45, 272
Rodd's Chemistry of the Carbon compounds, 163
Royal Society, 18
Royal Society of Chemistry, 13
'Rubber' Handbook, see *Handbook of Chemistry and Physics*
Russian literature, 12, 58, 128, 132
  guides, 40

Sadtler spectra, 131
Safety, see Health and safety
Scattering, Bradford, 32, 43
Science Abstracts, 46
Science Citation Index, 42, 56, 311
Search languages and protocols, 69
Searching, 23, 308–312
  CA, 37
  by chemical structures, 35, 76, 95
  older literature, 33, 309
  for physical data, 88–94, 120
SI units, 122
Societies, see Associations
Solubilities, 128, 129
Solvents, 172
Specialist Periodical Reports, 59, 133, 142, 174
Spectra, 90, 125, 126, 130, 185–188
Spectroscopy, 130, 184–188
Standards, 246, 287, 307
Statistics, economic, 224–229, 261, 292
Statutory Instruments, 291, 294
Structure determination 184–188 (see also Spectra)
Substructure searching, 75, 76, 79, 81, 95, 107
Symposia, see Conferences
Synthesis, 177–181, 266

Synthetic Methods of Organic Chemistry, Theilheimer, 82, 84

Tables of properties, inverted, 124
Technical writing, 317
Techniques of Chemistry, Weissberger, 133, 179, 306
Technique of Organic Chemistry, Weissberger, 179
Textbooks
  analytical, 142
  inorganic, 139–42
  physical, 132
Theilheimer's *Synthetic Methods of Organic Chemistry*, 268
Thermodynamic data, 91, 126, 127, 128
Theses, 12, 28, 29, 299
Toxic chemicals, 92, 93, 173, 234–238, 294, 295
TOXLINE, 75
Trade
  associations, 220
  literature, 126, 142
  names, 172, 214, 216, 220, 222, 223
  statistics, 224–229
Translations, 27, 239, 299

Ullmann's Encyclopaedia of Industrial Chemistry, 51
Ulrich's International Periodicals Directory, 21
Universities, 13, 28, 308
US Government publications, 300

Vendors, see Hosts, database
Videos, 134, 140, 316

Weissberger's *Techniques of Chemistry*, 133, 179, 306
Weissberger's *Technique of Organic Chemistry*, 179
Who's Who type guides, 308
World List of Scientific Periodicals, 21
World Patent Index, see WPI
Writing, 317

X-ray crystallography data, 129

*Zahlenwerte und Funktionen aus
  Physik, etc., Landolt-Boernsteins*,
  120, 121